MARINE
NATURAL PRODUCTS
CHEMISTRY

NATO CONFERENCE SERIES

I   Ecology
II   Systems Science
III   Human Factors
IV   Marine Sciences
V   Air—Sea Interactions

IV MARINE SCIENCES

Volume 1   Marine Natural Products Chemistry
            edited by D. J. Faulkner and W. H. Fenical

MARINE NATURAL PRODUCTS CHEMISTRY

Edited by

D. J. Faulkner and W. H. Fenical

Scripps Institution of Oceanography
University of California, San Diego
La Jolla, California

Published in coordination with NATO Scientific Affairs Division by

PLENUM PRESS · NEW YORK AND LONDON

Library of Congress Cataloging in Publication Data

Main entry under title:

Marine natural products chemistry.

  (NATO conference series: IV, Marine sciences; v. 1)
  "Sponsored by the NATO Special Program Panel on Marine Sciences."
  1. Marine biology—Congresses. 2. Biological products—Congresses. 3. Biological chemistry—Congresses. 4. Chemical oceanography—Congresses. 5. Marine pharmacology—Congresses. I. Faulkner, D. John, 1942-   II. Fenical, W. H. III. North Atlantic Treaty Organization. Special Program Panel on Marine Sciences. IV. Series.
  QH91.8.B5M37   574.92   76-58470
  ISBN-13: 978-1-4684-0804-1   e-ISBN-13: 978-1-4684-0802-7
  DOI: 10.1007/978-1-4684-0802-7

Proceedings of a conference on Marine Natural Products held in Jersey, Channel Islands, United Kingdom, October 12-17, 1976, sponsored by the NATO Special Program Panel on Marine Sciences

Preface

This volume contains the lectures presented at the NATO-sponsored conference on "Marine Natural Products" held in Jersey, Channel Islands, U. K., October 12-17, 1976. The intent of the organising committee was to encourage a dialogue between organic chemists who study the metabolites of marine organisms and biologists, ecologists, and pharmacologists who study the effects of these metabolites on other organisms. A feature of the conference was the three workshop sessions on chemotaxonomy, applications of marine natural products, and chemical communication.

The papers presented at the conference contain a mixture of original research in marine natural products and reviews of some of the more important subjects. The biologists were asked to present papers which could initiate new directions for marine natural products research. Their contributions to the meeting were warmly received by the chemists in the audience. We hope that this volume contains not only past and present research but a suggestion of future research trends.

The conference was first suggested by Dr. E. D. Goldberg. The organising committee, Drs. G. Blunden, D. J. Faulkner, W. Fenical and L. Minale, received constant help and encouragement from Dr. Andreas Rannestad and Dr. Tom Allan, the Executive Officers of the NATO Marine Science Program. On behalf of all participants, we wish to thank the Scientific Affairs Division of NATO for their generous financial support of the conference. The final draft of this symposium volume was typed by Mrs. Jean Dudley and Mrs. Theresa Koch, whose diligent efforts have ensured its rapid publication.

D.J.F.

W.H.F.

La Jolla, California
November, 1976

Contents

CONTENTS

FIELD OBSERVATION - PRELUDE TO DISCOVERY OF MARINE NATURAL PRODUCTS

Leon S. Ciereszko

Chemistry Department, The University of Oklahoma

Norman, Oklahoma 73019, U.S.A.

I feel that field work should be considered an appropriate activity of the natural products chemist. I feel that the observations the chemist can make in the field will help him make a better choice in selecting material to work up for "natural products."

Many marine stations accept visiting investigators and provide facilities for field studies. Some field studies can be made using a resort hotel or cottage as a base of operations. There are many varied shallow water environments easily accessible from shore or by small boat.

The tropics and subtropics are particularly attractive marine areas for field work. I am partial to coral reefs, mangrove forests or swamps, and sea grass meadows. These three types of habitats are major producers of organic matter and have a global distribution. Often, all three of these environments occur close together, so that observations and collections in all three can be made in small craft within one working day. All coral reefs, mangrove forests and sea grass communities are not the same. The organisms found in these habitats vary from one location to another, and the same genus or even species collected from different locations may yield different products having the same or similar functions.

The clear shallow waters of the habitats mentioned offer a large variety and abundance of organisms, dependent on the nature of the bottom, many of which are sessile or slow moving and, thus, easy to observe and to collect by gloved hand or hand tool.

Field observation as a prelude to the discovery of new natural products is not a novelty. M. Henze and Werner Bergmann based much of their chemical work on material they personally obtained at marine stations. Cornman has called attention to seaside screening as an aid in the search for leads to pharmacologically active substances in marine organisms. Andersen and Faulkner have reported on shipboard screening for antibacterial and antiyeast activities. Extensive surveys for organic halogen compounds and for antimicrobial activities in marine organisms were made during a cruise in the Gulf of California by workers from the University of Illinois (Hager et al.; Shaw et al.). Even identification of compounds in selected marine organisms by instrumental methods were accomplished on shipboard by Rinehart's group.

I feel that simple use of the senses and field experiments with modest means can lead to the discovery of marine natural products. Our work at the University of Oklahoma on the chemistry of marine organisms was founded on field observations (Ciereszko and Karns, 1973). My choice of octocorals as a group of organisms worth examining for "natural products" was based on a number of observations in the field. At Bermuda and at Bimini in the Bahamas I was struck by the abundance of gorgonians on suitable bottom, by their large size and apparent success in terms of freedom from predation and in competition for hard bottom. It was the pronounced "interesting" odour of Eunicea mammosa that prompted me to collect specimens of the gorgonian and to extract it for "terpenes." The initial extraction, in 1955, led to the crystallization of "eunicin," the first of a long series of cembranolides to be discovered in octocorals. Steam distillation of the extracts yielded sesquiterpene hydrocarbons.

Examination of other species of gorgonians yielded the various sesquiterpene hydrocarbons reported by W.W. Youngblood in his Ph.D. dissertation (1969). By 1960 I had found at least six different substances displaying strong carbonyl absorption in the infrared in the first six species of gorgonians I had encountered in the field, substances which in time turned out to include cembranolides, polymethylenebutenolides, seco-sterols related to gorgosterol and a halogenated diterpenoid. Collection of Euniceas of the species mammosa and succinea from additional locations has led to the discovery of still other cembranolides; jeunicin, eupalmerin acetate, cueunicin and cuenicin acetate, and most recently peunicin.

Work at Bermuda in 1962 led to the observation by Attaway and myself that water draining from an aquarium containing freshly collected Plexaura homomalla into a concrete tank containing Panulirus argus caused distress in these spiny lobsters. This observation and the one of the ejection of "brown clouds" by P. homomalla when disturbed under water led to the suggestion of

P. homomalla as an object for chemical study in our group and resulted in the discovery of the occurrence of prostaglandins in this gorgonian by Spraggins and Weinheimer.

The discovery of palytoxin in Palythoa, a zoanthid common in various coral reef areas, has been made independently at least three times. We collected Palythoa because we found it flourishing in what I consider a difficult environment and because we found that it recovered from rather serious damage without being eaten or overgrown.

Our concern for the fate of sterols in predator-prey pairs led to the discovery of two new C_{31} sterols, tentatively identified as 4-methylgorgosterol and 4-methylgorgostanol, as well as the new diterpenoid xenicin in the soft coral Xenia elongata.

Other examples of the role of field observation and experiment in the discovery of marine natural products could be given. Pigments are of obvious interest to the chemist. Odours are due to chemicals and provide their own guide to isolation. Texture, feel, such as sliminess, changes in the appearance of freshly collected material on standing or upon maceration are worth noting in the field. Interaction between neighbouring species may yield clues.

I feel that field work may be fun. It should be recognized, when it is done properly, as "hard" work, and as an important ingredient in the education of budding natural products chemists and a proper activity of professional natural products chemists.

RECENT DEVELOPMENTS IN TERPENOID AND STEROID CHEMISTRY OF ALCYONACEA

J.C. Braekman*

Collectif de Bio-écologie, Unité de Chimie Bio-organique

Université Libre de Bruxelles, Belgium

The subclass of the Octocorallia (or Alcyonaria) is usually divided into five orders: Stolonifera, Alcyonaceae, Gorgonacea, Pennatulacea and Telestacea.

Among these, the Alcyonacea (soft corals) and the Gorgonacea (sea fans) are specially important because of their contribution to the biomass of the tropical coral reefs. This fact is well illustrated in Table 1 where octocorallian species distribution is shown for four different coral reefs.

Table 1. Octocorallian species distribution.

	Alcyonacea	Gorgonacea	Pennatu-lacea	Stolo-nifera	Teles-tacea
West Indies[1]	3	85	5	0	7
Na Thrang[2] (Vietnam)	92	1	0	0	1
New Caledonia[3]	181	16	4	2	0
Madagascar[4]	150	65	7	3	1

* Chercheur qualifie du Fonds National de la Recherche Scientifique Belge.

This distribution clearly explains why most of the shallow-water Octocorallia which have been until now submitted to a chemical investigation belong to the orders Alcyonacea or Gorgonacea, by far the most abundant and diversified.

The current situation in the chemistry of the Alcyonacean has been recently reviewed by Tursch[5]. My purpose here will be to discuss new results obtained by the group of the University of Brussels in this field.

SESQUITERPENES

The structure and the absolute configuration of lemnacarnol (1), a sesquiterpene isolated from the soft coral Lemnalia carnosa, have been established by X-ray diffraction analysis[6].

Further chemical examination of Lemnalia and Paralemnalia species yielded three novel structurally related sesquiterpenes: lemnalactone (2), 2-desoxylemnacarnol (3) and 2-desoxy-12-oxo-lemnacarnol (4) (Table 2)[7].

Table 2. Distribution of derivatives (1) to (4) into different Lemnalia and Paralemnalia species.

	(1)	(2)	(3)	(4)
L. africana	-	-	+	-
L. carnosa	+	-	-	-
L. laevis	-	-	+	-
P. digitiformis	-	+	-	-
P. thyrsoides	-	-	+	+

These structures, first proposed on the basis of the comparison of their spectroscopic properties with those of (1), have now been unambiguously proved by chemical intercorrelation[7].

Compounds (3) and (4) were correlated with lemnalactone through the lithium aluminium hydride reduction product (5), while lemnalactone was correlated with lemnacarnol following the reaction path described for (1) to (8).

It is not my intention to discuss this scheme at length, but the following facts may be pointed out.

The two diols (5) and (6),obtained on lithium aluminium hydride reduction of either derivative (3) or (4), are epimeric at C-7. In diol (5) the existence of a strong intramolecular hydrogen bond indicated by IR dilution experiments and corroborated by the behaviour of the diols in tlc, (6) being more polar than (5), implies that the secondary hydroxyl group at C-7 is cis with regard to the 1-hydroxyisopropyl group at C-6 and thus equatorial. Consequently, in compound (6), the secondary hydroxyl group is necessarily trans and axial.

When either of the two diols is oxidized by Jones reagent, the only compound formed is 2-desoxylemnacarnol. This implies that the first step of the oxidation is the formation of a keto group at C-7, followed by cyclization to give a hemiketal group. The same reaction takes place when triol (7) is oxidized to (8). Since the 1-hydroxyisopropyl group at C-6 is α, the cyclization will produce preferentially the hemiketal having the hydroxyl group in the β position. Thus in 2-desoxylemnacarnol the stereochemistry at C-7 is the same as in lemnacarnol.

The sign of the rotatory power of diol (5) obtained either from (2), (3) or (4) is always negative. The same applies to enone (8) obtained either from (1) or (7). All the compounds thus have the same absolute configuration,which is that of lemnacarnol.

The only asymmetric carbon atom for which the relative configuration remained to be determined was the carbon atom C-7 of compound (4). Its configuration was obtained by X-ray diffraction analysis[8].

The absolute configuration of lemnacarnol and its companions is antipodal to that of the related nardosinane sesquiterpenes of plant origin[9]. This constitutes a further example of the remarkable antipodal relationship between sesquiterpenes from marine Coelenterates and their corresponding terrestrial forms. However, it is interesting to notice that we have recently isolated from the soft coral Cespitularia aff. subviridis three well known aromadendrane sesquiterpenes: (+)-palustrol[10], (+)-ledol[11], (-)-viridiflorol[11]. Whereas the optical rotations of palustrol and viridiflorol have opposite signs depending on their plant or animal origin, that of ledol is always positive (Table 3). As far as we know, this constitutes the first exception to the antipodal relationship mentioned above.

Table 3. Rotatory power of palustrol, ledol and viridiflorol following their origin.

	Plant Origin[12]	Animal Origin
Palustrol	-17°	$+14^{\circ}$ [10]
Ledol	$+ 5^{\circ}$	$+ 5^{\circ}$ [11]
Viridiflorol	$+ 4^{\circ}$	$- 5^{\circ}$ [11]

DITERPENES

Diterpenes are another group of terpenoids frequently encountered in the Alcyonacea. The structures of most of them are related to the cembrane skeleton[5].

From the soft coral Lobophytum crassum, we have isolated the highly oxygenated cembrane diterpene named crassolide $(C_{26}H_{34}O_9)$[13]. Crassolide was shown to contain three acetates, an α-methylene-γ-lactone, two methyl groups on trisubstituted double bonds and an epoxide having an α-methyl group. These functions account for nine of the ten unsaturations present in crassolide. The remaining unsaturation is attributed to a ring, leading to the reasonable hypothesis that crassolide is a monocyclic diterpene having a cembrane skeleton.

The structure (9) was deduced from NMR decoupling experiments (Table 4) and the relative position of the two trisubstituted double bonds from the results of ozonolysis experiments.

The triol (13), obtained by mild hydrolysis (MeOH/K_2CO_3) of (10),was left unchanged on treatment with sodium periodate, implying that an α-glycol is not present.

Table 4. NMR spectrum of (10), (11) and (12) $(CDCl_3/TMS/\delta$ ppm).

	Crassolide (10) (100 MHz)	Monoepoxicrassolide (11) (270 MHz)	Diepoxicrassolide (12) (270 MHz)
CH_3-C-12	1.46 s	1.43 s	1.31 s
$CH_3-C=C$	1.74 s	1.51 s	1.39 s
CH_3COO	1.86 s	1.77 s	1.48 s
	2.01 s	1.98 s	1.98 s
	2.07 s	2.06 s	2.06 s
	2.14 s	2.09 s	2.08 s
H_{13}	2.83 d	2.61 d	2.97 d
H_1	3.17 m	3.04 m	3.15 m
H_2	4.47 dd*	4.32 dd*	5.29 ddd
H_{20a}	5.79 d	5.83 d	5.78 d
H_{20b}	6.44 d	6.40 d	6.42 d
H_{14}	–	4.93 dd	4.51 dd
	4.86-5.42 m	2.54 dd (H_5 or H_9)	3.15 –] H_5+H_9
	$(H_5+H_7+H_9+H_{11}+H_{14})$	5.07 dd] H_5 or H_9	3.48 dd]
		5.12 dd] + H_7	4.97 dd] H_7+H_{11}
		5.35 dd] + H_{11}	5.12 dd]

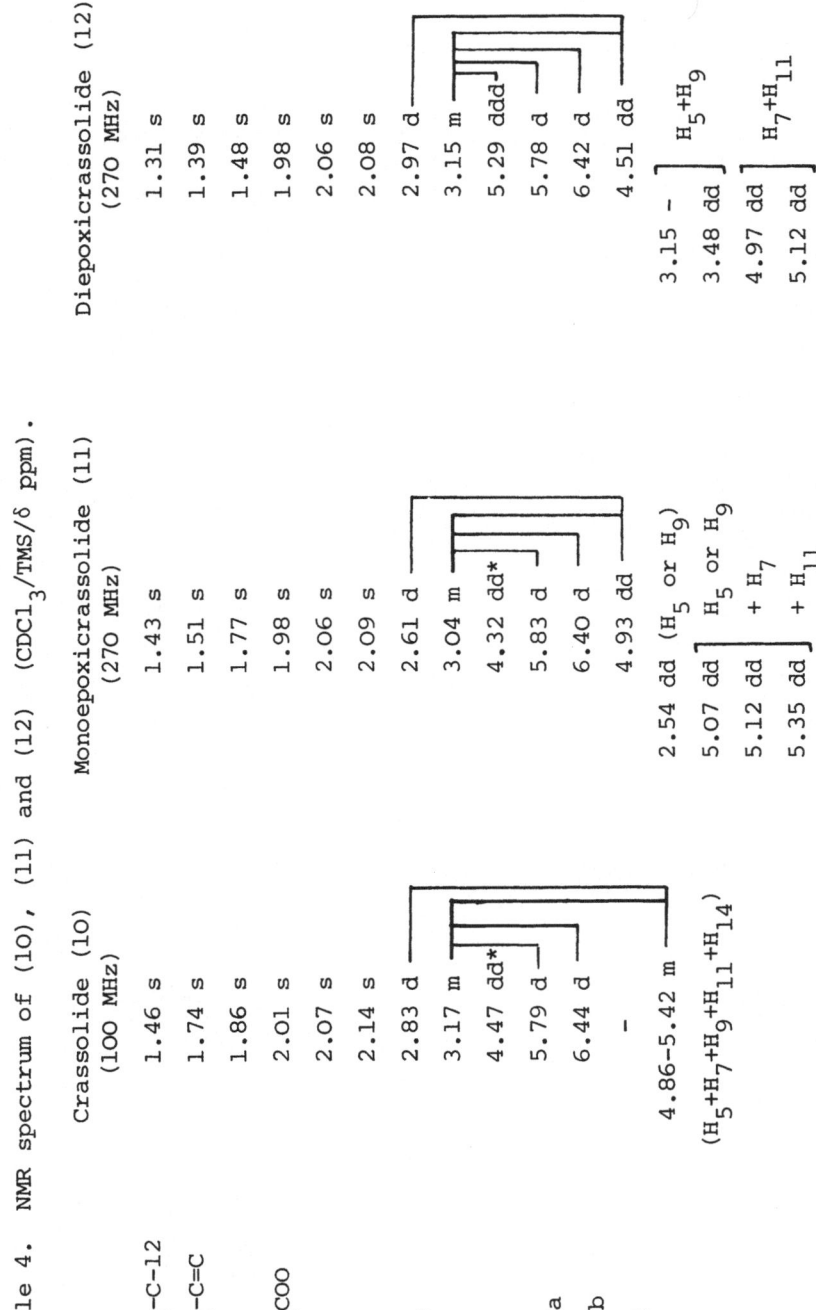

* H_2 appears as a dd and not as a ddd because the coupling constant between H_1 and H_2 is close to 0.

To locate the last two secondary acetyl groups on the cembrane skeleton, it was necessary to simplify the 5 ppm region of the NMR spectrum of crassolide so that the multiplicity of the α-protons could be observed.

Crassolide was therefore treated with m-chloroperbenzoic acid. The epoxidation reaction took place in two steps. First, a monoepoxide (M=506,(11))was formed, the NMR spectrum of which showed chemical shifts closely related to those of the starting compound. Decoupling experiments further confirmed sequence (9) (Table 4). At this stage the position of one of the acetate groups could be determined since there was no coupling between the proton at C-5 (or C-9) and one of the protons appearing at about 5 ppm. When the epoxidation reaction was carried out for a longer time, a diepoxide (M=522,(12)) was formed. In this case unexpected changes occurred in the NMR spectrum (Table 4), showing the existence of a coupling between the lactonic proton at 4.51 ppm and the doublet of the epoxide proton at C-13. Concurrently there was obviously no coupling between any of the α - acetoxy protons and any of the α-epoxy protons.

All these observations could only be explained if we admit
that during the second epoxidation, a transesterification similar
to that described in Figure 1 had occurred. Examples of this kind
of acyl migration, although not frequent, are well known[14]. In
conclusion, structure (10) agrees with all the observed spectral
and chemical data of crassolide.

POLYHYDROXYLATED STEROLS

Most of the marine naturally occurring polyhydroxylated
steroids so far described have been reported from starfish and
soft corals where they coexist with complex mixtures of the common
3-β-hydroxysterols.

A previous study of the chemical content of the soft coral
Litophyton viridis[15] had yielded the cembrane diterpenes (-)-
nephtenol and 2-hydroxynephtenol. Further work on this animal
has led to the isolation, amongst others, of two novel and interesting compounds.

On the basis of the spectral data it was deduced that the
more polar compound of the two was a 24-methylenecholestane
bearing a trisubstituted double bond, two secondary alcohols (one
of them probably at C-3) and, unexpectedly, one primary alcohol
(probably located at C-19).

Since only small amounts of this compound were available,
its structure was solved by single-crystal X-ray diffraction
analysis leading to structure (14). The less polar compound was
found to be the 7-monoacetate derivative of (14).

14

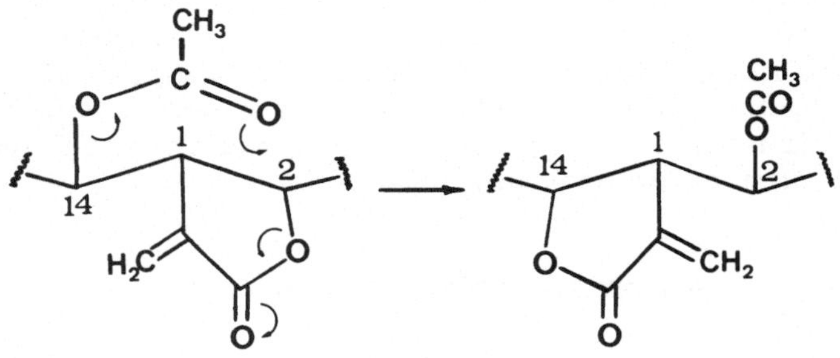

Figure 1.

Biogenetically, these derivatives are very interesting because they are potential precursors of the economically important
19-norcholesterols, compounds which have been encountered recently
by Djerassi et al.[16] in the gorgonian Plexaura homomalla.

In Table 5 we have listed all the polyhydroxylated sterols
isolated so far from soft corals. It is interesting to notice
that: (a) all are hydroxylated derivatives of 24-methyl or 24-
methylene-cholesterol; (b) these two sterols are by far the major
compounds of the sterol mixture present in soft corals living in
symbiosis with zooxanthellae[17]; (c) the sterol composition of the
entire animals is identical to that of the isolated zooxanthellae[17];
(d) the whole animal is able to incorporate labelled carbon dio-
xide into its sterol fraction[17]; (e) the absolute configuration at
C-24 of 24-methylcholesterol isolated from the soft coral
Capnella imbricata is S[17]. The same applies to lobosterol (24S-
methyl-cholestane-3β ,4β ,5β ,25-tetrol-6-one 25-monoacetate)
isolated from Lobophytum pauciflorum[18].

It is well known that 24-methylenecholesterol is abundant in
many unicellular algae[19a]. Moreover, it is also known that, in
general, algae produce sterols with the 24S configuration while
in most higher plants the sterols have the 24R configuration[19b].

Considering all these facts it seems reasonable to assume
that in the soft coral/zooxanthellae symbiosis, the algae are
responsible for the biosynthesis of the C-28 sterols and that the
zooxanthellae are the precursors of the polyhydroxylated sterols
and that hydroxylation may be a way for the soft corals to
metabolize an excess of C-28 sterols.

Table 5. Polyhydroxylated sterols isolated from soft corals.

Sterol	Source
24 ξ-methylcholestane-3β,5α,6β,12β,25-pentol 25-monoacetate	Sarcophyton elegans[20]
25-acetoxy-24 ξ-methyl-3β,5α,6β-trihydroxycholestane	Sarcophyton elegans[21]
25-hydroxy-24 ξ-methylcholesterol	Sinularia mayi[22]
	Lobophytum crassum[25]
24S-methylcholestane-3β,4β,5β,25-tetrol-6-one 25-monoacetate	Lobophytum pauciflorum[18]
24 ξ-methylcholestane-3β,5α,6β-triol	Sinularia dissecta[23]
	Lobophytum crassum[26]
24-methylenecholestane-3β,5α,6β-triol	Sinularia dissecta[23]
	Lobophytum crassum[26]
24 ξ-methylcholestane-3β,5α,6β-triol 6-monoacetate	Sinularia dissecta[23]
24-methylenecholestane-3β,5α,6β-triol 6-monoacetate	Sinularia dissecta[23]
24-methylenecholest-5-en-3β,7β,19-triol	Litophyton viridis[24]
24-methylenecholest-5-en-3β,7β,19-triol 7-monoacetate	Litophyton viridis[24]

REFERENCES

1. F. M. Bayer, <u>The Shallow-Water Octocorralia of the West-Indian Region</u>, ed., M. Nyhoff, The Hague, 1961.

2. A. Tixier-Durivault, Extrait des Cahiers du Pacifique, <u>14</u> (1970).

3. A. Tixier-Durivault, <u>Faune de Madegascar. XXI: Octocoralliaires</u>, O. R. S. T. O. M., Paris, 1966.

4. A. Tixier-Durivault, <u>Les Octocoralliaires de Nouvelle-Caledonie</u>, Laboratoire de Malacologie, Museum National d'Histoire Naturelle, Paris.

5. B. Tursch, Pure Appl. Chem., in press.

6. B. Tursch, M. Colin, D. Daloze, D. Losman, and R. Karlsson, Bull. Soc. Chim. Belg., <u>84</u>, 81 (1975); R. Karlsson and D. Losman, Acta Crystallogr., <u>B32</u>, 1614 (1976).

7. D. Daloze, J. C. Braekman, P. Georget, B. Tursch, and R. J. Wells, in press.

8. D. Losman, in press.

9. H. Hikino, Y. Hikino, S. Koakutsu, and T. Takemoto, Phytochem., <u>11</u>, 2097 (1972).

10. C. J. Cheer, D. H. Smith, C. Djerassi, B. Tursch, J. C. Braekman, and D. Daloze, Tetrahedron, <u>32</u>, 1807 (1976).

11. J. C. Braekman, D. Daloze, and B. Tursch, unpublished data.

12. G. Ourisson, S. Munavalli, and C. Ehret, <u>Selected Constants: Sesquiterpenoids</u>, Pergamon Press, 1966.

13. H. Dedeurwaerder, J. C. Braekman, D. Daloze, and B. Tursch, in press.

14. P. de Mayo, <u>Molecular Rearrangements</u>, Vol. 2, Interscience, 1964.

15. B. Tursch, J. C. Braekman, and D. Daloze, Bull. Soc. Chim. Belg., <u>84</u>, 767 (1975).

16. S. Popov, R. M. K. Carlson, A. M. Wegman, and C. Djerassi, in press.

17. M. Kaisin, J. C. Braekman, and B. Tursch, unpublished results.

18. B. Tursch, C. Hootele, M. Kaisin, D. Losman, and R. Karlsson, Steroids, 27, 137 (1976).

19. (a) M. Barbier, Introduction a l'Ecologie Chimique, Masson, 1976; (b) G. W. Patterson, Lipids, 6, 120 (1971).

20. J. M. Moldowan, W. L. Tan, and C. Djerassi, Steroids, 26, 107 (1975).

21. J. M. Moldowan, B. M. Tursch, and C. Djerassi, Steroids, 24, 387 (1974).

22. J. P. Engelbrecht, B. Tursch, and C. Djerassi, Steroids, 20, 121 (1972).

23. M. Bortolotto, J. C. Braekman, D. Daloze, and B. Tursch, Bull. Soc. Chim. Belg., 85, 27 (1976).

24. M. Bortolotto, J. C. Braekman, D. Daloze, D. Losman, and B. Tursch, Steroids, in press.

25. P. Wautelet, J. C. Braekman, D. Daloze, and B. Tursch, unpublished results.

26. M. Bortolotto, J. C. Braekman, D. Daloze, and B. Tursch, unpublished results.

CEMBRANE DERIVATIVES FROM SARCOPHYTUM GLAUCUM

Y. Kashman

Department of Chemistry, Tel-Aviv University

Tel-Aviv, Israel

In two previous publications[1,2] we described the isolation and structure elucidation of several new cembrane derivatives isolated from the soft coral Sarcophytum glaucum. The most prominent diterpene, sarcophine (1), which crystallizes from the crude petrol-ether extract and is believed to be one of the fish repellents which protect the coral against predators, was found to be accompanied by several other closely related diterpenes, of which the most abundant is its dihydrofuran analog (2). The relative quantities of compounds (1) and (2) change remarkably from batch to batch (0 to 4%, dry weight, of each one of them). The factors governing changes in the relative amounts are as yet unknown.

The structure of (1) was unequivocally established by an X-ray analysis[1]; however, its absolute configuration could not be deduced in a straight-forward manner. The usual method for such determinations is the utilization of CD measurements, which, in the case of (1), could be expected to disclose the chirality in the vicinity of the endo α,β-unsaturated lactone. We felt, however[1], that Snatzke[3] and Beecham's[4] empirical correlation between the sign of the n-π^* Cotton effect (C.E.) and the C=C-C=O chirality is ambiguous when

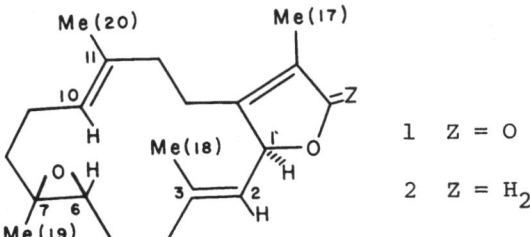

Figure 1. Compounds (1) and (2)

17

Table 1. CD data of (1), two of its derivatives, and isomers (3)
and (4)[+,1,2]

Compound	$n \to \pi^*$(λ_{nm}, $\Delta\varepsilon$)		$\pi-\pi^*$(λ_{nm}, $\Delta\varepsilon$)	
(1)	246	−6.5	223	+22.0
10,11-dihydro-(1)	245	−4.1	225	+13.3
2,3,10,11-tetrahydro-(1)	inflection		216	+ 6.4
(3)	268 inflection (0.6)		224	+ 6.0
(4)	+		233	+ 9.4

*
 Taken in MeOH

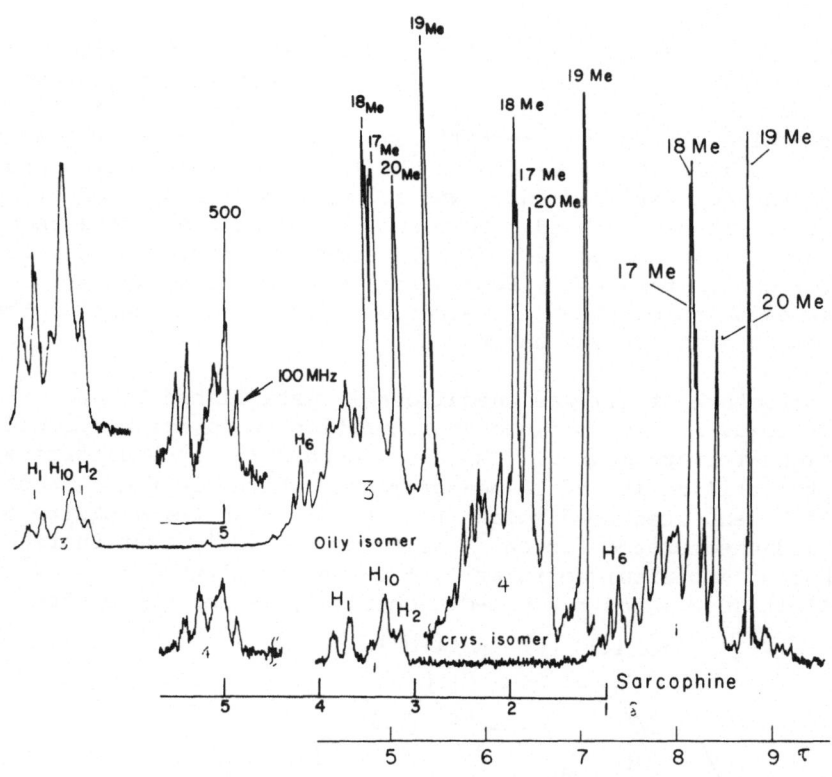

Figure 2. NMR spectra of (1), (3), and (4)

applied to butenolides, as the ring atoms of most butenolides are
coplanar (X-ray analysis). Indeed, from Table 1 it can be seen that
the measured n-π* C.E. of 2,3,10,11-tetrahydro-sarcophine is almost
zero and therefore the n-π* C.E. of (1) cannot be related directly
to the C=C-C=O chirality. We concluded that the C-1, γ-carbon,
chirality, rather than the C=C-C=O one, determines the sign of the
C.E. Furthermore, Uchida and Kuriyama came to the same conclusion[5].
On the basis of a series of suitable butenolides, they were able to
propose a new empirical rule which correlates the π-π* C.E. to the
γ-carbon chirality. According to this rule, the absolute configuration of (1) is as shown in Figure 1.

A challenging, as yet unsolved problem is the structure of two
isomers (one oily (3), the other crystalline (4))found accompanying
sarcophine in the soft coral. The isolation and the mutual interconversion of (3) and (4) have already been described[2]. The almost
identical NMR of (1), (3) and (4) (Figure 2) and the same IR and mass
spectra suggested only minute structural differences between the
three. The CD curves of (3) and (4) (Table 1), which, contrary to
(1) possess a positive n-π* C.E., may be interpreted as the epimeric
C-1 configuration; however, in the light of Kuriyama's rule, the
π-π* C.E. of the same sign in all three compounds casts doubt on this
suggestion, thereby leaving the structure problem a matter of debate.

Table 2. CMR spectra of (1) and (4)

	Sarcophine			Isomer (4)	
1	78.8	4	36.3*		
2	120.8	8	37.4*	2	+0.7
3	122.9	12	38.9		
6	61.4	5	25.3	6	-0.7
7	59.9	9	23.3	7	-0.7
10	125.0	13	27.4		
11	135.4	17	8.9		
14	162.3	18	15.5		
15	143.9	19	16.1		
16	173.0	20	17.2	all others	< ± 0.3

* Signals may be reversed

Table 3.   H-Relaxation time (T_1) of (1) and (4) (sec)

H	(1)	(4)	%
1	2.0	2.6	+30
10	1.9	2.1	+10
2	2.2	2.8	+27
6	2.5	3.0	+20
18	2.0	1.9	-10
17	2.2	2.3	+ 4
20	2.2	2.3	+ 4
19	1.7	2.0	+17

$(T-180^\circ-t-90^\circ)$, 28° 0.3 M in d_6-acetone

The ^{13}C NMR spectra of compounds (1) and (4) (as well as the ^{1}H NMR) indicates only very delicate changes in the C-1 and C-6,C-7 environments.   The identification of the various carbon signals in the CMR spectrum of (1) (Table 2) was based mainly on chemical shift considerations and comparison with the spectra of (2) and 10,11-epoxy sarcophine.

The relaxation times (T_1) of eight different, easily measured protons were measured by the inversion-recovery method.   The largest differences in T_1 (Table 3) were found for H_1 and H_2 and to some smaller extent for H_6 (a value of up to 20% is predicted for the error).   This result may strengthen the C-1 epimer differentiation between (1) and (4).

A similar problem exists with (5), the closely related isomer of (2); moreover, the situation with this pair was even more complicated because the compounds are unstable and lack CD-observable chromophore.   For the time being, we are trying to find chemical transformations delicate enough to aid in solving the problem.   In addition, more polar compounds[1] found in the crude coral extract may also assist in clarifying this matter.

REFERENCES

1.   J. Bernstein, U. Shmueli, E. Zadock, Y. Kashman, and I. Neeman, Tetrahedron, 30, 2817 (1974).

2.   Y. Kashman, E. Zadock, and I. Neeman, Tetrahedron, 30, 3615 (1974).

3.   G. Snatzke, Tetrahedron, 21, 413, 421, 439 (1965).

4. A. F. Beecham, Tetrahedron Letters, 1669 (1972).

5. I. Uchida and K. Kuriyama, Tetrahedron Letters, 3761 (1974).

THE CHEMISTRY OF SOME OPISTHOBRANCH MOLLUSCS

D. John Faulkner and Chris Ireland

Scripps Institution of Oceanography

La Jolla, California, U.S.A. 92093

During the course of our studies on marine natural products, we have come to regard the opisthobranch molluscs as a most reliable source of new and interesting metabolites. The opisthobranchs are not the primary source of these compounds, which are generally algal metabolites. We have shown that the sea hare Aplysia californica, collected in La Jolla, stores halogenated metabolites from Laurencia pacifica, Laurencia subopposita, Plocamium cartilagineum and Plocamium violaceum[1]. We have also isolated compounds which have not been found in local algae, but these compounds were closely related to the known algal metabolites[2]. We have found some evidence of chemical reactions occurring in the digestive gland, but we found no obvious catabolic products[3]. The occurrence of halogenated metabolites in the skin of A. californica suggests that these compounds may comprise a chemical defence system for an otherwise vulnerable organism.

We have collected a number of opisthobranch molluscs from the Gulf of California. The opisthobranch mollusc Dolabella californica was collected at the same location in April 1975 and April 1976. The digestive glands were excised and homogenised in acetone. The acetone extracts were subjected to chromatography on florisil to obtain a series of diterpenes. The first collection yielded six diterpenes, while the second collection gave twelve diterpenes, of which four were common to both collections. The results are summarised in Table 1. From the molecular formulae, the infrared spectra, and the pmr spectra, it was immediately apparent that the compounds were diterpene alcohols and the corresponding acetates. On the basis of pmr spectra, we have divided the compounds into four groups.

Table 1.   Diterpenes isolated from <u>Dolabella californica</u>

Collection			1975	1976
Number of animals			40	100
Total weight of digestive gland extracts			16 g	65 g
Compound No.	Molecular Formula	mp (^{o}C)	Wt (g)	Wt (g)
1	$C_{24}H_{38}O_4$	oil	0.25	1.69
2	$C_{22}H_{36}O_3$	78	0.5	6.0
3	$C_{20}H_{34}O_2$	152-3	-	0.4
4	$C_{26}H_{40}O_6$	oil	-	0.6
5	$C_{24}H_{38}O_5$	136-7	-	1.94
6	$C_{24}H_{38}O_5$		1.5	-
7	$C_{24}H_{38}O_5$		-	0.4
8	$C_{22}H_{36}O_4$		1.75	7.0
9	$C_{20}H_{34}O_3$	168-9	1.3	-
10	$C_{24}H_{38}O_5$	oil	-	0.5
11	$C_{22}H_{36}O_4$	153-4	-	1.5
12	$C_{22}H_{36}O_4$		-	1.0
13	$C_{20}H_{34}O_3$	157-8	-	1.5
14	$C_{20}H_{34}O$	oil	0.1	0.95
		Total recovery	5.4(34%)	23.5(36%)

On treatment with lithium aluminium hydride in dry ether, diacetate (1) and monoacetate (2) were converted into the diterpene diol (3). Although the structure and relative configuration of the monoacetate (2) was eventually determined by X-ray analysis[4], the spectral data clearly indicated that we were dealing with a diterpene having an unknown carbon skeleton. The pmr spectrum of (2) contained 6 methyl signals: a singlet at δ 2.05 due to the acetate, a singlet at 1.62 due to a vinyl methyl, two singlets at

1.25 and 1.18 assigned to methyl groups on carbon bearing oxygen,
a doublet (J = 7 Hz) at 0.94, and a singlet at 0.82 ppm. The low-
field region of the spectrum contained an α-acetoxy proton at
δ 4.81 (J = 9, 1, 1 Hz), a broad triplet at 5.10 (J = 7 Hz) due to
an olefinic proton, and two signals at 5.07 (d, J = 16 Hz) and
5.22 ppm (dd, J = 16, 9 Hz) due to a _trans_ disubstituted olefin.
The cmr spectrum confirmed the presence of the disubstituted and
trisubstituted olefinic bonds (δ 135.1, 130.9, 127.8 and 126.8
ppm) and indicated only two other tetrasubstituted carbon atoms,
one at 72.7 ppm bearing hydroxy and two methyl groups, the other
at 20.8 ppm bearing a methyl group at a ring junction. Irradiation at δ 2.32 in the pmr spectrum caused the double doublet at
5.22 to collapse to a doublet and the doublet at 0.94 to become a
sharp singlet, indicating that the _trans_ olefinic bond must be
positioned in a ring between a methine carbon atom bearing a
methyl group and a tetrasubstituted carbon atom bearing a methyl
group. Since these partial structures could not be accommodated
by any known bicyclic diterpene skeleton, we submitted the crystalline monoacetate (2) to Dr. J. Clardy for X-ray analysis. The
structure, shown in Figure 1, is in complete accord with all
spectral data. Thus compounds (1) and (3) must be the corresponding diacetate and diol, respectively.

     The next group of compounds, a triacetate (4), three diace-
tates (5) to (7), and a monoacetate (8), were all converted by
treatment with lithium aluminium hydride in ether into the triol
(9), which was also identified as a natural product. The major
constituent from both collections was the monoacetate (8). The
pmr spectrum of (8) was compared with that of monoacetate (2),
with which it shared many similar features. The pmr spectrum of

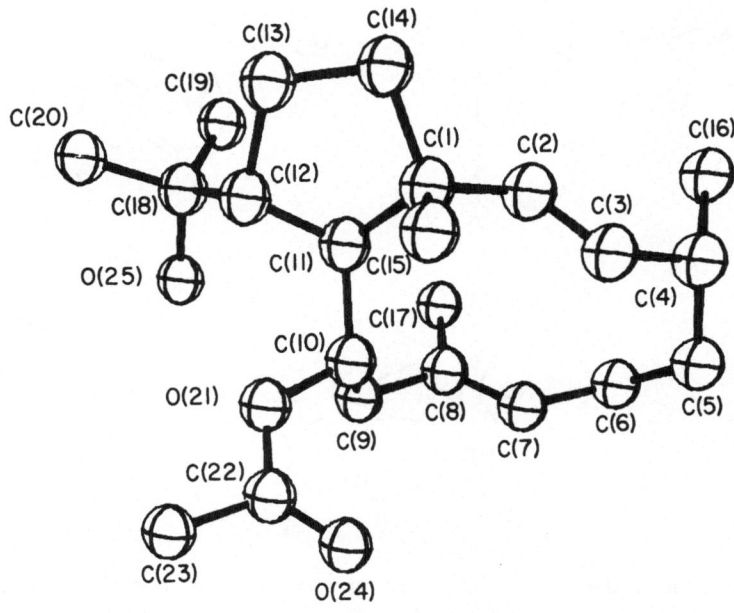

Figure 1. A computer generated perspective drawing of (2).
Hydrogens are not shown and no absolute stereochemistry is implied.

(8) contained 6 methyl signals: a singlet at δ 2.02 due to the
acetate, two singlets at 1.55 and 1.50 due to methyl groups on a
carbon atom bearing an acetate, a singlet at 1.20 due to a methyl
on a tetrasubstituted carbon atom bearing a hydroxyl, a singlet at
1.05, and a doublet (J = 7 Hz) at 0.95 ppm. The low-field region
of the spectrum contained an α-hydroxy proton at δ 3.94 with
doublets at 5.00 (J = 16 Hz) and 5.48 (J = 16 Hz) and a two-proton
multiplet at 5.18 ppm, all due to olefinic protons. In the pmr
spectrum of the corresponding triol (9), the four discrete olefinic signals at δ 4.92 (d, J = 16 Hz), 5.09 (dt, J = 16, 8, 8 Hz),
5.32 (dd, J = 16, 10 Hz) and 5.75 ppm (J = 16 Hz) were assigned to
two _trans_ disubstituted olefinic bonds on the basis of the observed
coupling constants. The major differences between the pmr spectra
of the diol (3) and the triol (9) could be explained by replacing
the trisubstituted olefinic bond in (3) by a _trans_ allylic tertiary alcohol moiety.

In order to confirm this assumption, the monoacetate (8) was
esterified with acetic anhydride in pyridine at room temperature
to obtain, after purification, the diacetate (5), identical to the
natural product. On treatment with phosphorus tribromide and
pyridine in hexane at -40°, the diacetate (5) gave the rearranged
allylic bromide (15) in ~80% yield. Reduction of the bromide
(15) with lithium in THF containing t-butanol gave the diol (3),
identical to the natural product, as the major product. Reduction
of the bromide (15) with lithium aluminium hydride in ether did
not result in the formation of the diol (3), suggesting that the
geometry of the medium-sized ring is such that steric hindrance
prevents a concerted displacement of bromide.

As a result of this experimental sequence, we were able to
define the structure and stereochemistry of the triol (9) with the
exception of the stereochemistry at C-8. The monoacetate (8) must
have the acetoxy group at C-18, since the two adjacent methyl
groups show pmr signals at δ 1.50 and 1.55 ppm. The three diacetates (5) to (7) and the triacetate (4) all have spectral data
completely in accord with the assigned structures.

A third group of compounds consist of a diacetate (10), two
monoacetates (11) and (12) and a triol (13). The acetates (10) to
(12) were all converted into the triol (13) by treatment with
lithium aluminium hydride in ether. The pmr spectra of the triol
(13) was similar to that of the diol (3), except in the low-field
region, which contained signals at δ 5.33 (dd, J = 16, 7 Hz) and
4.90 (d, J = 16 Hz) due to the trans disubstituted olefin, together with a doublet at 5.08 (J = 9 Hz) which was coupled to a
multiplet at 4.50 ppm. In the diacetate (10), the similar multiplet was located at 5.60 ppm, suggesting that the signals were
due to allylic α-hydroxy and α-acetoxy protons, respectively.
Comparison of the pmr spectra of (10) to (13) with that of the
bromide (15) led to the proposal that all five compounds had the
same carbon skeleton and substitution pattern. In order to confirm

this hypothesis, the bromide (15) was treated with silver acetate
in acetic acid at room temperature to obtain a mixture of two
triacetates which could not be separated by chromatography. Reduction of the crude reaction product with lithium aluminium hydride
in ether gave a 60:40 mixture of the triols (9) and (13), which
were separated and shown to be identical to the natural products.

The two products result from addition of acetate to an allylic
carbonium ion. Since the geometry of the olefinic bonds in both
products is trans and the acetate must approach from outside the
ring, there are two pairs of products which can arise, depending
on the alignment of the allylic carbonium ion with respect to the
medium-sized ring. If the vinyl methyl lies below the ring, the
triols (16) and (17) would be expected, whereas the triols (18)
and (19) would result from the alternative geometrical arrangement.
On the basis of examining molecular models and analysing pmr data,
we favour the former situation and propose structure (16) for
triol (9) and structure (17) for triol (13).

The final compound isolated was a mono-alcohol. On the basis
of analysis of the pmr spectrum, we have assigned the structure
(14). We have not confirmed this structural assignment.

Dolabella californica was not observed feeding and could only
be observed and collected at night. They were collected in an
area which was devoid of macroalgae, and they must therefore be
presumed to feed on a microalgal film. In general, the compounds
found in the digestive glands of opisthobranch molluscs have
reflected the dietary preference of the organism.

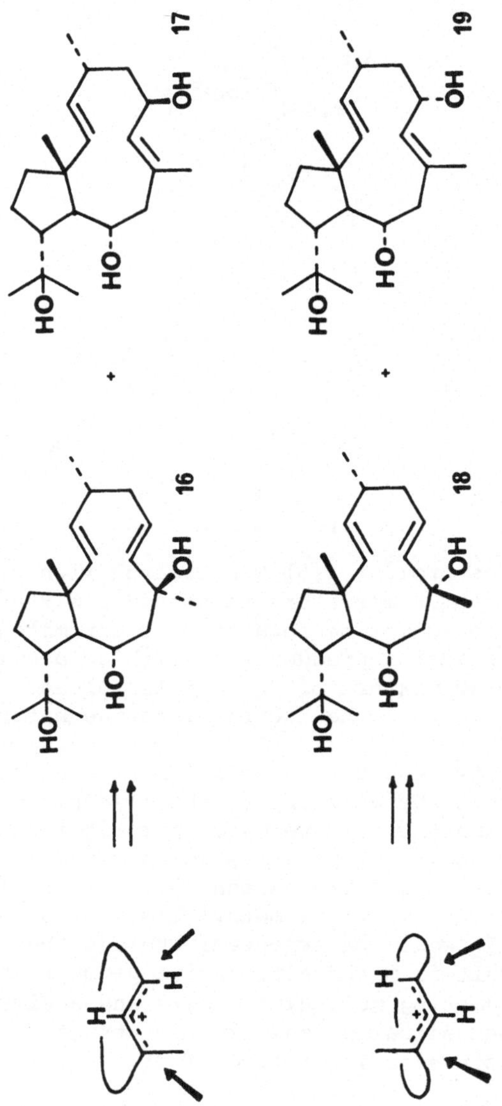
16
17
18
19

In addition to our published research on Aplysia californica, we have isolated johnstonol (20), a metabolite of Laurencia johnstonii, from A. californica collected from Isla Ildefonso in the Gulf of California. We also collected A. californica in a brackish pond on Isla Carmen, Gulf of California, and did not find any halogenated metabolites in the digestive gland. We have isolated pachydictyol A (21), a metabolite of Pachydictyon coriaceum, from A. vaccaria[5]. Minale et al.[6] have isolated dictyol A (22) and dictyol B (23), also found by Fattorusso et al.[7] in Dictyota dichotoma var. implexa, from A. depilans. Schmitz et al. have isolated dactylyne (24)[8] and three sesquiterpene ethers[9] from A. dactylomela, and Schulte and Fenical[10] have isolated all of these compounds from Laurencia poitei. Pettit et al.[11] have isolated dolatriol (25) and a monoacetate derivative (26) from Dolabella auricularia but do not comment on a possible dietary origin. Kato and Scheuer[12] have isolated aplysiatoxin (27) and debromoaplysiatoxin (28) from Stylocheilus longicauda, and Mynderse and Moore[13] have recently found that debromoaplysiatoxin (28) occurs in a blue-green alga, Lyngbya gracilis.

We have isolated aeroplysinin-2 (29) and two brominated dienones (30) and (31) from collections of Tylodina fungina. Since Tylodina feeds exclusively on sponges of the genus Verongia, it is not surprising that Tylodina contains Verongia metabolites and may use these compounds as deterrents to predation. Burreson et al.[14] have isolated an isonitrile, 9-isocyanopupukeanane (32), from the defensive secretion of the nudibranch Phyllidia varicosa and found that the source of this compound was the sponge Hymeniacidon sp.

These results suggest that the opisthobranch molluscs are among the most interesting marine organisms from the viewpoints of chemical communication and marine natural products chemistry. Their ability to concentrate the more interesting compounds from their diet is a most endearing capability[15].

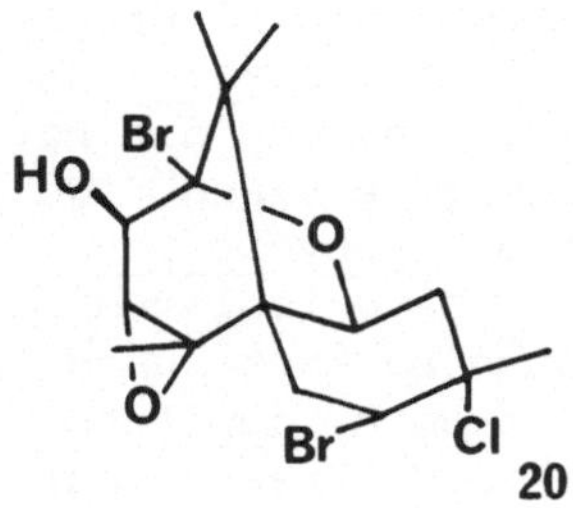

20

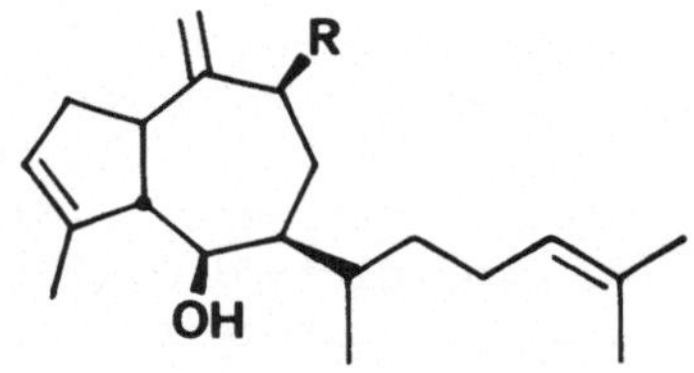

21 R = H
23 R = OH

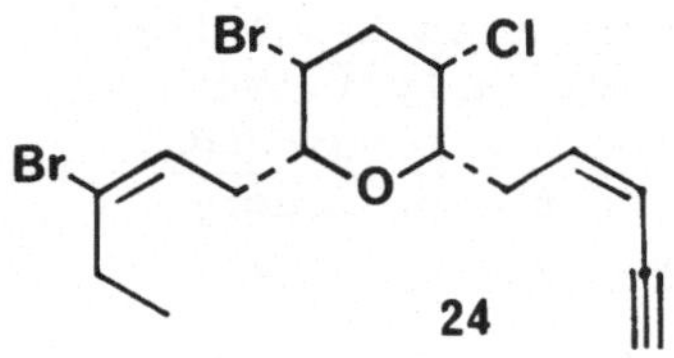

24

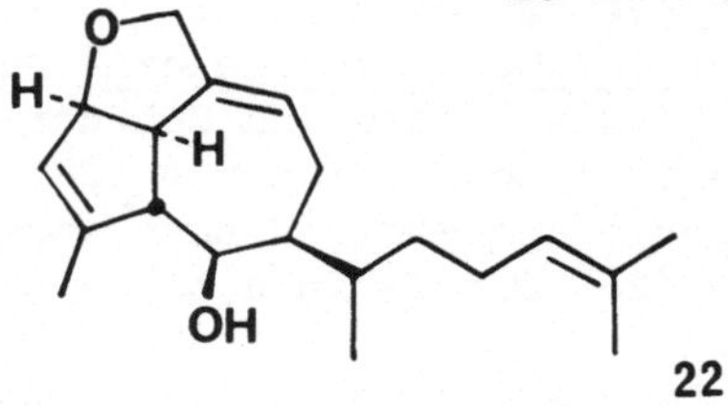

22

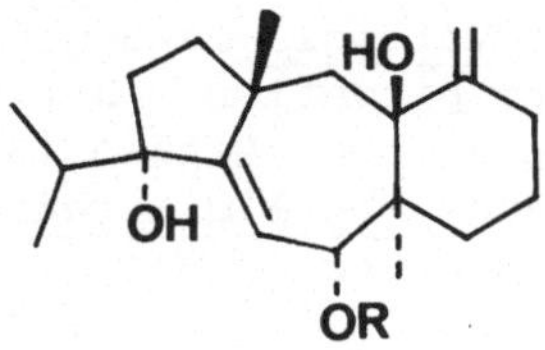

25 R = H
26 R = Ac

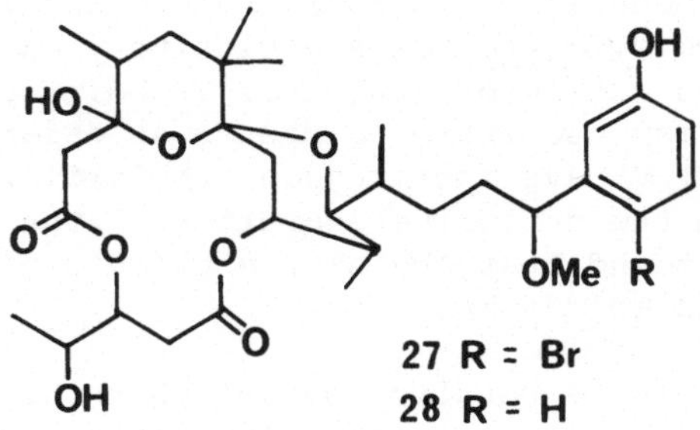

27 R = Br
28 R = H

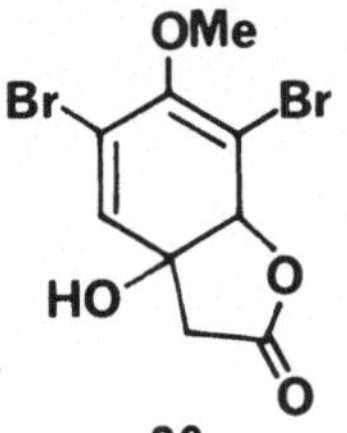

29

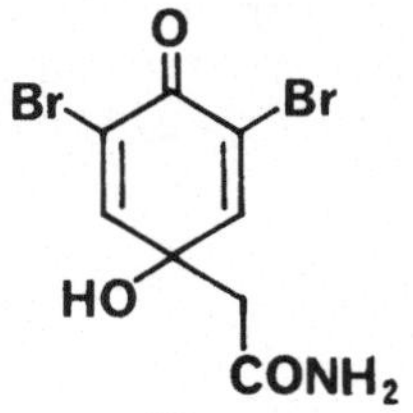

30

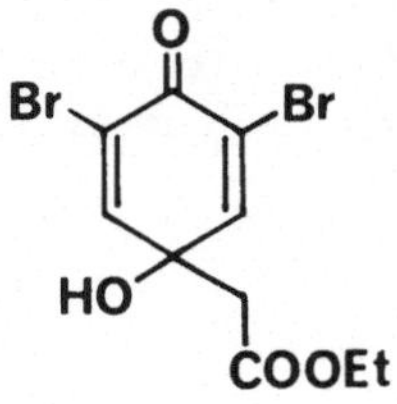

31

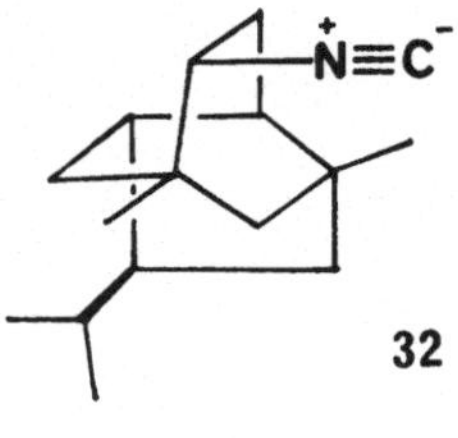

32

Table 2

Opisthobranch	Compound	Dietary source
Aplysia californica (La Jolla)	halogenated sesquiterpenes halogenated monoterpenes	Laurencia species Plocamium species
A. californica (Isla Ildefonso)	johnstonol	Laurencia johnstonii
A. californica (Isla Carmen)	no halogenated products	no marine algae available
A. vaccaria	pachydictyol A	Pachydictyon coriaceum
A. depilans	dictyol A, dictyol B	Dictyota dichotoma
A. dactylomela	dactylyne dactyloxene A, B, C	Laurencia poitei Laurencia poitei
Dolabella californica	diterpenes	unknown
D. auricularia	diterpenes (dolatriol)	unknown
Stylocheilus longicauda	aplysiatoxin debromoaplysiatoxin	Lyngbya gracilis
Tylodina fungina	aeroplysinin-2 brominated dienones	Verongia species Verongia species
Phyllidia varicosa	9-isocyanopupukeanane	Humeniacidon species

REFERENCES

1. M. O. Stallard and D. J. Faulkner, Comp. Biochem. Physiol., 49B, 25 (1974).

2. C. Ireland, M. O. Stallard, D. J. Faulkner, J. Finer, and J. Clardy, J. Org. Chem., 41, 2461 (1976).

3. M. O. Stallard and D. J. Faulkner, Comp. Biochem. Physiol. 49B, 37 (1974).

4. C. Ireland, D. J. Faulkner, J. Finer, and J. Clardy, J. Amer. Chem. Soc., 98, 4664 (1976).

5. D. J. Vanderah, unpublished observation.

6. L. Minale and R. Riccio, Tetrahedron Lett., 2711 (1976).

7. E. Fattorusso, S. Magno, L. Mayol, C. Santacroce, D. Sica, V. Amico, G. Oriente, M. Piatelli, and C. Tringali, J. C. S. Chem. Comm., 575 (1976).

8. F. J. McDonald, D. C. Campbell, D. J. Vanderah, F. J. Schmitz, D. M. Washecheck, J. E. Burks, and D. van der Helm, J. Org. Chem., 40, 665 (1975).

9. F. J. Schmitz and F. J. McDonald, Tetrahedron Lett., 2541 (1974).

10. G. Schulte and W. H. Fenical, unpublished observations.

11. G. R. Pettit, R. H. Ode, C. L. Herald, R. B. Von Dreele, and C. Michel, J. Amer. Chem. Soc., 98, 4677 (1976).

12. Y. Kato and P. J. Scheuer, J. Amer. Chem. Soc., 96, 2245 (1974).

13. J. S. Mynderse and R. E. Moore, personal communication.

14. B. J. Burreson, P. J. Scheuer, J. Finer, and J. Clardy, J. Amer. Chem. Soc., 97, 4763 (1975).

15. This research was supported by a grant from the National Science Foundation (GB-37227).

CONSTITUENTS OF THE HEMICHORDATE PTYCHODERA FLAVA LAYSANICA

Tatsuo Higa and Paul J. Scheuer

Department of Chemistry, 2545 The Mall

University of Hawaii, Honolulu, Hawaii

The phylum Hemichordata occupies an intermediate phyletic position between the echinoderms and the chordates. The morphological link to the chordates is a series of pharingeal clefts or gill slits, but to date no unique chemical relationship between the two phyla has been uncovered. Some members of one of the two classes of the hemichordates, the acorn worms or Enteropneusta, give off an offensive odour, which is often described as iodoform-like[1]. This is an intriguing observation since evolution of thyroid function with its iodine-containing hormone has never been traced to the hemichordates[2]. Previous attempts at uncovering thyroid-like activity in a hemichordate, where a likely biosynthetic path might begin with iodination of tyrosine[3], have included studies of iodine distribution in Balanoglossus gigas[4] and ^{131}I radiochromatography, which proved the presence of 3-iodotyrosine in Sacoglossus horsti[5].

Another interesting property of some acorn worms, in addition to their odourous secretion, is their bioluminescence[6]. Ashworth and Cormier[7], suspecting a possible relationship between bioluminescence and the odourous mucus of Balanoglossus biminiensis, isolated as a major metabolite of the animal 2,6-dibromophenol (1). These workers reported that (1) was responsible for the iodoform-like odour of B. biminiensis.

Ptychodera flava laysanica, an acorn worm that is widely distributed in the tropical Indo-Pacific, shares with other members of the class bioluminescence and an odour said to be reminiscent of iodoform and presumed to be a constituent of a defensive secretion[8]. When we examined an ethereal extract of the animal by mass spectrometry, we failed to detect iodoform or any other iodocarbon

compound, but saw evidence of a number of brominated compounds.
Subsequent separation showed the absence of 2,6-dibromophenol (1).
The simplest bromophenol that we isolated was 2,4,6-tribromophenol
(2), which was odourless, as were the more complex bromophenols
and their derivatives (vide infra). We succeeded in tracing the
odour of P. flava to a mixture of three haloindoles[9]. The principal constituent, present to the extent of 3×10^{-4}% of fresh
animal weight, was 3-chloroindole (3), which is unstable when exposed to the atmosphere. Two trace components were identified as
3-bromoindole (4) and 6-bromo-3-chloroindole (5). We have described a synthesis of compound (5)[10], which was not known prior to
our work.

    The animals (about 750 g including ingested sand) collected
at Paiko beach were macerated with acetone in a blender. The
brei was filtered, washed with acetone, and the combined acetone
was concentrated. A petroleum ether extract of the concentrate
was subjected to silicic acid chromatography. Elution with petroleum ether furnished 7 mg 2,4,6-tribromophenol (2), mp 91-94°,
after sublimation and recrystallization from carbon tetrachloride
and identified with a synthetic sample (mixed mp, ir, pmr). Petroleum ether containing 10% ether eluted 100 mg tetrabromohydroquinone (6), mp 254-255° (dec) after tlc purification and recrystallization from chloroform, identical (mp, ir, ms spectra)
with a sample prepared as described in the literature[11]. From
another collection, we isolated by methanol extraction of the
animals (about 800 g) collected at Kahala beach 241 mg tribromohydroquinone (7), mp 144°, after recrystallization from chloroform, identical (mp, ir spectrum) with a sample prepared from
hydroquinone and 3 equivalents of bromine in chloroform.

1 R = H
2 R = Br

3 R = H, X = Cl
4 R = H, X = Br
5 R = Br, X = Cl

Structural Formulas

6 R = Br

7 R = H

The methanol extraction also yielded small amounts of (6).

Elution of the silicic acid column with petroleum ether-ether (4:1) yielded after tlc and recrystallization from benzene 70 mg of fine white crystals, mp 240-242° (dec), of 2,4,5,2',3',5'-hexabromo-3,6,4'-trihydroxydiphenyl ether (8). The corresponding trimethoxy derivative (9) was prepared (diazomethane), white needles, mp 151.5-152°, after recrystallization from methanol-chloroform. The composition of (8) was established by mass spectra of (8) and (9) and by combustion data of (9). An important clue to the assigned structure was the high field resonance of the aromatic protons of (8) (δ 6.90, acetone-d_6) and of (9) (δ 6.59, CCl$_4$) as compared with tribromohydroquinone (6) (δ 7.27, acetone-d_6), its monomethyl ether (δ 7.30, acetone-d_6); 7.06, CDCl$_3$), and its di-methyl ether (δ 6.97, CCl$_4$). The sole aromatic proton must there-fore be <u>ortho</u> to the diphenyl ether linkage and must be shielded by the second aromatic ring, which is perpendicular to it because of the bulky <u>ortho</u> substituents[12]. Hydrogenolysis of (9) (10% Pd–C) yielded (10) identical (ir, pmr, ms spectra) with a sample prepared as described[13]. Attempted synthesis of (9) by demethylation of (10) to (11) and bromination led only to a pentabromoderivative, pre-sumably (12), transformable to its trimethyl ether (13), which showed in its pmr spectrum two aromatic protons resonating as a singlet at δ 7.01. The hexabromo ether (8), which has the <u>meta</u> bromine atoms already in place, could successfully be synthesized <u>via</u> (14), which was prepared from 2,5-dibromo-1,4-dimethoxybenzene and 2-bromo-4-methoxyphenol; demethylation of (14) and bromination of the free phenol yielded (8) identical (mp, ir) with the natural product. Synthetic trimethoxy (9) had identical mp, pmr, and mass spectra with (9) derived from natural (8).

The final fractions of the petroleum ether-ether (4:1) eluate, after solvent removal, tlc, and recrystallization from chloroform, yielded 12 mg of fine white crystals, mp 212-215°, to which we have assigned structure (15), 2,4,2',3',5'-pentabromo-3,6,4'-trihydroxydiphenyl ether, based on its mass and pmr spectra (δ 6.85, 7.34, both singlets) and those of its trimethoxy derivative (16), mp 148-152°, δ 3.80, 3.84, 3.86, 6.52, 7.14, all singlets.

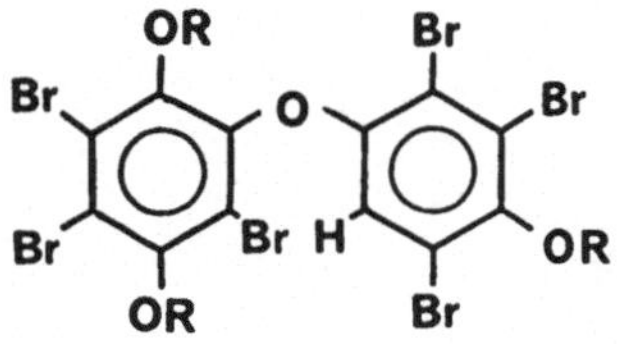

8   R = H
9   R = Me

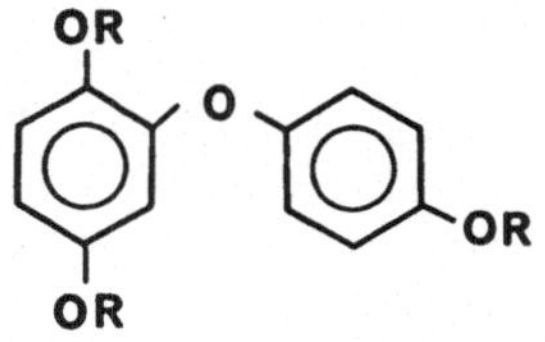

10 R = Me
11 R = H

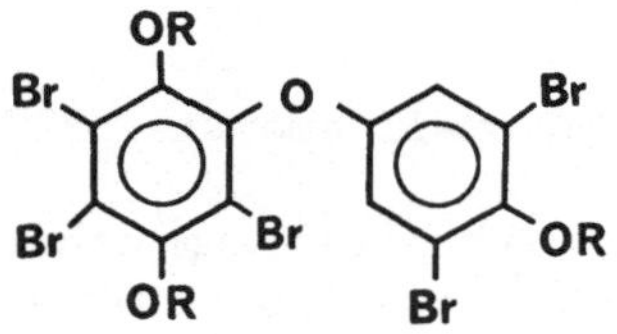

12 R = H
13 R = Me

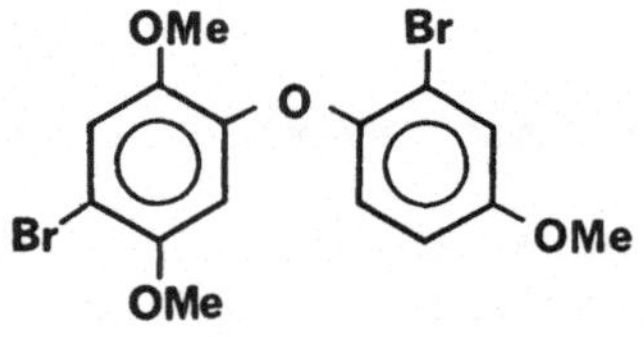

14

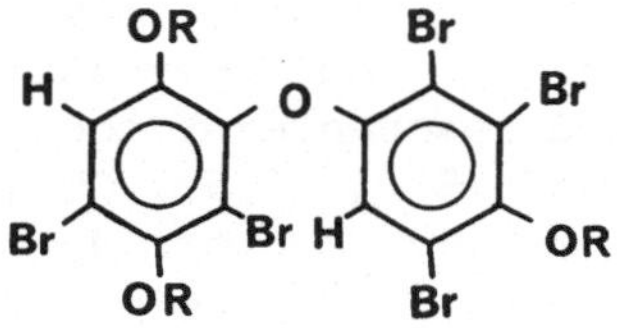

15 R = H
16 R = Me

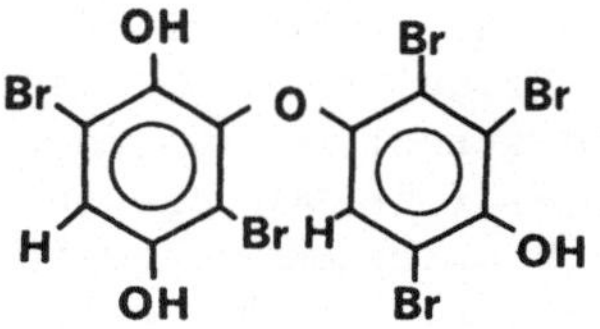

17

The pmr data, however, are equally compatible with structure (17).

Early mass spectral screening of ether extracts of the hemi-chordate had shown highly brominated compounds above m/e 900. A subsequent collection of animals yielded no (8) or (15), but elution with petroleum ether-ether (7:3) furnished 55 mg of 2,5-bis(4-hydroxy-2,3,5-tribromophenoxy)-3,6-dibromohydroquinone (18), mp 293–295° (dec), transformable to its tetramethyl ether (19), mp 248–250°, after tlc. The pmr spectrum indicated a highly symmetrical compound, and the mass spectrum proved the presence of eight bromine atoms. Hydrogenolysis of (19) furnished (20), mp 147–149°, which we synthesized from 2,5-dibromo-1,4-dimethoxybenzene with two equivalents of 4-methoxyphenol in the presence of cuprous oxide in refluxing collidine.

The high field nature (δ 6.63) of the two equivalent aromatic protons in (19) might suggest alternate formulation (21). We reject (21) on the following grounds. Since the two aromatic protons in the middle ring of (20) resonate at δ 6.64, we would expect corresponding signals of (22) at higher field than the

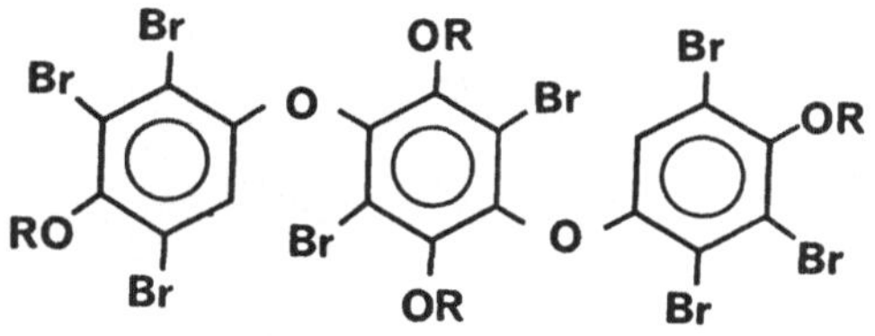

18 R = H
19 R = Me

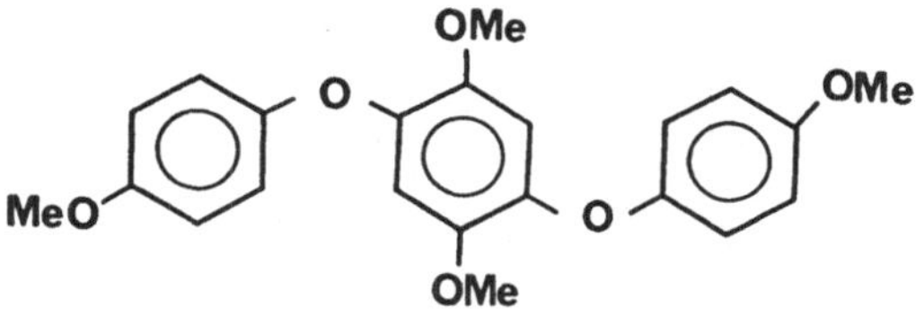

20

observed δ 6.63; biogenetic considerations favour (18); and our attempt to brominate the natural trimer resulted in complete recovery of starting material, which is unlikely in case of (21). Again, as in case of (8), the alternately formulated (21) would involve meta-bromination.

This trisphenyl ether (18) is an unprecedented natural product. Two diphenyl ethers, (23) and (24), presumably catechol derivatives, have been reported from the marine sponge Dysidea herbacea[14] and the unhalogenated diphenyl ether (25) from the brown alga Bifurcaria bifurcata[15]. Simple bromophenols, on the other hand, have been encountered frequently in red[16], occasionally in brown algae[17], and only rarely so far in other marine organisms[18].

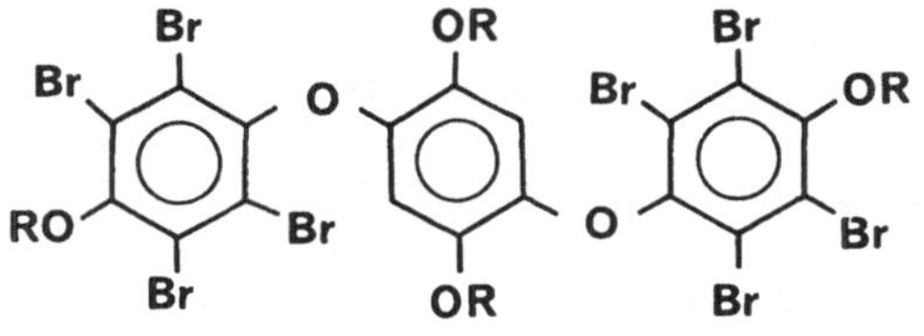

21 R = H
22 R = Me

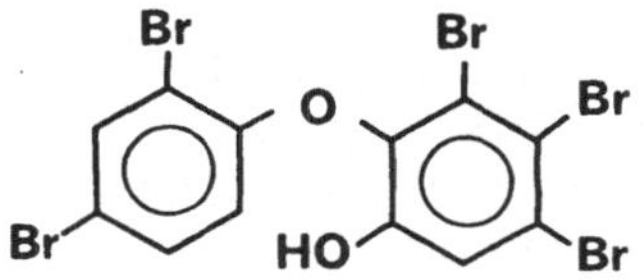

23

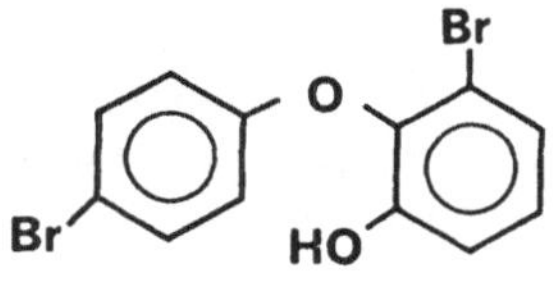

24

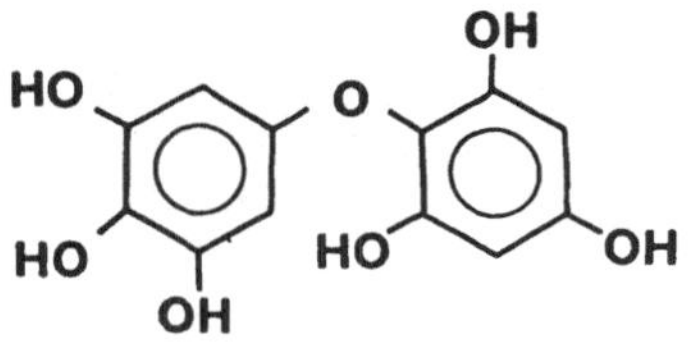

25

26

27

28

29

As has been mentioned, methanol extraction of <u>Ptychodera</u> furnished tri- (7) and tetrabromohydroquinone (6), and to our surprise four orange-red pigments. All four, (26) to (29), were known synthetic compounds and were identified by spectral comparisons. Pertinent data are summarized in the Table.

Presence of orange-red pigments in yellow animals was puzzling and, indeed, acetone rather than methanol extraction of the yellow animals failed to yield quinone pigments; we therefore suspected that all four pigments might be artefacts, derivable in principle from tribromo- and tetrabromohydroquinone, methanol, and an oxidizing agent. A likely oxidizing agent appeared to be a <u>Ptychodera</u> peroxidase since, as was mentioned, <u>Ptychodera</u> is bioluminescent, a property which it shares with other hemichordates, <u>inter alia</u>, <u>Balanoglossus biminiensis</u>. Dure and Cromier[23] had shown that the enzyme which catalyzes the bioluminescence of <u>B. biminiensis</u> is also capable of promoting the transformation of pyrogallol to purpurogallin. We therefore assumed that a similar situation might exist in <u>P. flava</u>. By a series of experiments we showed that a crude enzyme preparation is extractable into ether and that this preparation catalyzes the transformation of tetrabromohydroquinone to methoxytribromobenzoquinone. When we macerated the animals with acetone, the enzyme was apparently not extracted since we only isolated bromohydroquinones and their corresponding bis and tris-ethers (<u>vide supra</u>). We showed, however, that the oxidant once extracted remains effective in acetone solution. We also showed that methanol displacement of bromine in a bromobenzoquinone takes place, albeit poorly, in phosphate buffer without enzyme.

As the name <u>P. flava</u> suggests, these animals are yellow in colour. At one collection site, Paiko beach, the animals are distinctly green and virtually without their characteristic odour.

Table. Properties of the quinone pigments

No.	Compound	Yield, mg	Mp, °	λ_{max}^{MeOH} nm	$\nu_{C=O}^{KBr}$ cm^{-1}	Pmr δ	Ms	Color, Appearance	Ref.
26		90	171–172	207 305	1687 sh 1670 1667 1659	4.26 (s)	380 378 374 372	orange-red needles	(19)
27		33	155–157	203 307	1680 sh 1674 sh 1663 1655	4.28 (s)	330 328 326 324	orange-red crystals	(20)
28		10	182–183	205 307	1689 sh 1678 1659	4.21 (s)	328 326 324	orange-red plates	(21)
29		14	188–190	202 295	--	6.15 (s) 3.87 (s)	298 296 294	greenish-yellow prisms	(22)

We showed[24] that the pigments that are responsible for the abnormal
colour are deep-blue compounds, the well-known Murex dye Tyrian
Purple (30) and two new closely related congeners, (31) and (32).
This uncommon pigmentation, coupled with the nearly odourless
character of the Paiko animals, strongly suggested to us that in
the Paiko environment--a shallow beach that becomes nearly dry at
low tide--one of the odourous constituents, 6-bromo-3-chloroindole,
undergoes oxidative dimerization, perhaps under photochemical conditions. Support for this idea came from our isolation in the
Paiko animals of 5,7-dibromo-6-methoxyindole (33), which has no
colour or odour and which, by oxidative dimerization or by oxidative
coupling with 6-bromo-3-chloroindole, would be a rational precursor
of the two new bisindoxyl pigments.

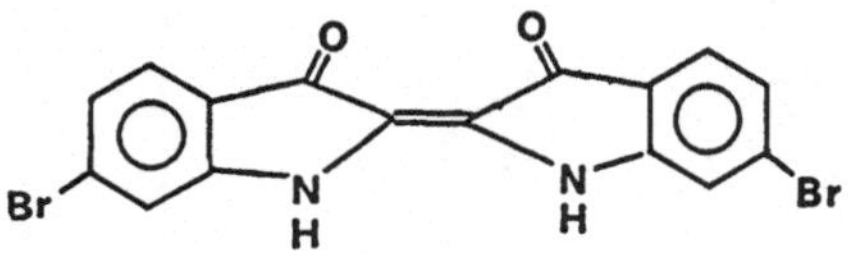

30

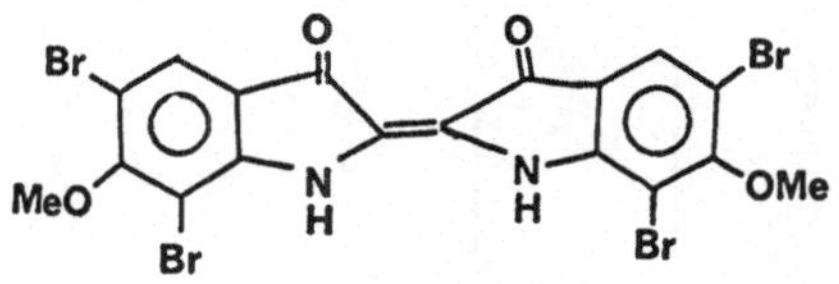

31

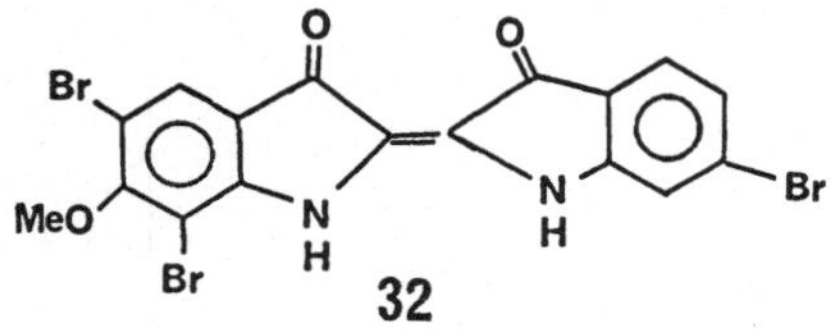

32

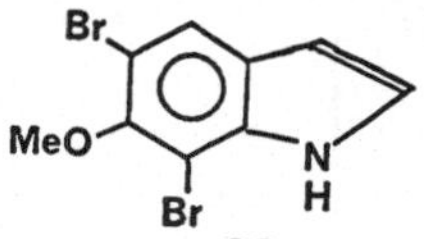

33

ACKNOWLEDGMENT

We thank the National Science Foundation for financial support.

REFERENCES

1.   L. H. Hyman, The Invertebrates, Vol. 5, McGraw-Hill, New York,
     1959, p. 143.

2.   I. M. Thomas, in The Evolution of Living Organisms, ed. G. W.
     Leeper, Melbourne University Press, 1962, p. 166.

3.   E. J. W. Barrington, Hormones and Evolution, The English
     Universities Press, London, 1964, p. 68.

4.   F. B. De Jorge, P. Sawaya, J. A. Petersen, and A. S. F. Ditadi,
     Science, 150, 1182 (1965).

5.   E. J. W. Barrington and A. Thorpe, Gen. Comp. Endocrinol., 3,
     166 (1963).

6.   P. B. Tett and M. G. Kelly, Oceanogr. Mar. Biol. Ann. Rev., 11, 89 (1973).

7.   R. B. Ashworth and M. J. Cormier, Science, 155, 1558 (1967).

8.   C. H. Edmondson, Reef and Shore Fauna of Hawaii, B. P. Bishop Museum, Honolulu, 1946, p. 317.

9.   T. Higa and P. J. Scheuer, Naturwissenschaften, 62, 395 (1975).

10. T. Higa and P. J. Scheuer, Heterocycles, 4, 231 (1976).

11. J. E. Kovacic, U. S. Patent 3,143.576 (1964); Chem. Abstr., 61, 11934e (1964).

12. G. Montaudo, P. Finocchiaro, E. Trivellone, F. Bottino, and P. Maravigna, Tetrahedron 27, 2125 (1971).

13. H. E. Ungnade and F. H. Otey, J. Org. Chem., 16, 70 (1951).

14. G. M. Sharma and B. Vig, Tetrahedron Lett., 1515 (1972).

15. K.-W. Glombitza and H.-U. Rösener, Phytochem., 13, 1245 (1974).

16. W. Fenical, J. Phycol., 11, 245 (1975).

17. M. Pedersén and L. Fries, Z. Pflanzenphysiol., 74, 272 (1975).

18. T. Higa and P. J. Scheuer, Tetrahedron, 31, 2379 (1975).

19. T. L. Davis and V. F. Harrington, J. Am. Chem. Soc., 56, 129 (1934).

20. O. Diels and R. Kassebart, Justus Liebigs Ann. Chem., 530, 51 (1937).

21. H. Davidge, A. G. Davies, J. Kenyon, and R. F. Mason, J. Chem. Soc., 4569 (1958).

22. J. M. Blatchly, R. J. S. Green, J. F. W. McOmie, and J. B. Searle, J. Chem. Soc. C, 1353 (1969).

23. L. S. Dure and M. J. Cormier, J. Biol. Chem., 239, 2351 (1964).

24. T. Higa and P. J. Scheuer, Heterocycles, 4, 227 (1976).

THE CHEMISTRY OF ADENOCHROME(S)

Giuseppe Prota, Shosuke Ito, and Giovanna Nardi

Stazione Zoologica di Napoli and Istituto di Chimica

Organica, Università, Napoli, Italy

During the past decade, a considerable variety of non-porphrin iron(III)-containing metabolites, known as siderochromes[1], has been isolated from bacteria, fungi, and actinomycetes. Typical representatives of such substances are ferrichrome[2] and enterobactin[3], which contain, respectively, hydroxamate and phenolate groups capable of coordination with a central ferric iron.

Although siderochromes are usually regarded as characteristic products of microbial sources, it has been known for a long time[4-6] that the common octopus possesses an iron-sequestering pigment, named adenochrome[7] (adeno = gland), found as intracellular red granules in the "glandular tissue" of the branchial heart.

Investigation of this unusual metabolite has proceeded very slowly and, until recently, little was known of its chemical nature beyond that revealed[7,8] by amphoteric properties, pH-dependent electronic spectra, and elemental analyses providing evidence for the presence of both nitrogen (11.8-12.9%) and sulphur (5.3-7.2%).

In this report we describe the results of a study which characterizes adenochrome as an inseparable mixture of closely related peptides derived from glycine and three novel phenolic amino acids, adenochromines A (1a), B (1b) and C (1c), accounting for the chelate formation with iron(III). This constitutes the main part of the present paper, which also includes preliminary experiments concerning the biosynthesis and distribution of adenochrome.

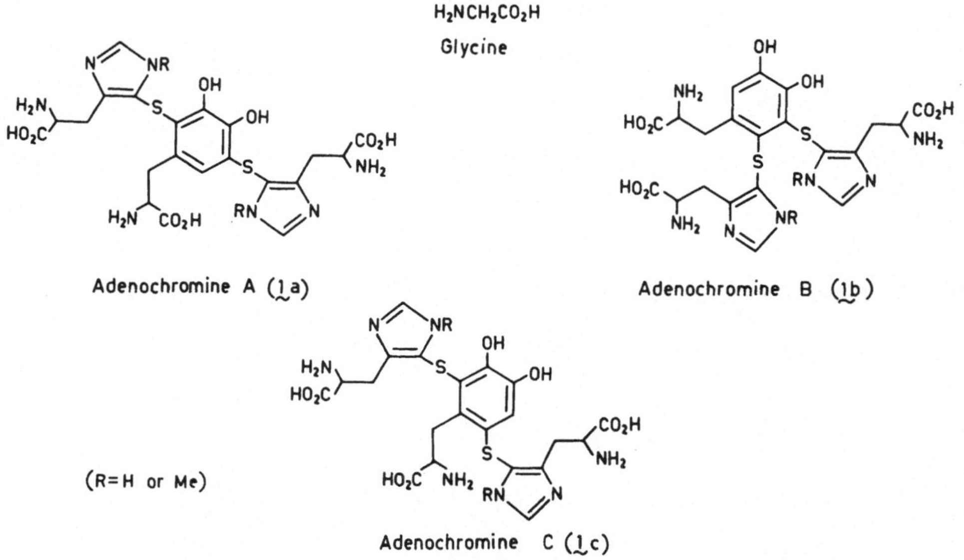

Figure 1.　Constituents of desferriadenochrome (DFA)

ISOLATION AND PROPERTIES OF ADENOCHROME

A procedure developed recently[8] for the isolation of the natural iron complex ferriadenochrome (FA) from branchial hearts of _Octopus vulgaris_ turned out to be unsatisfactory due to some unavoidable oxidation of the pigment during extraction and chromatographic purification at pH 8.3. Moreover, since branchial heart also contained large amounts of desferriadenochrome (DFA), alternative procedures involving the complete conversion of the material to either FA or DFA at some stage were examined. Preliminary experiments in both directions revealed that FA was more difficult to handle than DFA because of its insolubility at pH 3-8 and rapid oxidation even in acid solutions. Accordingly, adenochrome was isolated in the form of DFA, using the procedure outlined in Figure 2. This involved extraction of branchial hearts with 0.5 M $HClO_4$ containing 2% thioglycolic acid, to reduce Fe^{3+} to Fe^{2+}, and subsequent chromatography of the extract on Sephadex G-10, which allowed the separation of DFA from smaller molecules and salts. The crude DFA was then chromatographed on Sephadex LH-20, to remove proteins and some oxidized DFA, and eventually purified by ion-exchange chromatography on Dowex 50W-X8. Elution with 2M pyridine acetate buffer, pH 4.8, and treatment of the appropriate fractions with acetone gave DFA as an almost colourless amorphous powder, slightly soluble in water and extremely insoluble in organic solvents.

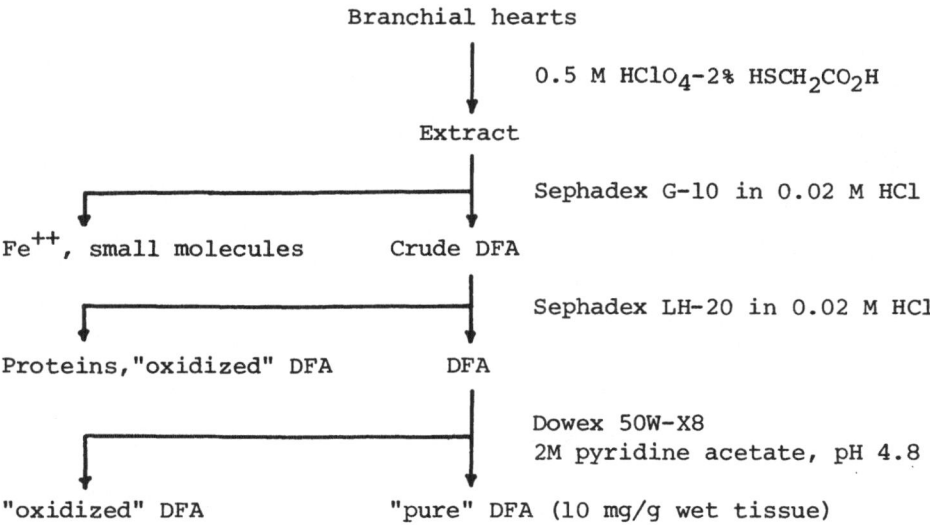

Figure 2.   Isolation of desferriadenochrome (DFA)

When examined on TLC and paper electrophoresis, DFA gave a single spot, detected with both ninhydrin (red-purple) and ferric chloride (green) reagents. However, in the course of degradative experiments it became evident that the isolated product was, in fact, a mixture of several compounds of similar structure. As obtained, DFA contained one molecule of acetic acid (nmr) and six molecules of water (drying) per monomer and gave elemental analyses consistent with a nitrogen/sulphur ratio of 9:2. The UV spectrum of DFA in 0.1 M HCl showed a well-defined peak at 306 nm which shifted to 320 nm at pH 10, suggesting the presence of catechol chromophore. On account of the complex nature of the structure(s), the nmr spectrum (in 2 M DCl) of DFA was not very informative, showing broadened signals in the aliphatic, aromatic, and hetero-aromatic regions.

ACID HYDROLYSIS OF DFA

Unlike normal peptides, hydrolysis of DFA under standard conditions (6 M HCl for 24 hours at 110°C) gave a very complex mixture of unusual amino acids, and their separation continues to be a major problem in this work. The only known amino acid obtained was glycine, while those remaining included adenochromines A, B, and C (1a, 1b, and 1c), secoadenochromines A, B, and C (2a, 2b, and 2c), and histidine-5-thiol disulphide (4), all partially methyl-ated at the N-1 position of the imidazole ring. Later, we found it

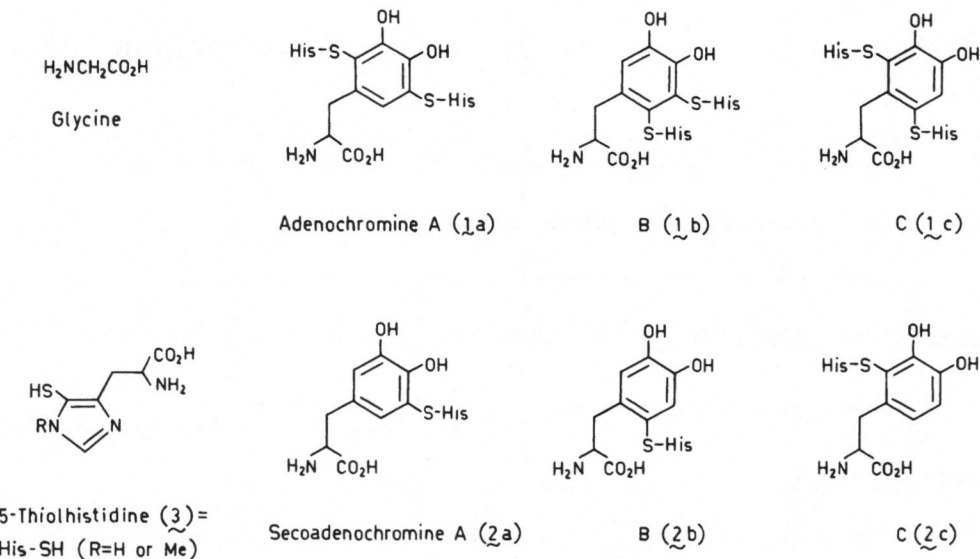

Figure 3.   Products of HCl hydrolysis of DFA

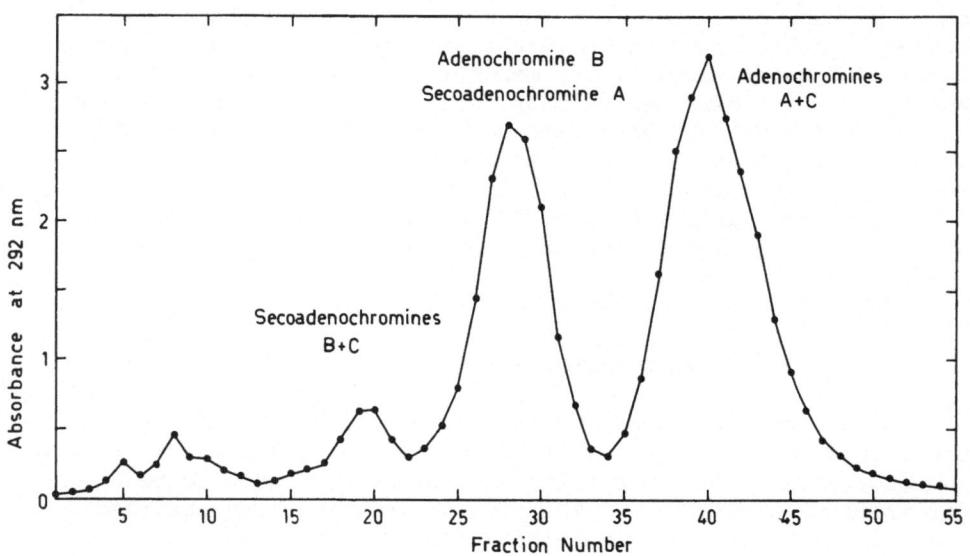

Figure 4.   Elution profile of the HCl hydrolysate of DFA from a
column of Dowex 50W-X2, developed with 3 M HCl

convenient to carry out the hydrolysis reaction with 6 M hydro-
chloric acid containing 4% thioglycolic acid, since these
conditions gave smaller amounts of by-products and the thiol (3)
that was formed in place of the disulphide (4) was easier to separate from the other products. Although the heterogeneity caused
by partial N-methylation hampered our efforts to separate the
adenochromines (1) and the secoadenochromines (2) into their
isomers, some separation could be achieved by careful column chromatography on Dowex 50W resin, coupled with paper chromatography
using n-propanol-1 M HCl (3:2, v/v). Thus, for example, fractionation of a hydrolysate of 250 mg of DFA afforded 101 mg of (1a) and
(1c) (ca. 3:1 ratio), 12.4 mg of (1b), 32.4 mg of (2a), and 6.5 mg
of (2b) and (2c) (ca. 4:1 ratio). Compounds (1c), (2b) and (2c)
could eventually be obtained in a practically pure form by means
of the chemical reaction which is described later.

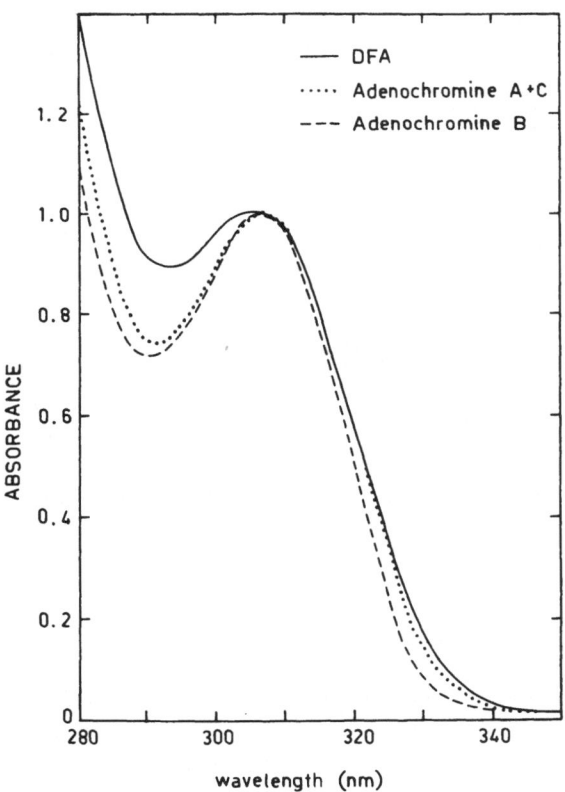

Figure 5. UV spectra of DFA and adenochromines

Table 1. Hydrolysis of DFA under mild and drastic conditions

Products	Yield[a]	
	3 M HCl, 20 hr[b]	6 M HCl, 120 hr
Adenochromine A+C (1a+c)	22.0 mg	20.6 mg
Adenochromine B (1b)[c]	8.3	1.6
Secoadenochromine A (2a)[c]	0.8	9.3
Secoadenochromines B+C (2b+c)[c]	---	1.7
5-Thiolhistidine (3)[d]	2.1	8.6
Dopa (5)	---	0.7

[a]In both experiments 63 mg of DFA were used.

[b]Hydrolysis of peptide bonds incomplete.

[c]Purified by paper chromatography ($\sim$80% recovery) with n-propanol-1 M HCl (3:2, v/v).

[d]Purified as the disulphide (4).

Among the products of hydrolysis, the adenochromines (1) exhibited UV spectra which closely paralleled that of DFA, suggesting that the secoadenochromines (2), as well as the thiol (3), might be artifacts. In order to clarify this aspect, hydrolysis of DFA with 6 M HCl-4% thioglycolic acid was examined under mild and drastic conditions. The results, summarized in Table 1, show that (2) and (3) were indeed formed from adenochromines (1),especially from isomer (1b), by a rather unusual acid cleavage of a thioether bond. Thus, it appeared likely that the actual amino acid constituents of DFA were only glycine and adenochromines A (1a), B (1b), and C (1c) in the ratio of ca. 3:2:1.

As hydrolysis of DFA with 6 M HCl was somewhat misleading, an alternative reaction was needed to determine the ratio of glycine to the adenochromines. The reductive hydrolysis with 57% hydriodic acid in the presence of red phosphorus (at 110°C for 48 hours) proved suitable for the quantitative cleavage of both thioether and peptide bonds. This hydrolysis of DFA afforded a much simpler reaction mixture (Figure 6), which was readily separated by column chromatography on Dowex 50W (eluent: 3 M HCl) to give glycine, histidine-5-thiol, and L-dopa in a molar ratio of 2:2:1. Hence, the ratio of glycine to adenochromines (1) was 2:1.

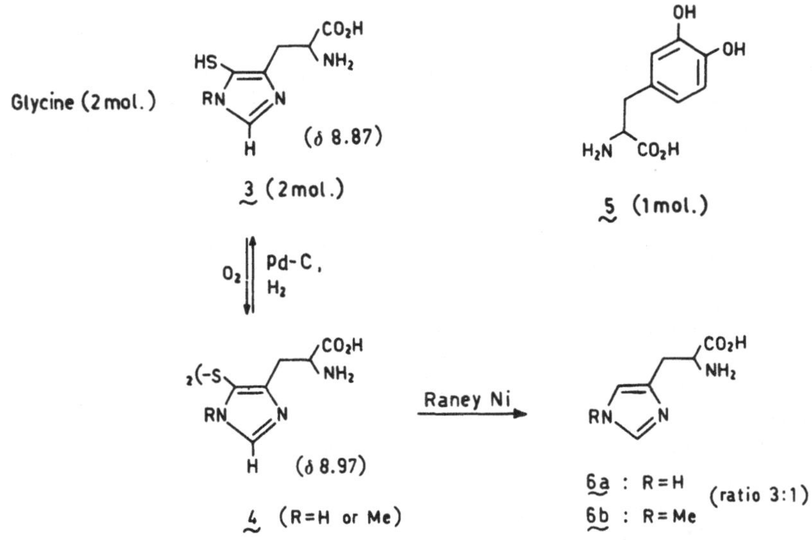

Figure 6. Products of HI hydrolysis of DFA

STRUCTURE OF THE THIOL (3)

The thiol amino acid (3), an amorphous powder, was available
in substantial amounts by hydriodic acid hydrolysis of DFA. On air
oxidation it gave the corresponding disulphide (4) as prisms from
6M HCl-acetone, $C_{12}H_{16}N_6O_4S_2 \cdot 2H_2O^9$, which could be reconverted
to the parent thiol by catalytic hydrogenation (Pd-C, 0.1M HCl).
Desulphuration of (4) with Raney-nickel, followed by paper electrophoresis at pH 6.5 of the reaction mixture, gave L-histidine and
1-methyl-histidine in ca. 3:1 ratio, reflecting the degree of
methylation. When compared with the nmr spectrum of histidine-2-
thiol (C-5 proton at δ 7.02), the spectrum of (3) $\left[\right.$ δ (2M DCl)
3.57 (2H, d, J = 7 Hz), 3.97 (ca. 0.8 H, S, N-Me), 4.57 (1H, t,
J = 7), and 8.87 (1H, s) $\left.\right]$ showed that it was the hitherto unknown
histidine-5-thiol, partially methylated at position 1.

STRUCTURES OF ADENOCHROMINES A, B, AND C

These characteristic constituents of adenochrome gave intense green colours with ferric chloride and exhibited very similar UV spectra (Figure 5) with an absorption maximum at 306 nm, shifted to 320 nm in alkali. Common features of the nmr spectra (in 2 M DCl) of the adenochromines were signals due to three $-CH_2CH(\overset{+}{N}H_3)$ COOH residues, one aromatic proton at ca. δ 7.0, and two low-field heteroaromatic protons around δ 8.8, typical for H-2 of an imidazole ring. The spectra also exhibited two N-Me singlets around δ 4.0 (each less than 1H), indicating that adenochromines were partially methylated. Moreover, on reductive hydrolysis with 57% hydriodic acid in the presence of red phosphorus (48 hours at 110°C) they gave, as expected, L-dopa and the thiol (3) in a molar ratio of 1:2. Accordingly, the adenochromines were regarded as isomers of the general structure (1), differing in the positions of the two 5-thiolhistidine residues on the dopa moiety.

In the early stage of the subsequent structural work, the two adenochromine fractions (see Figure 4) obtained by HCl hydrolysis of DFA were considered as corresponding to two of the three possible positional isomers, but later experiments revealed that the major fraction was, in fact, a mixture of two isomers (1a and 1c), thus creating an additional difficulty for the assignment of the relative substitution pattern. Although the adenochromines A and C were examined first, it seems preferable, for the sake of clarity, to begin with the evidence leading to the identification of adenochromine B as (1b).

When heated in 48% hydrobromic acid containing 4% thioglycolic acid, adenochromine B gave the thiol (3), dopa (5), and secoadeno-chromine A (63% yield), $C_{15}H_{18}N_4O_6S \cdot 3 HCl$[9], which was assigned structure (2a) on the basis of its nmr spectrum [δ (2 M DCl) 3.17 (2H, d, J = 6.5 Hz), 3.60 (2H, d, J = 7.5), 3.96 (ca. 0.8 H, s, N-Me), 4.37 (1H, t, J = 6.5), 4.49 (1H, t, J = 7.5), 6.70 and 6.84 (2H, ABq, J = 1.8), and 8.82 (1H, s)]. The assignment of the aromatic substitution as in (2a) was substantiated by the nmr spectra of two related compounds, 5-S-cysteinyldopa[10,11] (ABq, δ 6.93 and 7.01, J = 2.0) and 6-S-cysteinyldopa[11] (singlets, δ 6.94 and 7.20).

The formation of (2a) ruled out structure (1c) for adenochro-mine B, which might therefore be either (1a) or (1b). The latter structure (1b) was indicated, since hydrogenolysis of adenochromine B in 0.1 M HCl over PtO_2 (24 hours at room temperature) led to a different secoadenochromine B, identified as (2b), along with dopa (5) and histidine (6) (partially methylated at N-1).

On this basis it became evident that the remaining inseparable adenochromines A and C (ca. 3:1 ratio by nmr) were represented by (1a) and (1c) or vice versa.

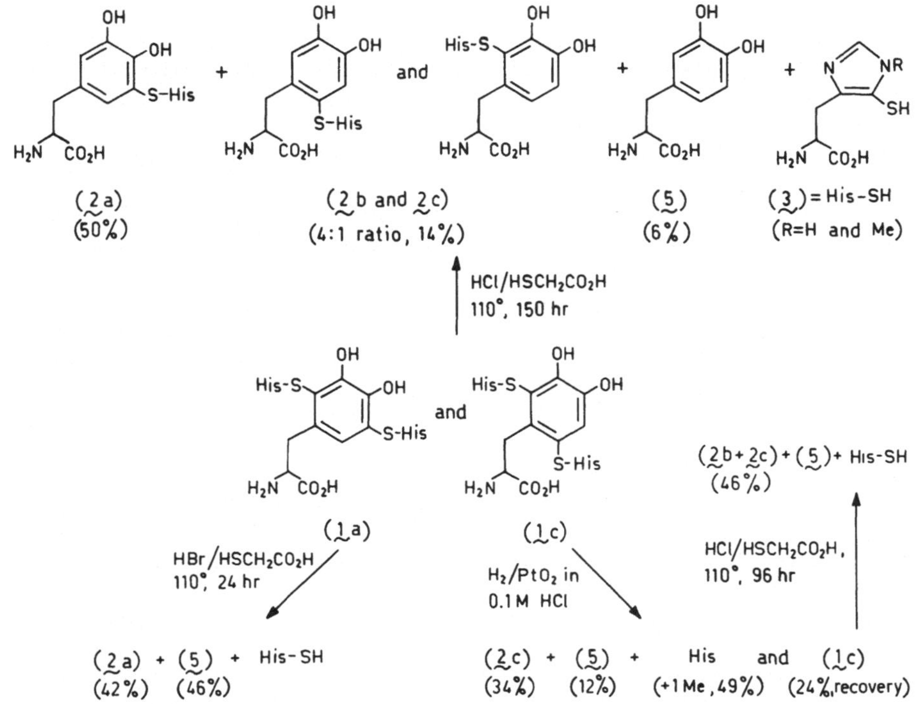

Figure 7.   Degradations of adenochromines A and C

Evidence suggesting (1a) as the major isomer of the mixture
was first obtained by the reaction of the adenochromines (Figure 7)
with 48% HBr-4% thioglycolic acid (24 hours at 110°C) which gave
the thiol (3), dopa (5), and secoadenochromine A (2a) in 42% yield.
In this experiment the accompanying isomer (1c) appeared relatively
more unstable to hydrobromic acid than (1a). Hydrolysis was
therefore carried out with 6 M HCl-4% thioglycolic acid. After
seven days at 110°C, about 45% of the starting material remained
unchanged, and fractionation of the reaction products on Dowex 50W
(eluent: 3 M HCl) gave secoadenochromine A (2a, 50%) and the
expected mixture of secoadenochromines B and C (2b and 2c, 14%) in
a ratio of ca. 4:1 (nmr).

In order to provide further characterisation of adenochromines
A and C, the mixture was subjected to hydrogenolysis over PtO_2
which afforded dopa (5), histidine (6) (partially methylated at
N-1), secoadenochromine C (2c), (ABq, δ 6.91 and 7.03, J = 8 Hz),
and a 24% recovery of starting material corresponding to the pure
adenochromine C (1c). This unexpected result was substantiated

by the nmr spectrum of the recovered product, which differed, albeit in very minor details, from that of the original mixture of (1a) and 1c). Moreover, hydrolysis of adenochromine C with 6 M HCl-4% thioglycolic acid gave no secoadenochromine A but a mixture of secoadenochromines B and C (46%).

Thus, by way of hydrogenolysis, a "chemical" separation of adenochromines A and C could be achieved as a consequence of the greater instability of the isomer (1a), which, under the conditions used, was selectively converted into secoadenochromine C. 2,5-S,S-Dicysteinyldopa, a related metabolite found in the eyes of fish[12] and, more recently, in the urine of patients with melanoma[13], behaves quite similarly to (1a) and, on treatment with PtO$_2$, is smoothly transformed to 2-S-cysteinyldopa[14] by selective removal of the cysteinyl residue at position 5.

MECHANISM OF ACID HYDROLYSIS OF ADENOCHROMINES

A dominant feature of the chemical behaviour of the adeno-chromines (1) is their tendency to undergo "hydrolysis" of the thioether bonds in 6 M HCl or, more effectively, in concentrated HBr, to give histidine-5-thiol (3), secoadenochromines (2), and dopa (5). Mechanistically, the reaction may be explained as an electrophilic elimination from a protonated species (Figure 8), of a sulphenyl ion, His-S$^+$, which is subsequently converted to the parent thiol by excess of thioglycolic acid. Therefore it seemed likely that under appropriate conditions the reaction might be reversible, since in strongly acidic solution disulphides are reported[15] to generate sulphenyl ions by the equilibrium reaction: R-S-S-R + H$^+$ $\rightleftarrows$ RS$^+$ + RSH. Indeed, heating of L-dopa (2 mmol) with the readily-available amino acid disulphide L-cystine (8 mmol) in 40% HBr for 6 hours afforded (Figure 9) 5-, 2-, and 6-S-cys-teinyldopa[10,11,14]. Under similar conditions (24 hours at 110°C), cystine reacted with tyrosine to give 3-S-cysteinyltyrosine (39%), a previously unknown amino acid.

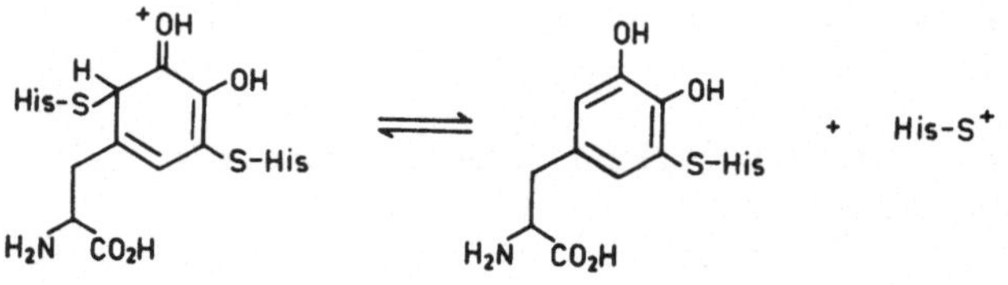

Figure 8. Hydrolytic cleavage of adenochromines

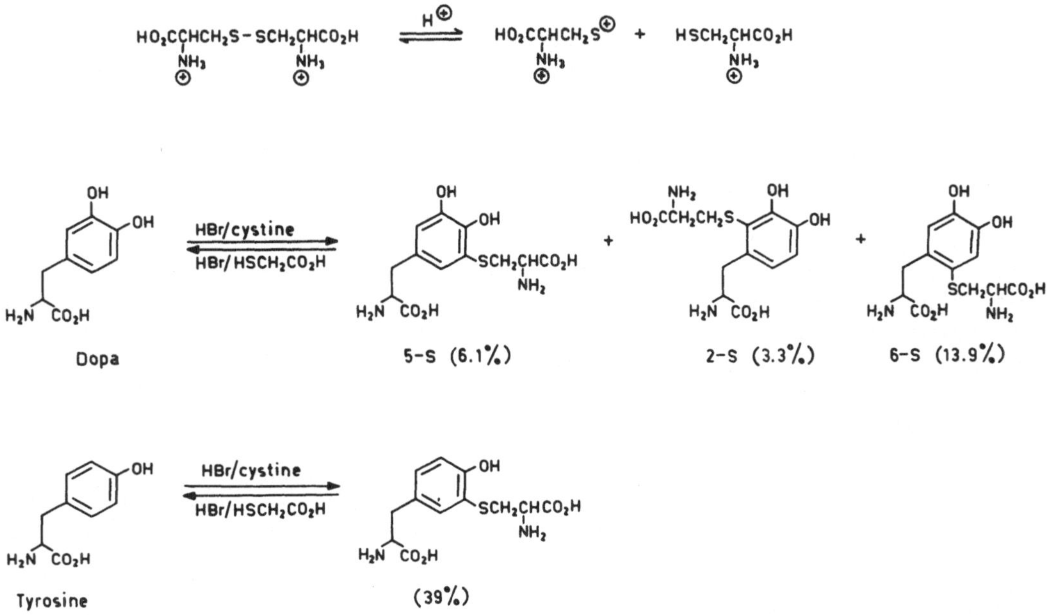

Figure 9. Substitution and elimination of cysteine sulphenyl ion

These reactions are of both theoretical and practical interest
in confirming the formation of cysteinyl sulphenyl ion in strongly
acidic solution[15] and in providing a new approach to the synthesis
of the cysteinyldopas and related compounds, which are the key
intermediates in the biosynthesis of the mammalian phaeomelanic
pigments[16].

BIOSYNTHESIS OF ADENOCHROMINES

A suggestion for the biosynthesis of the adenochromines (1)
follows from their close structural similarity to the cysteinyldopas[10,11,14], especially 2,5-S,S-dicysteinyldopa[11,12], which are
formed by addition of cysteine to dopaquinone arising _via_ dopa by
tyrosinase oxidation of tyrosine[16]. Therefore, a similar reaction
between histidine-5-thiol (3) and dopaquinone would account for
the biosynthesis of adenochromines (1). In fact, when a solution
of dopa (1 mmol) and (3) (2 mmol) in phosphate buffer, pH 6.8, was

oxidized with oxygen in the presence of mushroom tyrosinase, a
smooth reaction took place, as revealed by UV spectroscopy. Fractionation of the reaction mixture on Dowex 50W and paper chromatography led to the isolation of secoadenochromines (2a) (57%), (2b),
and (2c) (30%, ca. 2:1 ratio), adenochromine (1a) (3%), and a
trace amount of (1b). Moreover, enzymic oxidation of (2a) in the
presence of (3) under similar conditons gave a 10% yield of (1a),
together with some (1b).

     In the light of these preliminary experiments, the formation
of adenochromines in octopus might be regarded as the result of
a deviation of the normal eumelanin pathway involving a non-
enzymic reaction between dopaquinone and histidine-5-thiol (or its
1-Me homologue). The analogy of such a process with phaeomelanin
biosynthesis is remarkable, especially when compared with the early
stages (Figure 10) leading to the formation of 5- and 2-S-cysteinyl-
dopa, as well as 2,5-dicysteinyldopa. From the viewpoint of
positional reactivity of dopaquinone in the two processes, it is
relevant that, unlike cysteine[11], histidine-5-thiol also gives the
addition product at the C-6 position, which is quite compatible
with the presence of substantial amounts of the 5,6-disubstituted
isomer (1b) among the adenochromines. Moreover, since the in vitro
oxidation of dopa with mushroom tyrosinase in the presence of
histidine-5-thiol gave mainly the monosubstituted adducts (2), it
seems reasonable to believe that the formation of the adenochromines
(1) in vivo requires somewhat different physiological conditions,
probably involving the intervention of a specific enzyme.

CONCLUDING REMARKS

     The results so far obtained provide evidence that adenochrome
is a complex mixture of closely related peptides, consisting of
two moles of glycine and one mole of adenochromines (1), which are
the amino acids responsible for the iron(III)-binding properties of
DFA and are represented by the three isomers A (1a), B (1b) and C
(1c) in 3:2:1 ratio. Moreover, one-fourth of the imidazole N-1
positions of these adenochromines are methylated, with the conse-
quence that the number of actual constituents of the mixture is
markedly increased. Although the fundamental aspects of the
structure of adenochrome have been largely clarified, many problems
remain to be studied. One of them concerns the position of attachment of the glycine units to adenochromines (1). Preliminary
experiments favour a structural situation in which a glycylglycine
unit is linked through a peptide bond to the amino group of the
dopa moiety of adenochromines, but the main uncertainty in
formulating a tentative structure concerns the molecular size of
DFA, which may well not be monomeric.

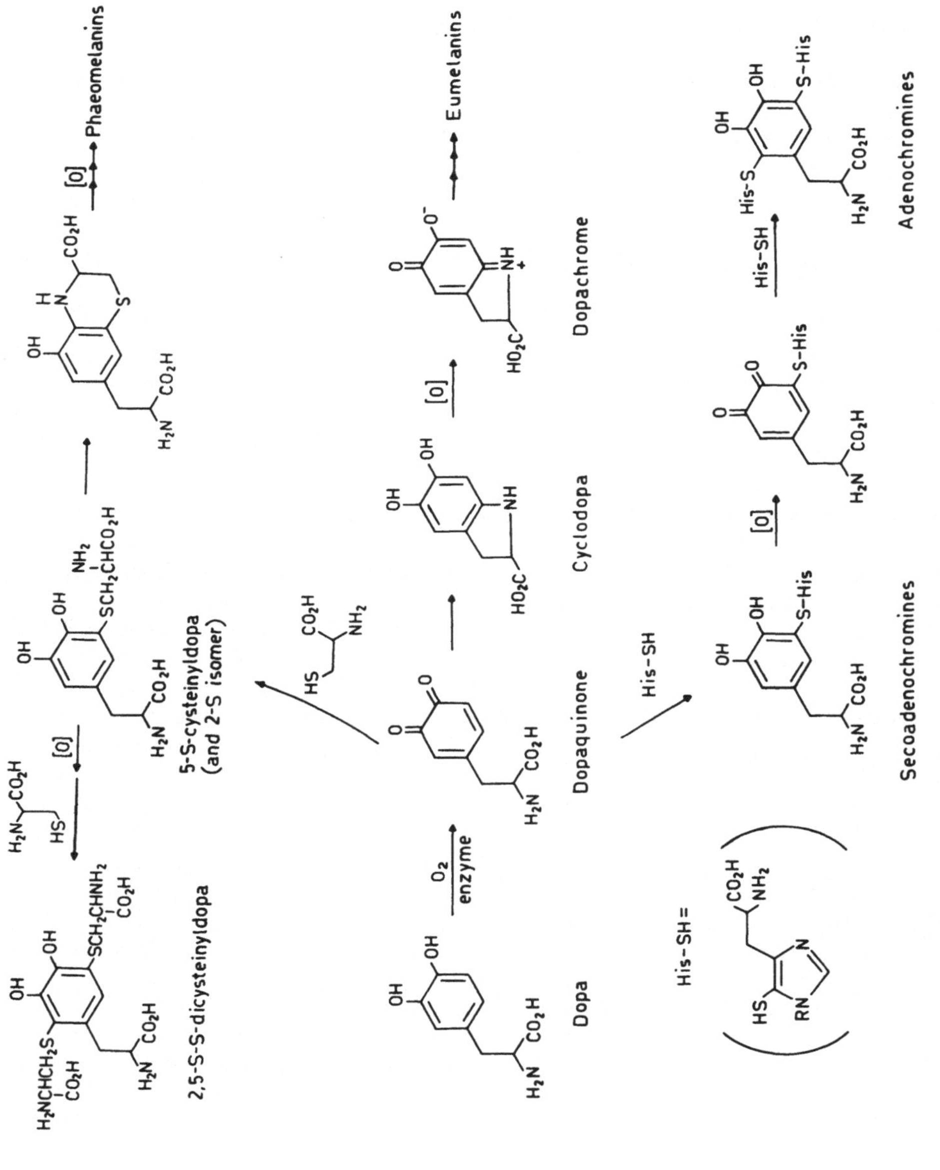

Figure 10. Biosynthetic relationship between adenochromines and melanin pigments

Table 2.   Adenochrome distribution in <u>Octopus</u> <u>vulgaris</u>

Source	Adenochrome Content
Branchial heart	12 mg/g wet wt.
White body	(1.7)*
Ovary	0.52
Amoebocytes	0.24
Gill	0.14
Hepatopancreas	+
Branchial gland	-
Kidney	-
Skin	-

*In a form of a possible precursor with the general structure
DFA-Asp$_2$

Another interesting aspect of biochemical significance is
related to the localization of the site of biosynthesis and to the
mechanism of distribution of adenochrome in octopus. As shown in
Table 2, the same adenochrome mixture, characteristic of the
branchial heart, is found also in ovary, gills, hepatopancreas and
amoebocytes, the latter being the circulating cells of the blood.

The occurrence in the white bodies of a DFA analogue is also
intriguing. As revealed by degradative studies, this is composed
of adenochromines (la - c), glycine, and aspartic acid in a molar
ratio 1:2:2. On partial hydrolysis under mild conditions[17]
(0.03 M HCl, 16 hours at 105°C), it is smoothly converted into
a shortened peptide, identical in all respects to DFA, by loss of
the two aspartic acid residues. Therefore, this particular form
of DFA, found exclusively in the white bodies, might represent a
possible biosynthetic precursor of the adenochrome distributed in
branchial hearts and other organs. Such a possibility would be
compatible with biological studies[18] suggesting that the white
bodies are the site of origin of the amoebocytes, which therefore
may act as carriers of DFA from the white bodies throughout the
organism[18].

ACKNOWLEDGMENT

The authors wish to thank Mr. V. Saggiomo for helpful
technical assistance.

REFERENCES

1. For a review, see J. B. Nielands, Inorganic Biochemistry, ed.
   G. Eichhorn, Elsevier, Amsterdam, 1972, p. 167.

2. S. Roger and J. B. Nielands, Biochem. $\underline{3}$, 1850 (1964).

3. J. R. Pollack and J. B. Nielands, Biochem. Biophys. Res.
   Comm., $\underline{38}$, 989 (1970).

4. Cuenot, Gonet et Bruntz, Arch. Zool. exp. Gen., $\underline{9}$ (4) n. 3,
   XLIX-LIII (1908).

5. J. Turchini, Arch. Morph. gén. exp., $\underline{18}$, 7 (1923).

6. Z. M. Bacq and M. Leiner, Z. vergl. Physiol., $\underline{22}$, 434 (1935).

7. D. L. Fox and D. M. Updegraff, Arch. Biochem., $\underline{1}$, 339 (1943).

8. G. Nardi and H. Steinberg, Comp. Biochem. Physiol., $\underline{48B}$, 453
   (1974).

9. The analyses were in agreement with the values calculated by
   taking into account the 3:1 ratio of NH and N-Me homologues.
   However, for clarity, the molecular formulae refer only to
   the NH homologues.

10. G. Prota, G. Scherillo, and R. A. Nicolaus, Gazzetta, $\underline{98}$, 495
    (1968).

11. S. Ito and G. Prota, Experientia, in press.

12. S. Ito and J. A. C. Nicol, Tetrahedron Lett., 3287 (1975);
    Biochem. J., in press.

13. G. Prota, H. Rorsman, A. M. Rosengren and E. Rosengren,
    Experientia, in press.

14. E. Fattorusso, L. Minale, S. De Stefano, G. Cimino, and
    R. A. Nicolaus, Gazzetta, $\underline{99}$, 969 (1969).

15. R. E. Benesch and R. Benesch, J. Am. Chem. Soc., $\underline{80}$, 1666 (1958).

16. For reviews, see (a) G. Prota in Pigmentation: Genesis and
    Biologic Control, ed. V. Riley, Appleton-Century-Crofts,
    New York, 1972, p. 615; (b) R. H. Thomson, Angew. Chem. Internat.
    Edit., $\underline{13}$, 305 (1974); (c) G. Prota and R. H. Thomson,
    Endeavour, $\underline{35}$, 32 (1976).

17. J. Schultz, Methods in Enzymology, $\underline{11}$, 255 (1967).

18. R. R. Cowden, J. Invert. Path., $\underline{19}$, 113 (1972).

A SURVEY OF SESQUITERPENOIDS FROM MARINE SPONGES

G. Cimino

Laboratorio per la Chimica di Molecole di Interesse

Biologico del C.N.R., Arco Felice, Napoli, Italy[*]

During the last four years there have appeared an ever-increasing number of papers describing the isolation and elucidation of the structures of many sesquiterpenoids occurring in marine sponges. It now seems that these very primitive animals contain the greatest variety of sesquiterpenoids in the animal kingdom. Almost fifty new sesquiterpenoids from marine sponges have now been isolated and characterized. In this paper I will present a broad survey of these compounds, which can be subdivided into three main groups: (a) furan sesquiterpenes, (b) sesquiterpenoid hydroquinones, and (c) sesquiterpenoid isonitriles.

Table 1 lists the sponges which have been reported to contain sesquiterpenoids. As you can see, these compounds have been isolated from eleven species belonging to four orders of the class Demospongiae.

FURAN SESQUITERPENES

Furan rings are present in the structures of the majority of the sesquiterpenes so far reported from sponges, even though they have been isolated from only three species, two Dictyoceratida and the unrelated Microciona toxystila. Among these, Disidea pallescens has proved to be the major source of furan sesquiterpenes. As can be seen in Figure 1, ten new compounds have been found in this sponge. Three of them, pallescensin-1 (1), -2 (2), and -3 (3), have a monocyclofarnesyl skeleton, while, for the seven pallescensins A-G (4-10), different cyclization patterns have led to closely related structural types having a 2,3-disubstituted furan ring and two additional rings.

Table 1. Sponges in which (a) furan sesquiterpenes, (b) sesquiterpenoid hydroquinones, and (c) sesquiterpenoid isonitriles have been reported.

Furan Sesquiterpenes	Sesquiterpenoid Hydroquinones	Sesquiterpenoid isonitriles
Order Dictyoceratida	Order Dictyoceratida	Order Axinellida
Family Aplisillidae	Family Disideidae	Family Axinellidae

Pleraplysilla spinifera
  Sample 1
  Sample 2
	Disidea pallescens
	Disidea avara
	Australian Disidea sp.
		Axinella cannabina
		Acanthella acuta

Family Disideidae	Family Spongidae	Order Halichondrida
Disidea pallescens	Stelospongia canalis (yellow-coloured form)	Family Halichondrida
	Stelospongia canalis (orange-coloured form)	Halichondria sp.
Order Poecilosclerida	Order Halichondrida	Family Hymeniacidonidae
Family Clathrüdae	Family Halichondrida	Hymeniacidon sp.
Microciona toxystila	Halichondria panicea	

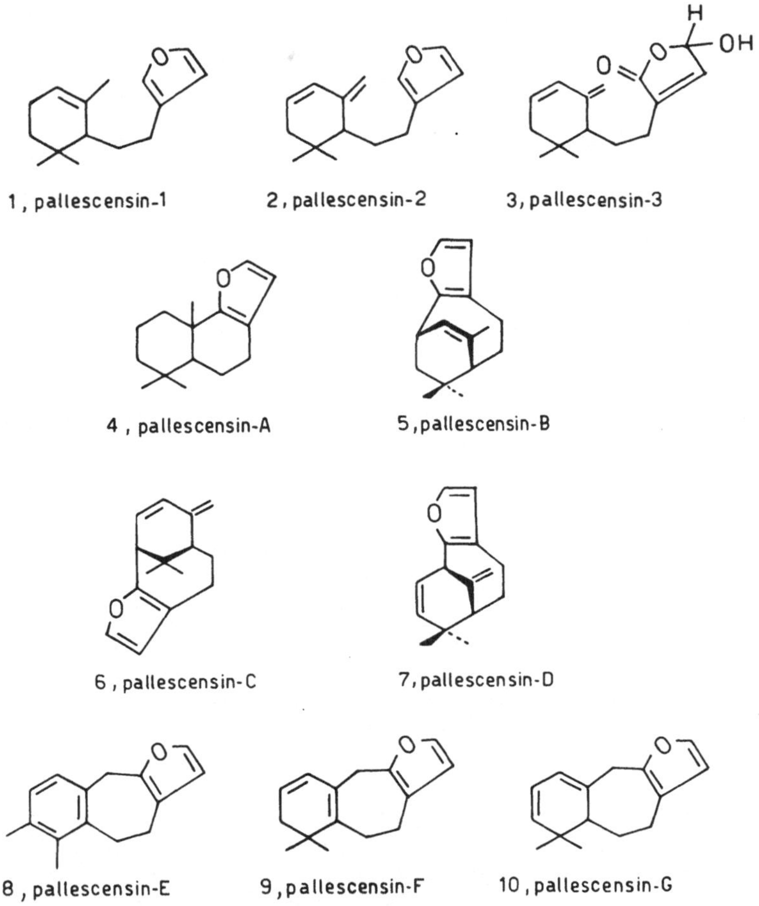

1 , pallescensin-1   2 , pallescensin-2   3 , pallescensin-3

4 , pallescensin-A   5 , pallescensin-B

6 , pallescensin-C   7 , pallescensin-D

8 , pallescensin-E   9 , pallescensin-F   10 , pallescensin-G

Figure 1. Furan sesquiterpenes from the sponge _Disidea pallescens_

Most of these molecules, which were present in the sponge in small amounts, were sensitive to light, air, and heat. Because of this, an extensive chemical investigation was impossible and the structural assignments were made mainly on spectral grounds, biogenetic considerations, and the chemical interrelations between them.

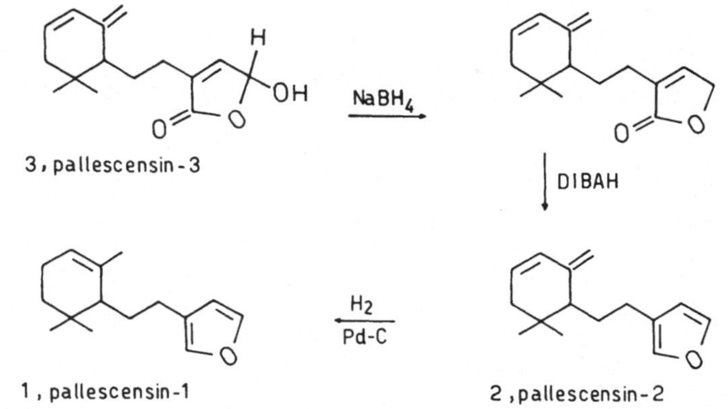

Figure 2. Chemical interrelations among pallescensins 1-3

The structures of pallescensins 1-3, proposed on the basis of spectral analysis, were confirmed by chemical interconversions[1], as shown in Figure 2. Pallescensin-3 (3), in which the furan ring is modified as a γ-hydroxy-α,β-butenolide, yielded, by a two-step reduction sequence, pallescensin-2 (2), which was hydrogenated to obtain pallescensin-1 (1).

It is worth noting that despite the abundance of sesquiterpenes in nature and the variety of their structures, few compounds possessing the simple monocyclofarnesyl skeleton have been found in nature until recently[2]. Other than in pallescensins 1-3, the simple monocyclofarnesyl skeleton has been found in only four other marine sesquiterpenes, the structures of which are shown in Figure 3. Two of them, the methyl ester of the trans-monocyclofarnesic acid (11)[3] and microcionin-3 (12)[4], have been isolated from the sponges Halichondria panicea and Microciona toxystila. The remaining two, α-snyderol (13) and β-snyderol (14), have been described by Howard and Fenical[5] from two species of the marine red alga Laurencia. Very recently, a simple synthesis of β-snyderol was recorded by Gonzales et al.[6].

The tricyclic structure (4), possessing a new farnesyl skeleton, has been proposed for pallescensin-A on the basis of spectral analysis and chemical interrelation with pallescensin-1 (1)[7], which, on treatment with boron trifluoride etherate, was converted into (4).

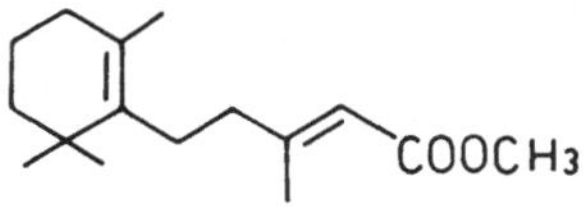

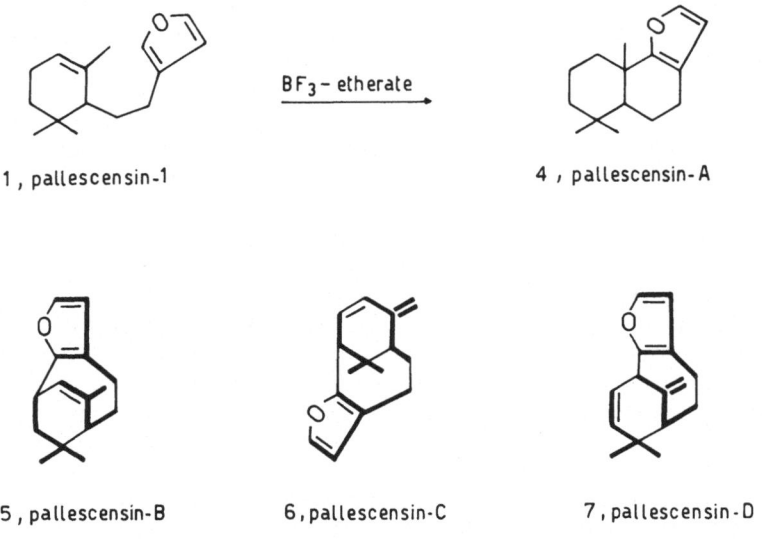

11 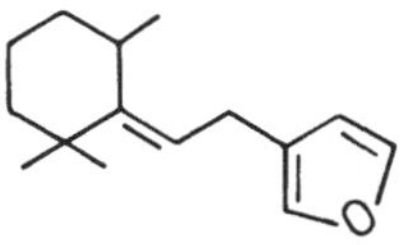   12, microcionin-3

from sponges

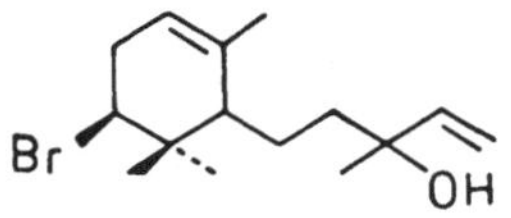

13, ⊀-snyderol 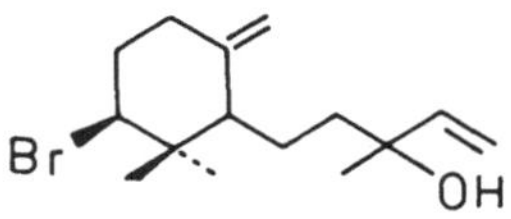   14, β-snyderol

from algae

Figure 3.   Marine compounds possessing the same monocyclofarnesyl skeleton of the pallescensins 1-3

1, pallescensin-1   BF₃ – etherate →   4 , pallescensin-A

5, pallescensin-B   6, pallescensin-C   7, pallescensin-D

Figure 4.   Structures of pallescensins A–D

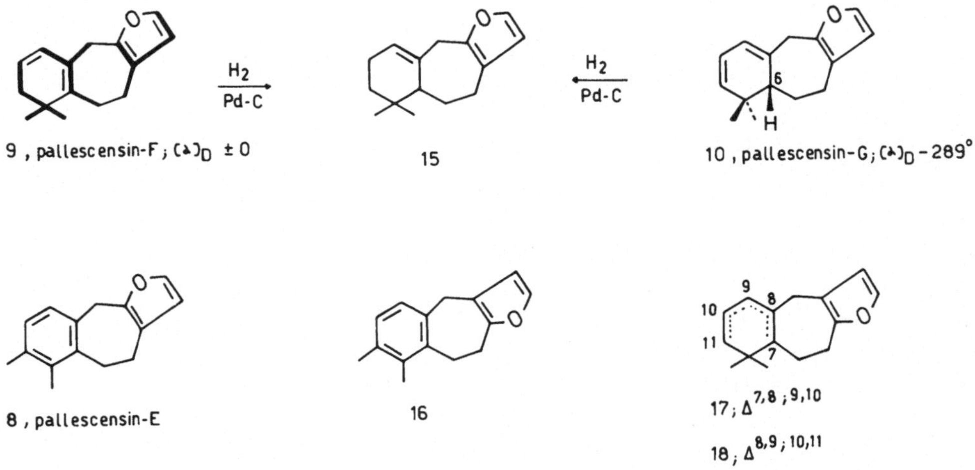

9 , pallescensin-F ; $(\lambda)_D \pm 0$

15

10 , pallescensin-G ; $(\lambda)_D - 289°$

8 , pallescensin-E

16

17 ; $\Delta^{7,8} ; ^{9,10}$

18 ; $\Delta^{8,9} ; ^{10,11}$

Figure 5.   Structures of pallescensins E-G

Pallescensins B-D (5-7)[7] also possess carbon skeletons not previously encountered. The cagelike structures of these compounds were mainly determined by spectral analyses of high resolution nmr spectra since, fortunately, almost all the protons resonate as well-separated signals in the nmr.

Structures of the three remaining pallescensins E-G (8-10)[8] were proposed on the basis of a combination of spectral data and chemical interrelation between them as shown in Figure 5.

Hydrogenation of the optically inactive pallescensin-F (9) yielded a 1,4-dihydroderivative (15) which proved to be identical to the major dihydroderivative of pallescensin-G (10). We also suggested that pallescensin-G had the absolute stereochemistry shown in (10). The chirality of the sole asymmetric center C-6 was proposed to be R, on the basis of the negative Cotton effect observed in CD spectrum, suggesting a left-handed helix conformation for the cisoid diene chromophore, coupled with nmr data indicating a quasi-axial orientation for H-6. The last component of this group, pallescensin-E (8), possesses a rearranged benzenoid skeleton which could arise from the co-occurring pallescensins F-G by a 1,2 methyl migration followed by dehydrogenation (Figure 6). Analysis of spectral and chemical data failed to distinguish (8-10) from their isomers (16-18). Decision in favour of the former was made on biogenetic grounds.

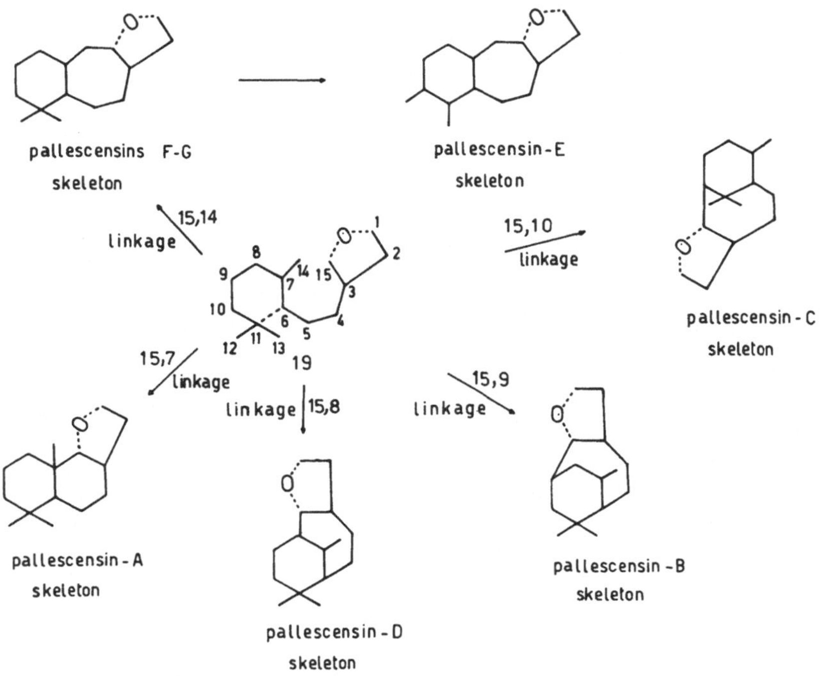

Figure 6. Possible biogenetic scheme for the formation of pallescensins A-G

    Figure 6 illustrates possible biosynthetic pathways to pallescensins A-G involving the intermediacy of a furanoid mono-cyclofarnesyl derivative (19), from which linkages between C-15 and C-7, -8, -9, -10, and -14 could lead formally to the formation of pallescensins A-G. The presence of the monocyclofarnesyl compounds pallescensins 1-3, which coexist in the same sponge, supports this hypothesis. It should be noted that only the spiro-skeleton arising from bond formation between C-15 and C-6 is absent among the sesquiterpenes from Disidea pallescens. However, bicyclic spiro-sesquiterpenes are common metabolites in many species of red algae of the genus Laurencia[9]. In particular, spirolaurenone (21), found in Laurencia glandulifera by Irie's group[10], could be considered formally derived from a monocyclofarnesyl precursor and subsequent linkage formation between C-15 and C-6 (20).

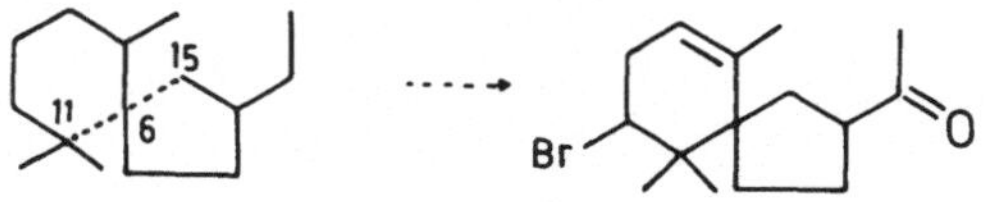

20   21, spirolaurenone

Another group of six closely related sesquiterpenes has been found in the sponge <u>Microciona toxystila</u> (order Poecilosclerida)[4]. Their structures are listed in Figure 7. All these compounds except the above-mentioned microcionin-3 (12) belong to the structural type (22), which, although new among marine sesquiter-penoids, has terrestrial representatives in ascochlorin and its analogs, sesquiterpenoid phenols isolated from the fungus <u>Aschochyta viciae</u>[11]. The structures of the microcionins, proposed on the basis of spectral analyses, were confirmed by chemical transformations, as illustrated in Figure 7. On ozonolysis, microcionin-3 (12) gave 2,2,6-trimethylcyclohexanone. Furthermore, the two double-bond isomers, microcionins-2 (23) and -4 (24), were

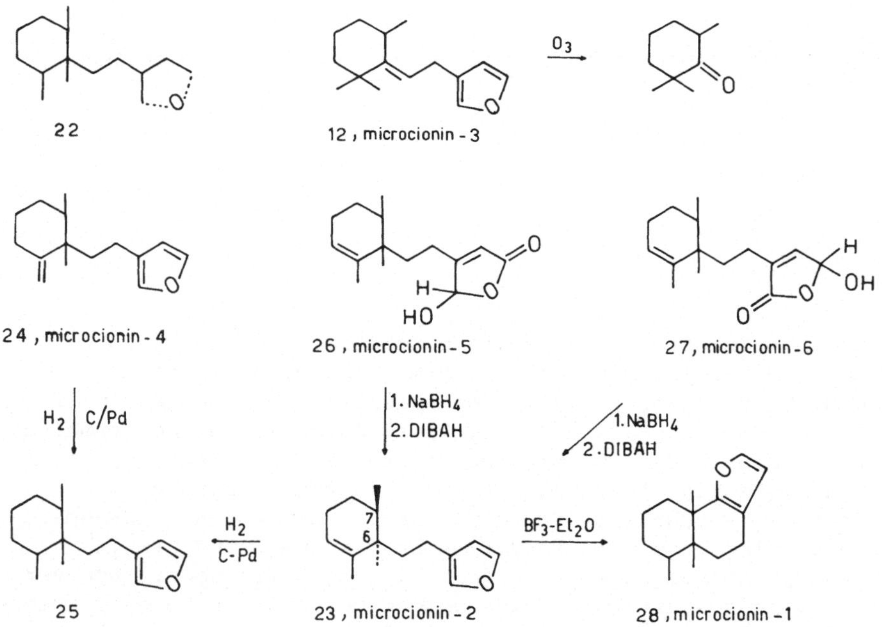

Figure 7. Furan sesquiterpenes from the sponge <u>Microciona toxystila</u>

interrelated by hydrogenation, which gave the same dihydroderivative (25). The isomeric microcionins-5 (26) and -6 (27), which
have the furan ring modified as a γ-hydroxy-α,β-butenolide, were
related by two-step reduction to microcionin-2 (23). The latter,
by treatment with boron trifluoride etherate, furnished the tricyclic microcionin-1 (28). For microcionin-2 (23), the relative
stereochemistry at C-6 and C-7 with the two methyl groups trans
to each other was derived from a careful study of Eu-induced shifts
made on the two stereoisomeric epoxides obtained by peracid
oxidation.

The co-occurrence of microcionins -1, -2, -4, -5, and -6,
having a rearranged monocyclofarnesyl skeleton, along with
microcionin-3, led us to suggest that the whole group should arise
from a common biosynthetic precursor, the ion (29), from which, as
is summarized in Figure 8, microcionin-3 could be derived by loss
of the hydrogen at C-5, while microcionins -1, -2, -4, -5, and -6
could be formed by migration of a methyl group from C-11 to C-6
and subsequent oxidations.

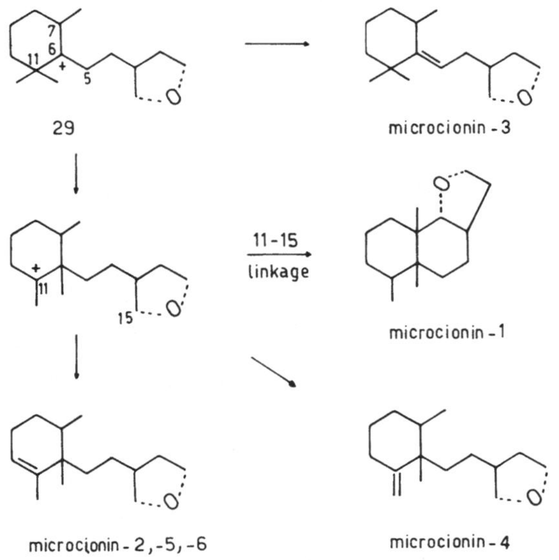

Figure 8. Possible biosynthetic pathways of microcionin 1-6.

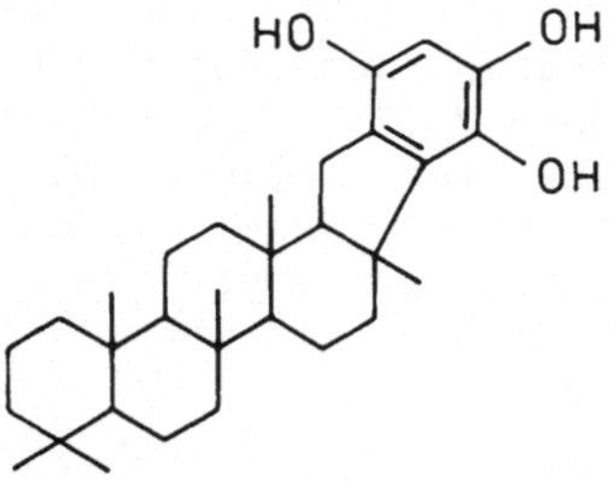

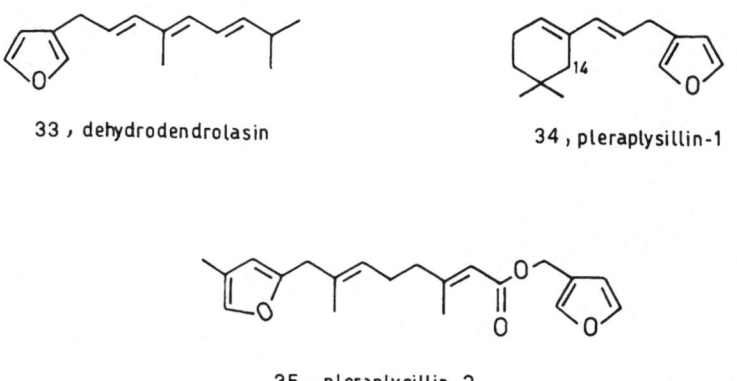

30, disidein

from Disidea pallescens

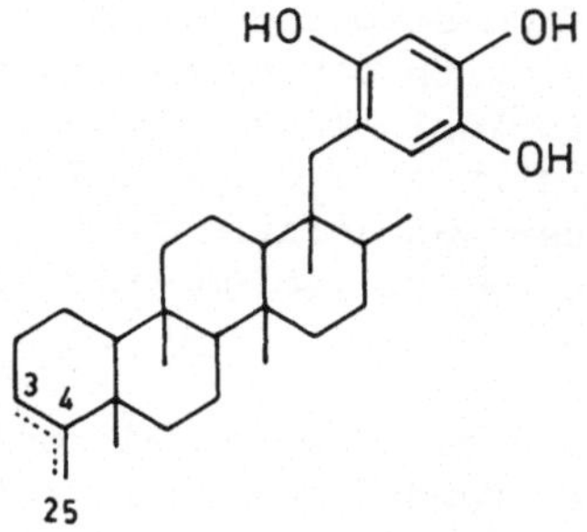

31, Δ³,⁴

32, Δ⁴,²⁵

from Microciona toxystila

Figure 9.   Sesterterpenoid   hydroxyhydroquinones from sponges

33 , dehydrodendrolasin

34 , pleraplysillin-1

35 , pleraplysillin-2

Figure 10.   Furan sesquiterpenes from Pleraplysilla spinifera
(sample 1)

It is worth noting that the sponges Disidea pallescens and
Microciona toxystila show a very similar pattern of secondary
metabolites, although they belong to two different orders. A
similar chemical pattern is also observed for the more polar metabolites. Among these, disidein (30)[12], a pentacyclic sesterterpene
linked to a hydroxyhydroquinone ring, is the most abundant compound
found in Disidea pallescens, while from the extracts of Microciona
toxystila two isomeric sesterterpenes (31) and (32) having
rearranged skeletons linked to a hydroxyhydroquinone residue have
recently been isolated by our group (Figure 9). In the light of
these findings, we think that the classification of Microciona
toxystila as belonging to the order Poecilosclerida is questionable.

The final furanosesquiterpenes to be considered are the unusual
constituents of the sponge Pleraplysilla spinifera. Two subspecies
of this sponge, one of which grows on the marine coelenterate
Paramuricea camaleon, are present in the Bay of Naples. The two
sponge types, according to the authoritative opinion of Professor
Sara (University of Genoa), are identical, from the standpoint of
a spicule analysis, even though they show slight morphological
differences. We have found two different sesquiterpenoid patterns
in the two samples. The structures of the sesquiterpenes isolated
from the first sample[13,14] are listed in Figure 10.

Dehydrodendrolasin (33), which is present in very large amounts
(ca. 5% of the dry sponge), is closely related to dendrolasin, the
defensive-odour substance of the ant Dendrolasius fuliginosus
which was discovered by Quilico's group in 1953[15]. Pleraplysillin-1
(34) belongs to a new type of sesquiterpene with a unique carbon
skeleton. In fact, it seems to arise from a very unusual C-C
cyclization involving the lateral C-14 methyl group of the poly-
isoprene chain. The third component of this sponge is the linear
ester pleraplysillin-2 (35)[14].

The second sample of Pleraplysilla spinifera, which grows on
the coelenterate Paramuricea camaleon, contains (Figure 11) two
very unstable polycyclic compounds, spiniferin-1 (36) and -2 (37)
or (38), along with longifolin (39)[16], a linear furanosesquiterpene
previously characterized by Hayashi et al.[17] from a terrestrial
source, the plant Actinodaphne longifolia. The elucidation of the
structure of spiniferin-2 was greatly aided by a detailed nmr
study and by an oxidative degradation which yielded a dicarboxylic
acid (40) characterized as the dimethyl ester. Unfortunately, the
collected data are equally compatible with structure (37) or with
its isomer (38). At present, additional proof in favour of one or
the other is lacking.

Two alternative structures (41) and (42) have also been proposed for spiniferin-1, the major (0.32% of the dry animal) and
most unstable furanoid component of this sponge. The presence in

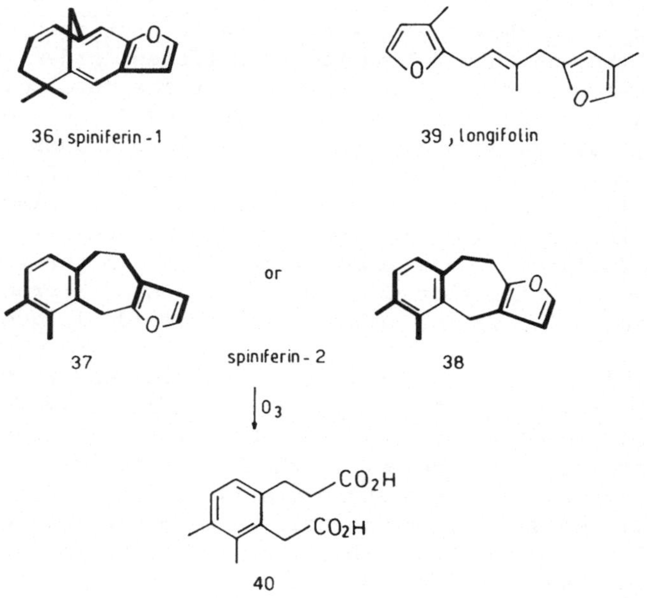

36, spiniferin-1   39, longifolin

37   or   38
              spiniferin-2

$\downarrow O_3$

40

Figure 11.   Furan sesquiterpenes from <u>Pleraplysilla</u> <u>spinifera</u>
(sample 2)

the nmr spectrum of a proton resonating as a doublet at δ 0.75 led
to an incorrect conclusion because we assigned it to a cyclopropyl
hydrogen.   This was the key argument which confused us for some
time.   Very recently, a study of this molecule by ^{13}C nmr, along
with some additional chemical proof and biogenetic considerations,
allowed us to propose the structure (36), which more closely
corresponded to all the collected data.   The ^{13}C nmr spectral data
are given in Figure 12.   There are signals for ten sp[2] carbons and
only five sp[3] carbons.   This finding clearly demonstrated that the
proposed alternative cyclopropane structures (41) and (42) were
both incorrect because they should have eight sp[2] and seven sp[3]
carbons.   It is interesting to note that the signal at 34.0 ppm,
assigned to C-6, in the off-resonance decoupled spectrum appeared
as a pair of doublets.   This is because the two hydrogens at C-6
are extremely magnetically non-equivalent[18], displaying signals

^{13}C n.m.r.

C-atom	ppm	
15	153.2	s
1	140.9	d
5 or 7	131.5	s
8	130.3	d
5 or 7	127.4	s
9	124.9	d
3	118.4	s
4 or 14	112.4	d
2	109.9	d
4 or 14	109.4	d
10	44.3	t
11	39.6	s
6	34.0	dd
13	30.7	q
12	28.2	q

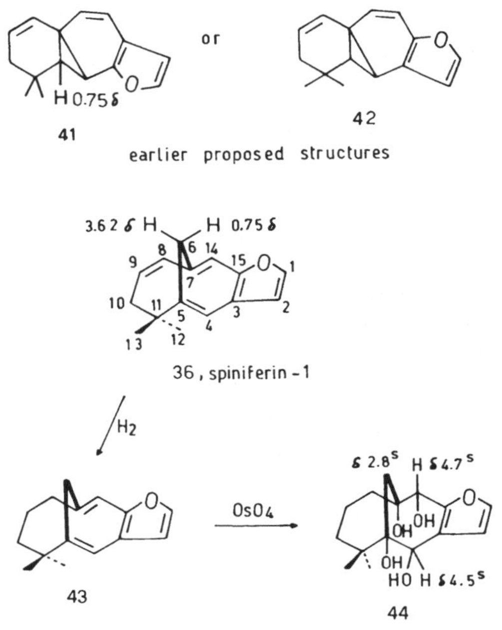

or

earlier proposed structures

36, spiniferin-1

Figure 12. Structure of spiniferin-1

at δ 0.75 and 3.62 in the proton nmr spectrum. Support for the proposed structure came from the very informative nmr spectrum of the tetrol (44), obtained from osmium tetroxide treatment of the dihydroderivative (43). Two singlets at δ 4.5 (1H) and 4.7 (1H) were easily assigned to two isolated CH-OH groups, while the presence of an isolated methylene was indicated by a two-proton singlet at δ 2.8. Decoupling experiments, showing interactions between the CH-OH signals and the furanoid protons, added confirmatory evidence.

In Figure 13 I would like to propose a simple scheme to explain the generation of the wide variety of cyclic furan sesquiterpenes found in sponges. In this scheme, spiniferin-2 (37) or (38) is missing because of the uncertainty regarding its structure. The ion (45) might be the precursor of all the furan sesquiterpenes which have been described, leading, through different cyclizations involving linkages between C-11 and C-14, C-5, and C-6, to the structural types (46), (47), and (19). Structure (46) represents the skeleton of pleraplysillin-1 (34). The carbon skeleton of

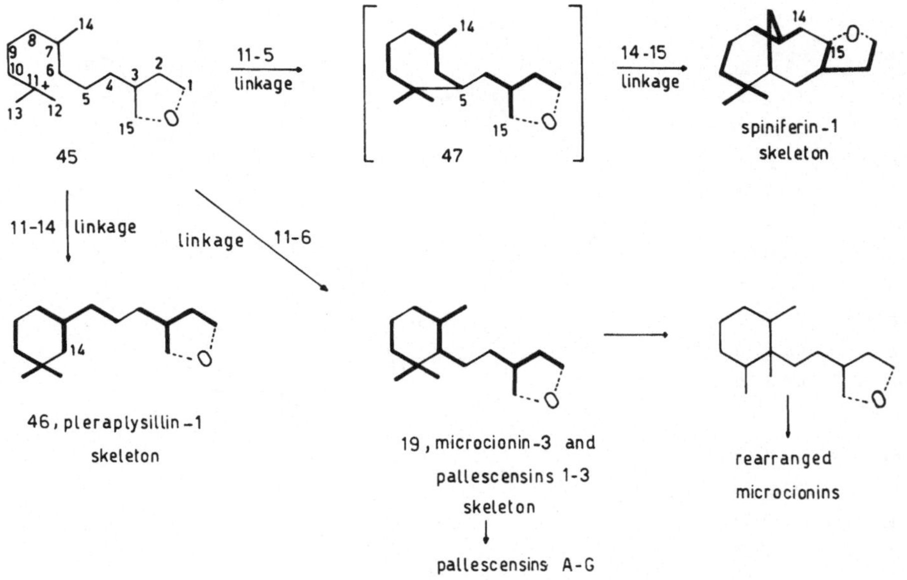

Figure 13. Possible biogenetic scheme for the formation of the whole group of cyclic furan sesquiterpenoids found in sponges

spiniferin-1 can be envisaged as being derived from (47) through further cyclization involving carbons 14 and 15. Finally, the rearranged microcionins and pallescensins A-G could be derived from the monocyclofarnesyl skeleton (19), which is found in the structures of pallescensins 1-3 and microcionin-3.

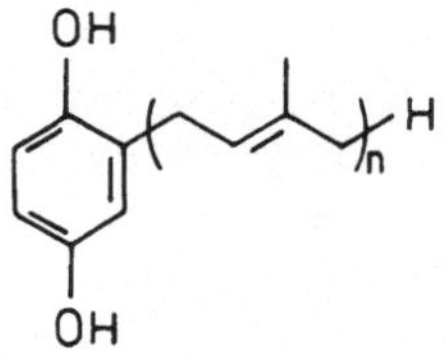

50, n = 2, 4, 5, 6, 7, 8

Figure 14. Linear polyprenyl benzoquinols isolated from sponges

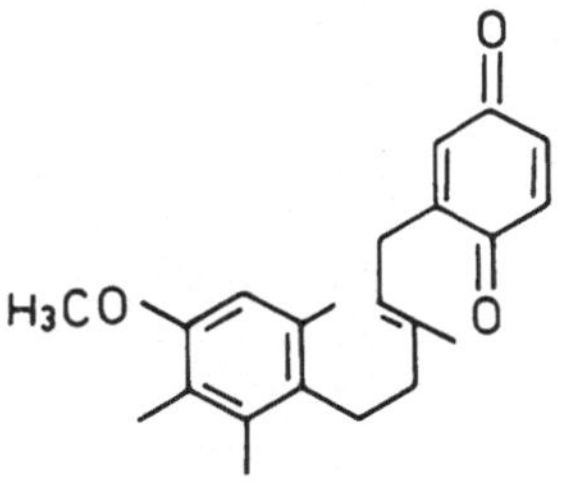

51, panicein-A

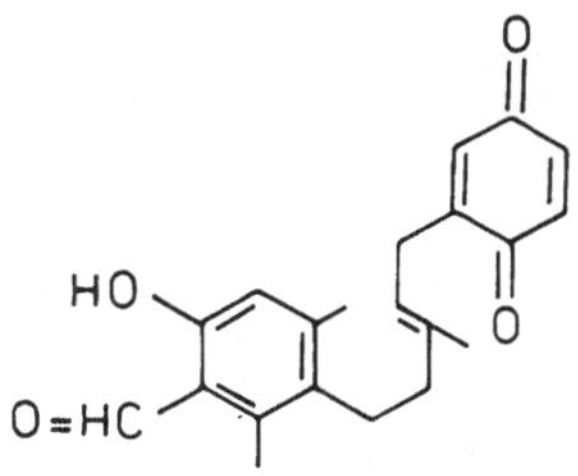

52, panicein-B1

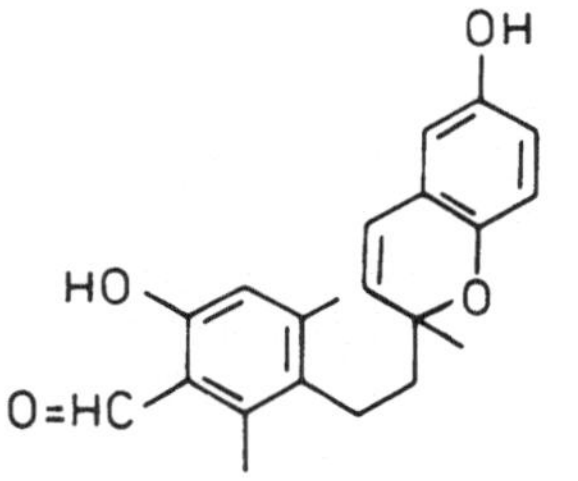

53, panicein-B2

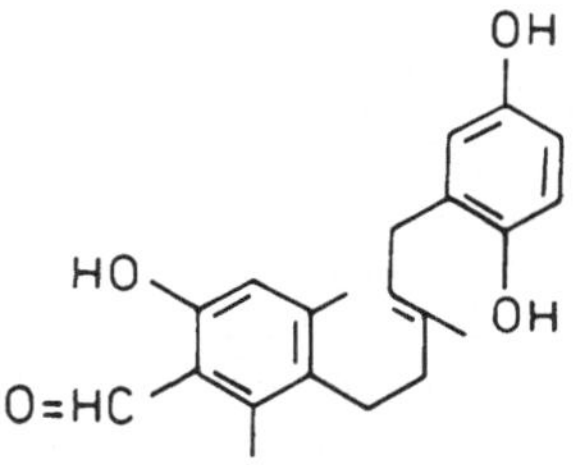

54, panicein-B3

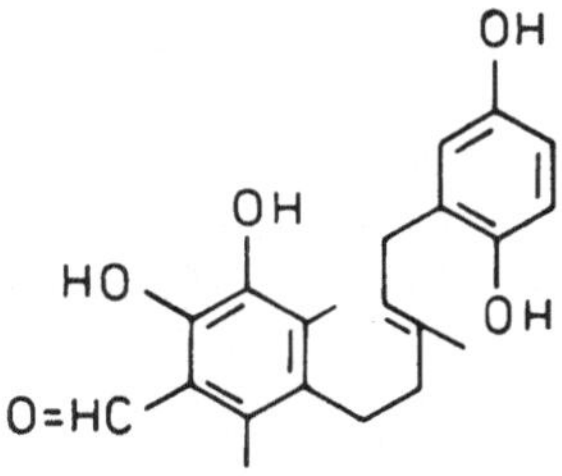

55

or

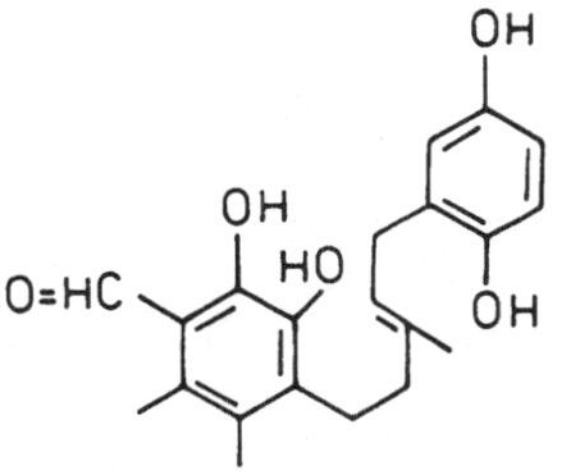

56

panicein-C

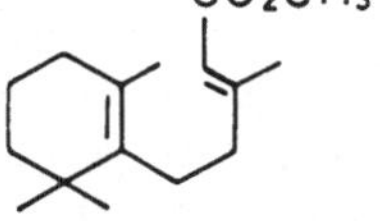

11

Figure 15. Sesquiterpenoids from Halichondria panicea

SESQUITERPENOID HYDROQUINONES

Compounds of mixed biogenesis, arising from the combination
of a part of sesquiterpenic origin with another part having a
different biogenetic origin, are not unusual in nature. Many of
them, such as the mould metabolites grifolin[19], tauranin[20] and
siccanin[21], and also zonarol (48) and isozonarol (49)[22], isolated
by Fenical and his colleagues from a marine organism, the brown
seaweed Dictyopteris zonarioides (syn. D. undulata), have been
shown to possess biocidal properties.

Sponges have also proved to be a rich source of compounds
having a quinone or quinol ring linked to variously cyclized sesquiterpenoid moieties. In spite of these findings, it is interesting to note that, as can be seen in Figure 14, the linear
triprenyl-1,4-benzoquinol is the missing member of a series of
linear polyprenyl benzoquinols isolated from Mediterranean[12,23a,b]
and Australian[24] sponges containing from two to eight isoprenic
units (50).

Until now, sesquiterpenoid benzoquinols have been found in
six species (Table 1), all belonging to the order Dictioceratida
with the exception of Halichondria panicea. The latter has produced five closely related compounds, panicein-A, -B_1, -B_2, -B_3, and
-C(51-56)[25] (Figure 15), all containing an aromatic sesquiterpenoid
moiety linked to a hydroquinol or quinone system except the
chromenol, panicein-B_2 (53), which could be an artifact of
extraction.

The structures of panicein -A, -B_1 and -B_2 were readily deduced
by chemical interrelation with panicein-B_3, along with spectral
evidence. For panicein-C we had to propose two alternative structures (55) and (56), on spectral and chemical grounds, even though
structure (55), with the carbon substitution pattern in the aromatic
terpenoid ring identical to those in the other paniceins, should be
favoured. It is worth noting that the uncommon feature of an
aromatic ring in a terpenoid moiety has already been encountered
in some arylcarotenoids found in the same sponge (syn. Reniera
japonica) by Yamaguchi's group[26a-e] and also among the constituents
of the Hawaiian alga Laurencia nidifica[27], from which Sun and his
colleagues have recently isolated an aromatic sesquiterpenoid
alcohol closely related to the snyderols (13) and (14). The biogenetic derivation of these aromatic groups is a matter for
conjecture, but it is likely that a 1,2 methyl migration, followed
by oxidation, occurs in a β-end group of the carotenoids or in a
monocyclofarnesyl intermediate. This hypothesis is supported by
the presence of methyl trans-monocyclofarnesate (11), which, as is
noted above, coexists with the paniceins in Halichondria panicea.

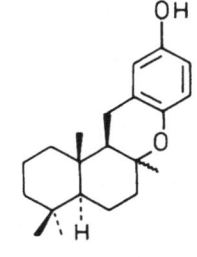

57, <u>ent</u>-chromazonarol

from <u>D.pallescens</u>

58, chromazonarol

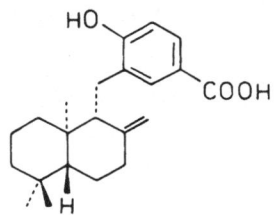

59, zonaroic acid

48, zonarol

from <u>D.zonarioides</u>

49, isozonarol

Figure 16. Sesquiterpenoid phenols from the sponge <u>Disidea</u> <u>pallescens</u> and the alga <u>Dictyopteris</u> <u>zonarioides</u>

The chemical versatility of the sponge Disidea pallescens, which is a rich source of furan sesquiterpenes, was made apparent by the isolation of a chroman-sesquiterpenoid, the dextrorotatory ent-chromazonarol (57)[28] with a drimane skeleton (Figure 16). It is interesting that the levorotatory antipodal isomer chromazonarol (58), with the rare iresane skeleton[29], has been found in the brown alga Dictyopteris undulata[30], along with its fungicidal phenolic isomers, zonarol (48) and isozonarol (49)[22] and zonaroic acid (59)[31]. The absolute stereochemistry of the whole group of compounds was recently established by our research group in conjunction with Fenical and Sims[31] by chemical interrelation with ambrein and manool, diterpenes of known stereochemistry. In the marine environment, few antipodal metabolites have been isolated from different organisms[9], and their biosynthesis is of considerable interest.

A further sesquiterpenoid, 1,4-benzoquinol, having a rearranged drimane skeleton, avarol (60), has been isolated from the Mediterranean sponge Disidea avara[32] by our research group in Naples. Shortly after, the same compound was reported by Baker[24] from an Australian Disidea sp. The structure of avarol (60) was deduced from chemical transformations and spectroscopic data. The relevant chemical data are summarized in Figure 17. Strong evidence for

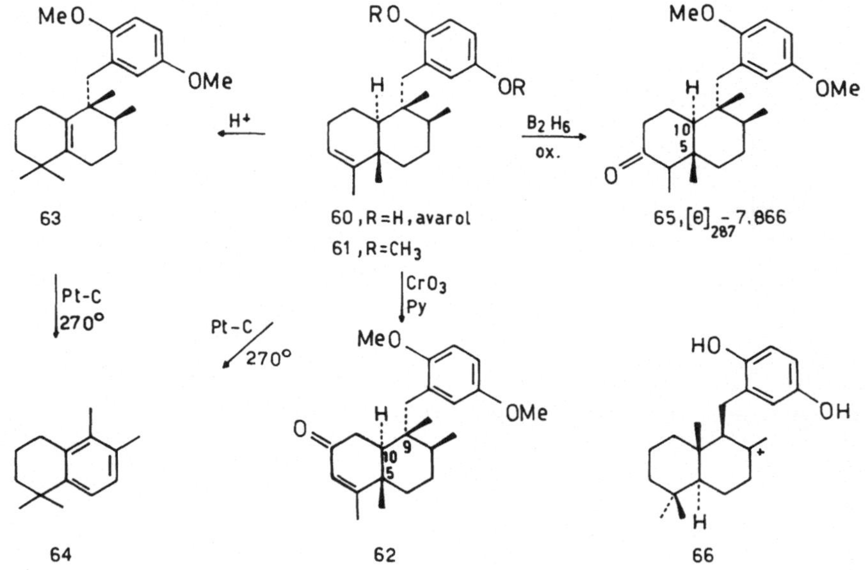

Figure 17. Structure of avarol

the proposed structure has been derived from CrO_3-pyridine complex oxidation of avarol dimethyl ether (61), which yielded the enone (62), the nmr of which clearly indicated that C-10 is tertiary and C-5 and C-9 quaternary.

Furthermore, dehydrogenation of both avarol and its acid catalyzed rearranged product (63), affording 1,2,5,6-tetramethyl-naphthalene and 1,2,5-trimethylnaphthalene, together with a greater amount of tetralin (64), demonstrated conclusively the structure (60) for avarol (without stereochemical implications). The stereo-chemistry of avarol has recently[33] been assigned with the aid of a Eu-induced-shift study on the diastereoisomeric epoxides, obtained by peracid oxidation, and by comparing the ^{13}C nmr spectra of both avarol dimethyl ether and its dihydroderivative with those of a series of cis- and trans-clerodane diterpenoids. Finally, the strong negative Cotton effect shown by the ketone (65) ($[\theta]_{277}$ - 7.866) led to the assignment of the 5β,10α absolute configura-tion. The occurrence of avarol (60), with a rearranged drimane skeleton, and ent-chromazonarol (57), with a drimane skeleton, in two sponges of the same genus is remarkable. It suggests a common biogenetic origin from an intermediate cation, such as (66), from which avarol could derive by a friedo rearrangement and deproto-nation. A further example of co-occurrence of sesquiterpenoids having both rearranged and unrearranged skeletons has recently been encountered by Baker[24]. He reported that the Australian sponge Stelospongia canalis occurs in an orange-coloured form and also in a yellow-coloured one. From the orange-coloured form, the Roche research group characterized four novel quinones (67-70) which were shown to be absent in the yellow-coloured form, which contains the hydroxyquinone (71), featuring a rearranged bicyclofarnesyl skeleton (Figure 18).

SESQUITERPENOID ISONITRILES

The final sesquiterpenoids to be considered are the interest-ing compounds bearing the isonitrile function, very rare in nature, which are listed in Figure 19. Up to 1973, only one naturally-occurring isonitrile, the mould metabolite xanthocil-lin[34a,b], had been reported. Since then, six sesquiterpenoid isonitriles have been isolated from four different sponge species (Table 1). All of them except acanthellin-1 (85) and 9-iso-cyanopupukeanane (84) were accompanied by the corresponding formamides and isothiocyanates. The relationship between these compounds was proved by converting the isonitriles to the isothio-cyanates and to the formamides. These findings prompted both Fattorusso's and Scheuer's groups to propose that a formamide could be the biogenetic precursor of the rare isonitrile function. This suggestion seems particularly relevant when we consider that in the earlier xanthocillin case all attempts to discover the bio-synthetic origin of the isonitrile function had failed[35].

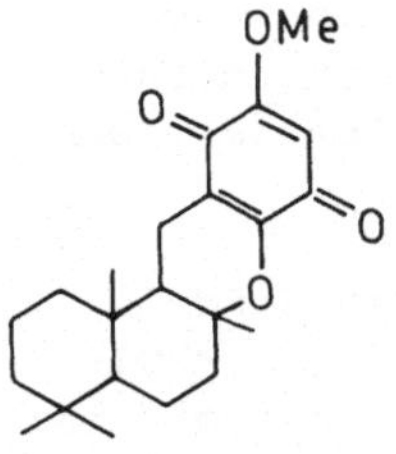

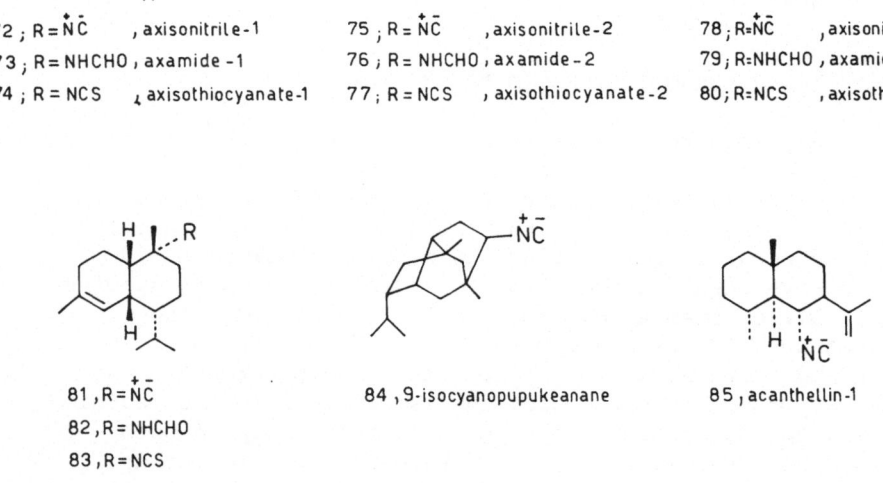

67 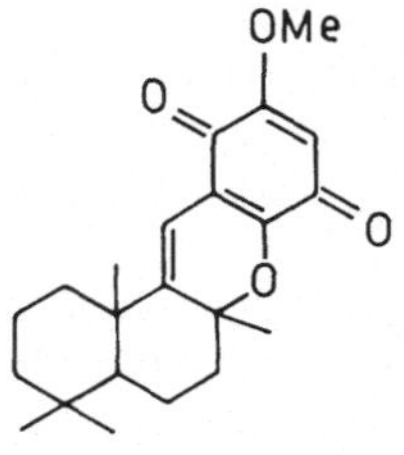   68 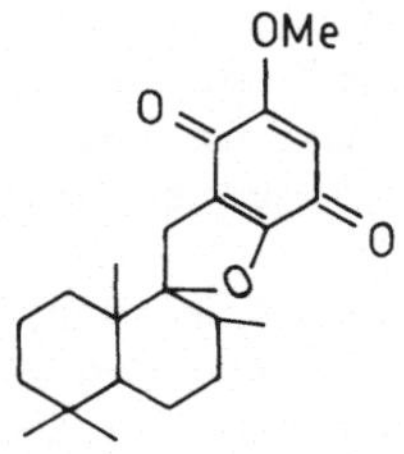   69

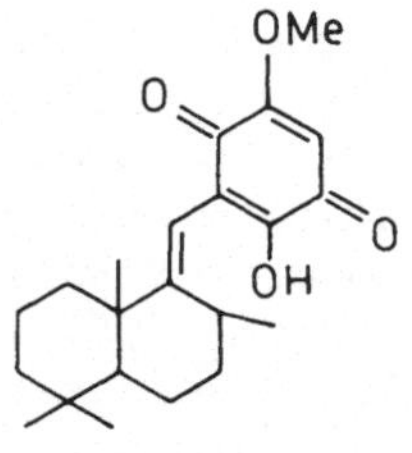

70 (from the orange c. form) 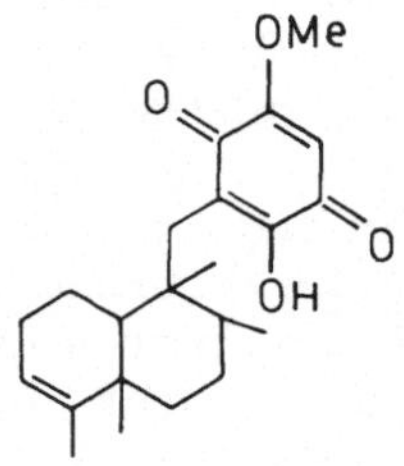   71 (from the yellow coloured form)

Figure 18. Sesquiterpenoids from the Australian sponge
Stelospongia canalis

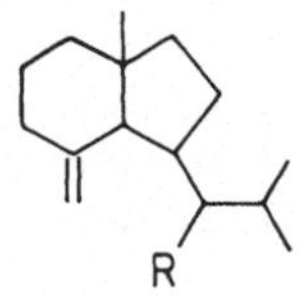

72 ; R = N̊C̄   , axisonitrile-1   75 ; R = N̊C̄   , axisonitrile-2   78 ; R=N̊C̄   , axisonitrile-3
73 ; R= NHCHO , axamide-1   76 ; R= NHCHO , axamide-2   79; R=NHCHO , axamide-3
74 ; R = NCS   , axisothiocyanate-1 77; R = NCS   , axisothiocyanate-2 80; R=NCS   , axisothiocyanate-3

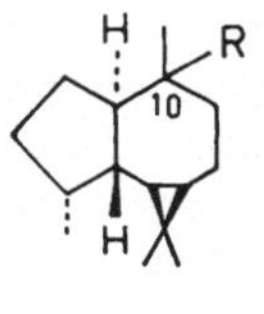

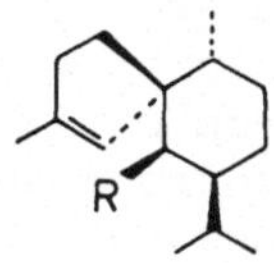

81 , R = N̊C̄   84 , 9-isocyanopupukeanane   85 , acanthellin-1
82 , R = NHCHO
83 , R = NCS

Figure 19. Sesquiterpenoid isonitriles, amides and isothiocyanates
from sponges

Fattorusso's group isolated the axisonitrile, axisothiocyanate, and axamide series from the sponge <u>Axinella</u> <u>cannabina</u>[36,37,38,39]. The structure of axisonitrile-1 (72)[36] was carefully and convincingly elucidated by chemical transformations and spectral determinations. The skeletal type of axisonitrile-1, which is probably derived in nature by ring contraction from a eudesmane precursor[40], has a further representative in oppositol[41], a brominated sesquiterpene alcohol found in the red alga <u>Laurencia</u> <u>subopposita</u>. The structure of axisonitrile-2 (75)[37], was secured by Birch reduction to a hydrocarbon identical to the hydrogenation product of aromadendrene. Since both reductions afforded a mixture of the two epimers at C-10, the authors, on the basis of this finding, were able to place the isonitrile function at C-10, thus establishing the stereochemistry except for C-10. The structure of the crystalline axisonitrile-3 (78) was secured by X-ray diffraction[39].

Acanthellin-1 (85) was isolated from the sponge <u>Acanthella</u> <u>acuta</u>[42], a species also belonging to the order Axinellida. The 4-epi-eudesmane structure, proposed mainly on the basis of a careful spectral analysis, was unambiguously secured by glc comparison

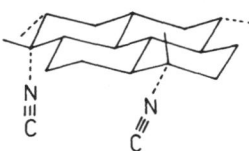

86, R = N̄C̄
87, R = NHCHO   from <u>Halichondria</u> sp.
88, R = NCS

89 from <u>Adocia</u> sp.

Figure 20.   Non-sesquiterpenoid isonitriles from sponges

of the hydrocarbon derived by Birch reduction with an authentic
sample of 4-epieudesmane.

Scheuer's group established the carbon framework, the site of
functionalization, and the absolute stereochemistry of the amorphane sesquiterpenoid group (81-83) found in a sponge of the genus
Halichondria[43].

The same group also discovered 9-isocyanopupukeanane (84)[44],
a rearranged sesquiterpenoid isonitrile lethal to fish and crustaceans, in an off-white sponge, Hymeniacidon sp., and also in its
browser, the nudibranch Phyllidia varicosa. The authors pointed
out that this defensive substance protects the delicate shell-less
mollusk from its predators and, at the same time, is the allomone
of the browser-prey relationship.

Two more non-sesquiterpenoid isonitriles have been reported
from sponges. The Halichondria species[43] was shown to contain, in
addition to the three above-mentioned amorphane sesquiterpenoids,
three linear diterpenoids bearing isonitrile (86), formamide (87),
and isothiocyanate (88) functions at C-3 (Figure 20), while Baker[24]
has recently announced the isolation from a Queensland sponge,
Adocia sp., of the first reported diisonitrile from a marine organism, the unsymmetrically-substituted hexadehydropyrene (89).

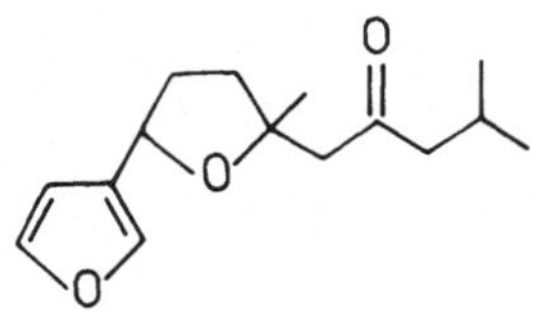

90 , ipomeamarone

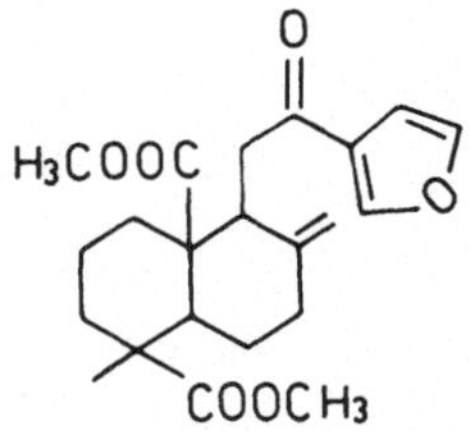

91 , dimethyl sciadonate

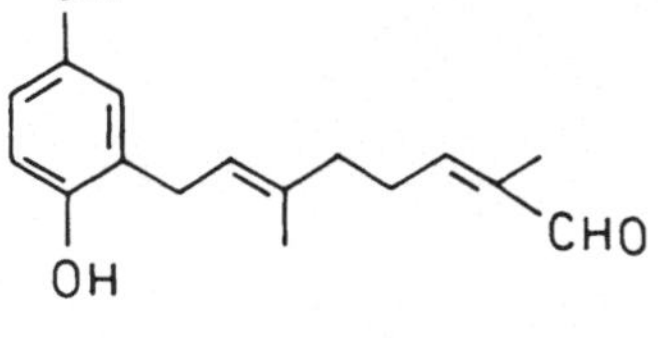

92 , alliodorin

In conclusion, we do not yet know the function of the sesquiterpenoids discovered in sponges. The biological properties of
similar terpenoids suggest that they could be not merely metabolic
"waste products" and could have important biological functions.
For instance, furanoterpenoids such as ipomeamarone (90)[45a,b,c]
are known to belong to phytoalexins and participate, by their
antibiotic activities[46], in the defence reaction of sweet potatoes
against penetrating fungi. Besides this, some furanoterpenes such
as dimethyl sciadonate (91) are of significance as growth inhibitors
for silk worm larvae[47]. It is also worth mentioning that Stevens
and Jurd reported the isolation of alliodorin (92)[48], a quinol
terpenoid compound from <u>Cordia alliodora</u>, a Panamanian tree which
possesses considerable resistance to attack by marine organisms[49],
termites[50], and terrestrial fungi[51].

It therefore seems to us that sesquiterpenoids in sponges
could play a role in chemical defence against marine predators,
as has been shown in the case of 9-isocyanopupukeanane (84).
However, a deeper understanding of these compounds, gained by
means of a closer study of their biological properties, is needed.
This will be made easier by a wider collaboration between organic
chemists, physiologists, and experts in other natural disciplines.

ACKNOWLEDGEMENTS

It is my privilege to thank the members of our research team,
Professor L. Minale and Drs. S. De Stefano, A. Guerriero, and
E. Trivellone, for their contributions to the work described in
this paper. The cooperation of the Zoological Station (Naples)
in the collection of sponges is acknowledged.

REFERENCES

1. G. Cimino, S. De Stefano, A. Guerriero, and L. Minale, Tetrahedron Lett., 1417 (1975).

2. V. Herout, in <u>Aspects of Terpenoid Chemistry and Biochemistry</u>,
   ed. T. W. Goodwin, Academic Press, London, 1971.

3. G. Cimino, S. De Stefano, and L. Minale, Experientia <u>29</u>, 1063
   (1973).

4. G. Cimino, S. De Stefano, A. Guerriero, and L. Minale, Tetrahedron Lett., 3723 (1975).

5. B. M. Howard and W. Fenical, Tetrahedron Lett., 41 (1976).

6. A. G. Gonzáles, J. D. Martin, C. Perez, and M. A. Ramírez, Tetrahedron Lett., 137 (1976).

7. G. Cimino, S. De Stefano, A. Guerriero, and L. Minale, Tetrahedron Lett., 1425 (1975).

8. G. Cimino, S. De Stefano, A. Guerriero, and L. Minale, Tetrahedron Lett., 1421 (1975).

9. P. J. Scheuer, Chemistry of Marine Natural Products, Academic Press, New York and London, 1973.

10. M. Suzuki, E. Kurosawa, and T. Irie, Tetrahedron Lett., 4995 (1970).

11. H. Sasaki, T. Hosokawa, Y. Nawata, and K. Ando, Agri. Biol. Chem. 38, 1463 (1974).

12. G. Cimino, P. De Luca, S. De Stefano, and L. Minale, Tetrahedron 31, 271 (1975).

13. G. Cimino, S. De Stefano, L. Minale, and E. Trivellone, Tetrahedron 28, 4761 (1972).

14. G. Cimino, S. De Stefano, and L. Minale, Experientia 30, 846 (1974).

15. A. Quilico, F. Piozzi, and M. Pavan, Tetrahedron 1, 177 (1957).

16. G. Cimino, S. De Stefano, L. Minale, and E. Trivellone, Tetrahedron Lett., 3727 (1976).

17. H. Hayashi, H. Komal, S. Eguchi, M. Nakayama, S. Hayashi, and T. Sakao, Chem. Indus., 572 (1972).

18. E. Wenkert, D. W. Cochram, E. W. Hagaman, F. M. Schell, N. Neuss, A. S. Katner, P. Potier, C. Kan, M. Plat, M. Koch, H. Mehri, J. Poisson, N. Kunesh, and Y. Rolland, J. Amer. Chem. Soc. 95, 4990 (1973).

19. T. Gato, H. Kakisawa, and Y. Hirata, Tetrahedron 19, 2079 (1963).

20. K. Kakisawa, K. Nakanishi, and H. Nishikawa, Chem. Pharm. Bull. Japan 12, 796 (1964).

21. K. Hirai, K. T. Suzuki, and S. Nozoe, Tetrahedron 27, 6057 (1971).

22. W. Fenical, J. J. Sims, D. Squatrito, R. M. Wing, and P.
     Radlick, J. Org. Chem. 38, 2388 (1973).

23. (a) G. Cimino, S. De Stefano, and L. Minale, Tetrahedron 28,
     1315 (1972); (b) G. Cimino, S. De Stefano, and L. Minale,
     Experientia 28, 1401 (1972).

24. J. T. Baker, Marine Natural Products Int. Symposium, Aberdeen,
     September, 1975.

25. G. Cimino, S. De Stefano, and L. Minale, Tetrahedron 29, 2565
     (1973).

26. (a) M. Yamaguchi, Bull. Chem. Soc. Japan 30, 111 (1957); (b)
     M. Yamaguchi, Bull. Chem. Soc. Japan 30, 1979 (1957); (c)
     M. Yamaguchi, Bull. Chem. Soc. Japan 31, 51 (1958); (d)
     T. Hamasaki, N. Okukado, and M. Yamaguchi, Bull. Chem. Soc.
     Japan 46, 1884 (1973).

27. H. H. Sun, S. M. Waraszkiewicz, and K. L. Erickson, Tetrahedron
     Lett., 585 (1976).

28. G. Cimino, S. De Stefano, and L. Minale, Experientia 31, 1117
     (1975).

29. G. Rucher, Ang. Chem. Intern. Edit. 12, 783 (1973).

30. W. Fenical and O. McConnell, Experientia 31, 1001 (1975).

31. G. Cimino, S. De Stefano, W. Fenical, L. Minale, and J. J. Sims,
     Experientia 31, 1250 (1975).

32. L. Minale, R. Riccio, and G. Sodano, Tetrahedron Lett., 3401
     (1974).

33. S. De Rosa, L. Minale, R. Riccio, and G. Sodano, J.C.S. Perkin
     I, 1408 (1976).

34. (a) I. Hagedorn and H. Tonjes, Pharmazit. 12, 567 (1957);
     (b) I. Hagedorn and H. Tonjes, Chem. Abs. 52, 6362 (1958).

35. H. Achenbach and H. Griseback, Z. Naturforsch. B 20, 137
     (1965).

36. F. Cafieri, E. Fattorusso, S. Magno, C. Santacroce, and D.
     Sica, Tetrahedron 29, 4259 (1973).

37. E. Fattorusso, S. Magno, L. Mayol, C. Santacroce, and D. Sica,
     Tetrahedron 30, 3911 (1974).

38.   E. Fattorusso, S. Magno, L. Mayol, C. Santacroce, and D. Sica,
      Tetrahedron 31, 3911 (1975).

39.   B. Di Blasio, E. Fattorusso, S. Magno, L. Mayol, C. Pedone,
      C. Santacroce, and D. Sica, Tetrahedron 32, 473 (1976).

40.   T. Mooney, in Terpenoids and Steroids, V. 5, The Chemical
      Society, London, 1975.

41.   S. S. Hall, D. J. Faulkner, J. Fayos, and J. Clardy, J. Amer.
      Chem. Soc. 95, 7187 (1973).

42.   L. Minale, R. Riccio, and G. Sodano, Tetrahedron 30, 1341
      (1974).

43.   B. J. Burreson, C. Christophersen, and P. J. Scheuer, Tetra-
      hedron 31, 2015 (1975).

44.   B. J. Burreson, P. J. Scheuer, J. Finer, and J. Clardy,
      J. Amer. Chem. Soc. 97, 4763 (1975).

45.   (a) M. Hiura, Report from GIFU Agri. Coll. 50, 1 (1943);   (b)
      T. Kubota and T. Matsuura, J. Chem. Soc. Japan 24, 101, 197,
      248, 668 (1953); (c) T. Akazawa, Arch. Biochem. Biophys. 90,
      82 (1960).

46.   I. Uritani, Ann. Rev. Phytopathol. 9, 211 (1971).

47.   C. Chang, A. Isogai, T. Kamikado, S. Murakoshi, A. Sakurai,
      and S. Tamura, Agr. Biol. Chem. 39, 1167 (1975).

48.   K. L. Stevens and L. Jurd, Tetrahedron 32, 665 (1976).

49.   C. R. Southwell and J. Bultman, Biotropica 3, 81 (1971).

50.   T. C. Schefer and C. G. Duncan, Tropical Woods 92, 1 (1947).

51.   J. P. Perry and J. Martinez Lima, J. Forestry 62, 398 (1964).

*Laboratorio per la Chimica di Molecole di Interesse Biologico del
 C.N.R. - Via Toiano, 2, Arco Felice, Napoli, Italy.

NON-CONVENTIONAL STEROLS OF MARINE ORIGIN

L. Minale and G. Sodano

Laboratorio per la Chimica di Molecole di Interesse Bio-

logico del C.N.R and Istituto di Chimica dell'Universita[*]

Our knowledge of sterols and sterol metabolism in marine invertebrates has increased very rapidly in the last few years. for many years the known carbon range of sterols extended from C_{27} to C_{29}, and the carbon variation occurred exclusively in the side chain at C_{24}[1]. Only recently have biogenetically interesting C_{26} and C_{30} sterols, C_{27} and C_{29} sterols with side chains involving new alkylation patterns, and, more notably, sterols with modified tetracyclic nuclei been reported from marine invertebrates. In this paper we will present a survey of these uncommon sterols, termed in the title "non-conventional".

At the present time there is no full explanation of the biosynthetic origin of these sterols, but we will attempt to focus attention on this problem by discussion of the few results obtained using tracer techniques and, also, by speculation based on the common structural features recognized in different molecules and their distribution.

Some of these sterols are widely distributed and occur as a low percentage (1-2%) of sterol mixtures; for these sterols, distribution through the food chain can be suspected, since many species are incapable of de novo sterol synthesis. Other sterols, with distributions restricted to certain classes, genera, or species, are present in the sterol mixtures in very high percentages; for these sterols, the participation of the invertebrate in the biosynthesis or in the modification of sterols taken up from the diet is most likely.

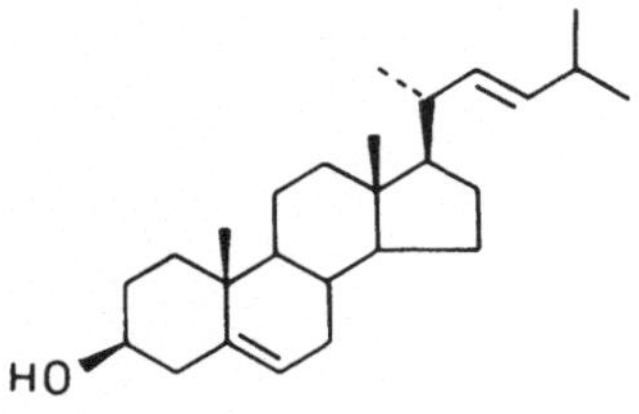

Figure 1. C_{26} sterols from marine sources.

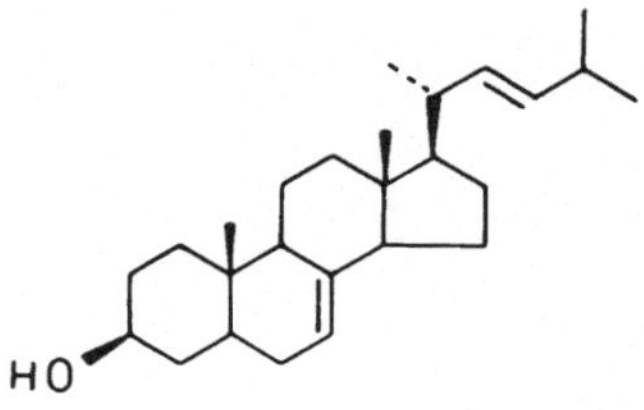

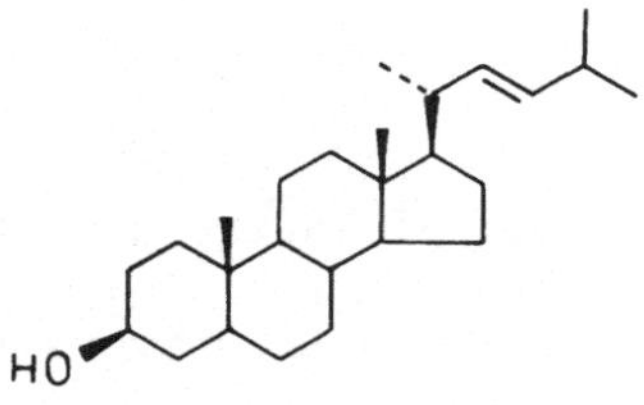

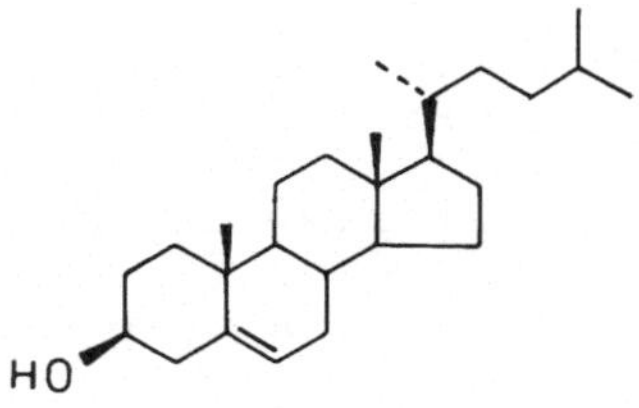

The sterols are usually grouped according to carbon content, and in this paper we will discuss the C_{26} sterols, the 27-norergo-stane-type sterols, the sterols having side chains modified by addition of "extra" carbon atoms at biogenetically unprecedented positions, and the uncommon sponge-derived sterols, including both those having unprecedented side-chain alkylation patterns and those with modified tetracyclic nuclei, in that order.

C_{26}-STEROLS

Figure 1 lists the structures of the C_{26} sterols so far iso-lated; the first of these unprecedented sterols with a C-7 side chain, 22-trans-24-norcholesta-5,22-diene-3β-ol (1), or, if you prefer, 22-trans-26,27-bisnorergosta-5,22-dien-3β-ol, was isolated by Idler et al.[2] from the mollusc (bivalve) Placopecten magellanicus, where it occurs as a very minor component. The same C_{26} sterol was also detected by the same authors in eighteen other species of molluscs belonging to the classes Pelecypoda and Gastropoda and in one species of phylum Molluscoidae, in a low percentage, ranging from 0.1 to ca. 6% of their total sterol content[3]. The structure was established by spectral techniques, especially by nmr, which clearly indicated a terminal isopropyl group, and was confirmed by partial synthesis, first reported by Fryberg et al.[4] and later by Barbier's group[5] in Gif-sur-Yvette (France). The synthesis eliminated the uncertainties associated with the stereochemistry at C-20, which was shown to be 20 R, as found in cholesterol, and with the configuration of the 22-double bond, which was shown to be trans.

The $\Delta^{7,22}$ C_{26} sterol (2) was first reported by Barbier's group from the tunicate Halocynthia roretzi[6], where it occurs along with the $\Delta^{5,22}$-isomer (1) and the dihydro-derivative (3). At the same time, (2) was detected in several asteroids and in the holothurian Stichopus japonicus by Kobayashi et al.[7]. The latter authors, who described the new sterol as a crystalline material from Asterias amurensis[8], have named it asterosterol, in view of its fairly wide distribution in asteroids. The Δ^7 nuclear unsaturation in asterosterol is not unexpected, in view of the predominant occurrence in both asteroids (sea stars) and holothurians (sea cucumbers) of the rare Δ^7 sterols, which are thought to arise by modification of the appropriate dietary Δ^5 sterols[9]. Very recently, the biotransformation of Δ^5 sterols into Δ^7 sterols was demonstrated by Smith and Goad[10] as occurring in at least two species, Asteria rubens and Solaster papposus. The Japanese workers have suggested that asterosterol (2) may be formed by asteroids from digested $\Delta^{5,22}$-C_{26} sterol (1)[8].

The structure of asterosterol was also confirmed by synthesis, which was announced almost simultaneously by Kobayashi et al.[11] and Boll[12], who later reported the isolation of asterosterol from the marine sponge Halichondria panicea.

C_{26} sterols have also been isolated from two other sponge species by Erdmann and Thomson[13].

The structure of the C_{26} sterol (3) was confirmed by synthesis by Metayer and Barbier[14], thus confirming the 20 R configuration and the trans stereochemistry of the 22-double bond.

The fourth member of this group has been recently detected as a very minor component in the scallop Placopecten magellanicus by Idler et al.[15].

The C_{26} sterols appear to be widespread in marine invertebrates, and very probably they are ubiquitous. During a recent gas chromatograph-mass spectrometer examination of the sterol mixtures of twenty-five different sponge species, representative of eleven families and seven orders, the ubiquitous occurrence of a minor component of short retention time was noted[16]; Goad et al.[9,17,18] enlarged the number of species of Echinodermata in which the C_{26} sterols have been detected, demonstrating the presence of the

Table 1. Marine animal phyla in which C_{26} sterols have been detected.

Phylum	Class
Porifera	Desmospongiae
Mollusca	Gastropoda
	Pelecypoda
Annelida	Polychoeta
Tunicata	Ascideacea
Coelenterata	Anthozoa
Brachiopoda	
Molluscoidea	
Echinodermata	
	Asteroidea
	Holothuroidea
	Ophiuroidea
	Echinoidea

$\Delta^{5,22}$-C_{26} sterol as a minor component (2%) in two species of ophiuroids (Ophiuroidea) and one of echinoids (Echinoidea); in addition, the asteroids and holothuroids they examined contained low levels of the $\Delta^{7,22}$-C_{26} sterol. More recently, Voogt et al.[19] have detected, by glc, C_{26} sterols in three sea anemones (Anthozoa, Coelenterata), and Voogt[20] and Kobayashi et al.[21] have reported the presence of C_{26} sterols in some Annelida. So, to date, the C_{26} sterols have been reported in seven marine phyla, the annelides, the coelenterates, the echinoderms, the molluscs, the molluscoids, the tunicates and the sponges (Table I). The unique C_{26} skeleton and the ubiquitous occurrence of these compounds pose questions concerning both the type of organism in which they originate and the manner of their biosynthesis. The discovery by Barbier's group of a C_{26} sterol in a marine phytoplankton suggested that this may be the common origin of all the C_{26} marine animal sterols[23], but these authors could not detect radioactive $\Delta^{5,22}$-C_{26} sterols (1) in the red alga Rhodymenia palmata after administration of labelled acetate, mevalonate and methionine[23]; radioactivities from acetate and mevalonate were recovered in the C_{27}-C_{29} sterols, while radio-activity from ^{14}C-methyl methionine was found, as expected, in the C_{28} and C_{29} sterols.

The authors gave different explanations of their results, but they concluded that the problem of the biological origin of the C_{26} sterols remained unsettled at that stage and that the main question to be solved was the discovery of the type of organism which could, in fact, synthesize the C_{26} skeleton. While evidence on the origin and biogenesis of these interesting compounds is lacking, many authors have speculated on their formation. The idea we prefer is that the C_{26} sterols arise by demethylation of a

24-methylated sterol. This seems well supported by the recent discovery by Kobayashi and Mitsuhashi[24-26] of a novel group of sterols, which appear to be widely distributed among marine phyla, having an unprecedented 27-norergostane-type side chain. Furthermore, the ability of organisms to dealkylate sterols is not surprising, in view of the evidence recently accumulated demonstrating that many arthropod species dealkylate dietary C_{28} and C_{29} sterols[27,28].

THE 27-NORERGOSTANE-TYPE STEROLS

Three members of this group have been isolated to date: 27-norergosta-7,22-dien-3β-ol (5), named amuresterol, its Δ^5-isomer 27-norergosta-5,22-dien-3β-ol (6), named occelasterol, and 27-norergosta-22-en-3β-ol (7), named patinosterol (Figure 2).

The structures of the first two members of this group were established and confirmed by partial syntheses by Kobayashi and Mitsuhashi in 1974. They found amuresterol (5) in six species of asteroids[24]. Occelasterol (6), first extracted from the annelid Pseudopotamilla occelata, was found[25] to be present in a large number of invertebrates belonging to different phyla (Table 2). Very recently, patinosterol (7) has been isolated by the same authors from the scallop Patinopecten yessoensis[26], the sterol fraction of which contains over twenty components. All three

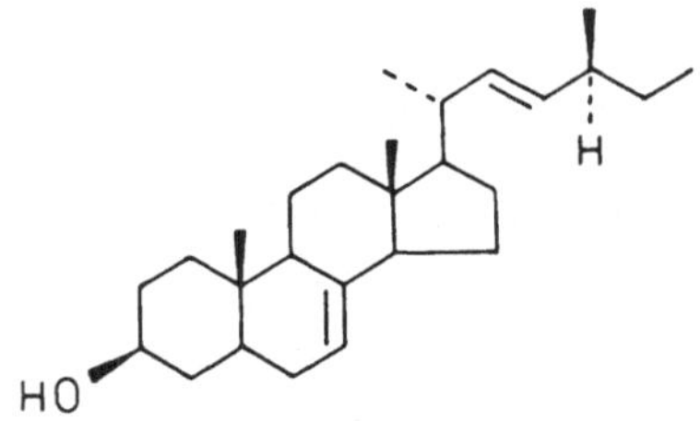

5, amuresterol

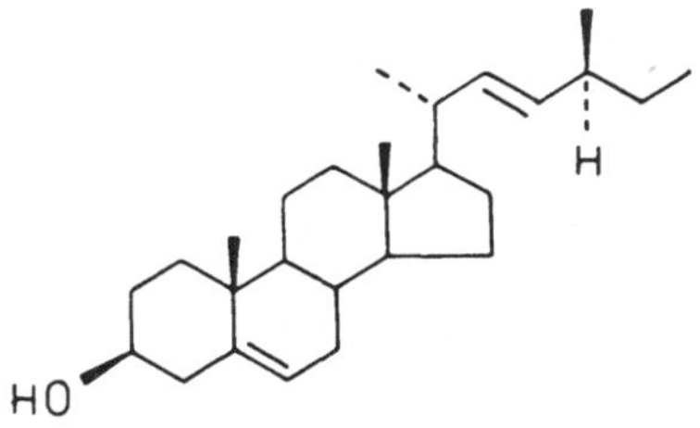

6, occelasterol

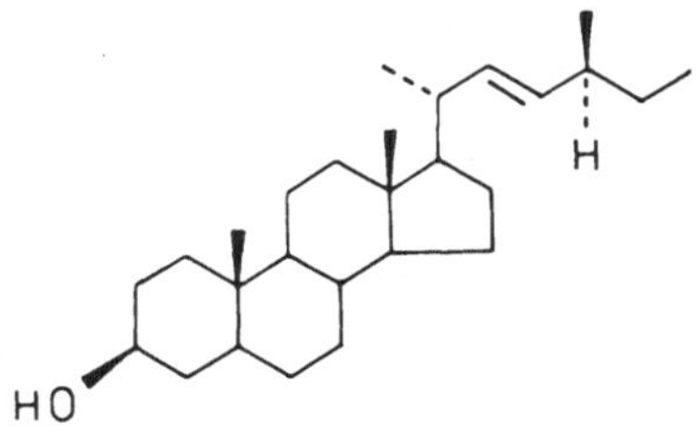

7, patinosterol

Figure 2. The 27-norergostane-type sterols from marine sources.

Table 2.   Marine animal phyla in which 27-norergostane-type sterols have been detected.

Phylum	Class
Coelenterata	Anthozoa
Annelida	Polychaeta
Tunicata	Ascidiacea
Mollusca	Bivalvia
	Gastropoda
Echinodermata	Asteroidea
	Holothuroidea
	Echinoidea

sterols, which are present in low percentages (1-2%), closely resemble the conventional Δ^{22}-trans-C_{27}-sterol analogues in ir and ms properties; however, they show slightly shorter retention time in glc than the corresponding conventional C_{27} sterols, and their acetates are slightly less polar on argentation chromatography.

Furthermore, the same authors also observed[25] that the sterol from marine invertebrates observed by many workers in the past, thought to be 22-cis-cholesta-5,22-dien-3β-ol from its glc retention time and mass spectral data, might, in fact, be occelasterol (6). This observation further enlarges the distribution of these 27-norergostane-type sterols in marine sources (Table 2).

As far as the biosynthetic origin of these sterols is concerned, three hypotheses can be formulated on structural grounds, taking into account the fact that mussels[29], sea anemones[29,30], and crustaceans[29,31], in which these sterols have been detected, are known to be unable to synthesize sterols de novo and that these sterols must, therefore, be accumulated via the food chain:

(1) by conversion from the corresponding dietary Δ^{22}-C_{27} sterol by an internal methyl migration;

(2) by methylation of a C_{26} sterol; or

(3) by demethylation of an ergostane-type sterol.

Since the sterol fraction of some diatoms (diatoms constitute the major biomass in marine water) was recently reported to be entirely composed of 22-trans-(24S)-24-methylcholesta-5,22-dien-3β-ol (the 24α isomer of brassicasterol, which was also isolated from P. occelata), and since occelasterol (6) has the same stereochemistry at C-24, Kobayashi and Mitsuhashi[25] conclude that the coincidence of the configuration at C-24 seems to suggest that

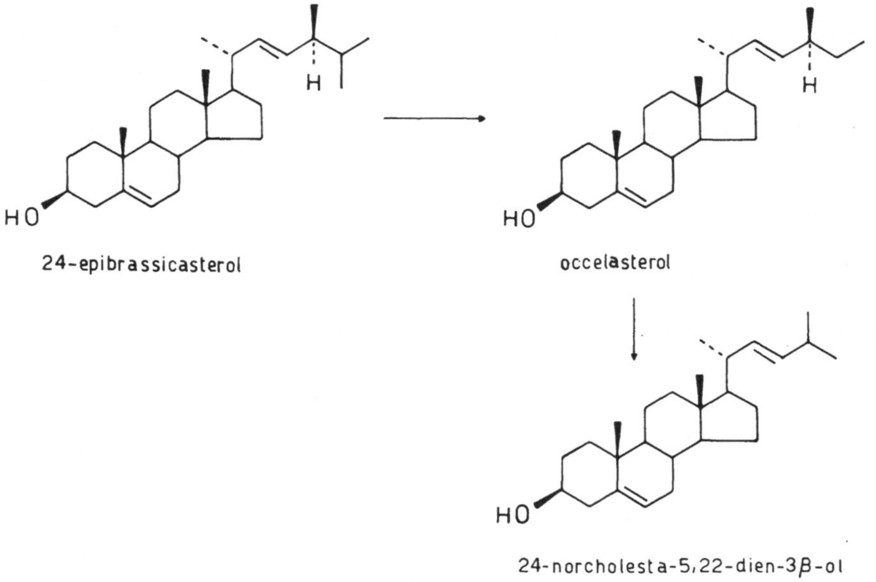

24-epibrassicasterol   occelasterol

24-norcholesta-5,22-dien-3β-ol

Figure 3. Possible biogenetic sequence to C_{26} and 27-norergostane-type sterols according to Kobayashi and Mitsuhashi[26].

occelasterol (6) was derived by demethylation from diatom sterol and later distributed and accumulated in marine invertebrates via the food chain. The possible biogenetic sequence to 26-norergo-stane-type and C_{26} sterols suggested by Kobayashi and Mitsuhashi[26] is summarized in Figure 3. The Δ^{22} and $\Delta^{7,22}$-sterols might be formed in different ways, directly from the corresponding $\Delta^{5,22}$-sterols or by demethylation of the Δ^{22} and $\Delta^{7,22}$-precursors. The authors also noted that C_{26} and 27-norergostane-type sterols always occur together and in association with a large amount of 24(S)-24-methyl sterols. Clearly, much work is required before any conclu-sions can be drawn about the origin of these unprecedented sterols, but, as in the case of the C_{26} sterols, which are probably C_{26} demethylated forms of occelasterol (6), the main problem to be solved is the discovery of the type of organism which fabricates the 27-norergostane skeleton and which could be the same as that which synthesizes the 26,27-bisnorergostane skeleton.

We have discussed the "non-conventional" sterols having side chains modified by the apparent loss of one or two carbon atoms from a normal C_9 side chain. Now let us consider the "non-conventional" sterols having side chains modified by the addition of "extra" carbon atoms at biogenetically unprecedented positions (e.g. C-22, C-23 and C-26) of a normal C_9 side chain. These seem to have a limited distribution among the marine phyla, in contrast to the wide occurrence of the above nor-sterols.

Table 3.   Occurrence of 22,23-cyclopropane-containing sterols.

Phylum	Class	Species	Reference
Coelenterata	Anthozoa	Gorgonia flabellum	40
		G. ventilina	40
		Plexaura sp.	51
		P. flexuosa	37
		Pseudopterogorgia americana	41
		Nephthea sp.	52
		Palythoa tuberculosa	53
Echinodermata	Asteroidea	Acanthaster planci	34,35

STEROLS WITH SIDE CHAINS MODIFIED BY ADDITION OF
EXTRA CARBON ATOMS AT POSITIONS OTHER THAN C-24

At the present time, two types of C_{30} sterols have been discovered in marine invertebrates: a 24-propylidene-substituted
sterol and those having a cyclopropane ring in the side chain at
the position 22,23.

(Z)-24-propylidenecholest-5-en-3β-ol (8) was first extracted
by the Idler group in Canada from the scallop Placopecten magellanicus[32], in which this sterol accounted for about 1% of the
sterol mixture. Later, Sheikh and Djerassi[33] identified this
sterol in the sponge Tethya aurantia as a minor component and also
described the synthesis. Unidentified C_{30} sterols have been
noticed[16] in other sponges.

All sterols reported to date which carry a cyclopropane moiety
at position 22,23 in the side chain (Figure 4) occur in soft corals
(Table 3), with the exception of acanthasterol (10), which was
isolated from the asteroid Acanthaster planci[34,35]. But because
Acanthaster planci feeds on soft corals, and in view of the abil-
ity[9,10] of asteroids to transform ingested Δ^5 sterols to Δ^7 sterols,
it is very probable that acanthasterol in the sea star originates
in ingested gorgosterol (9).

8

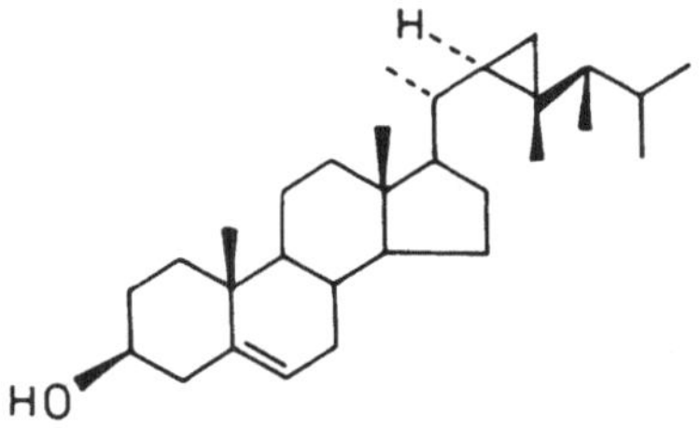

9 gorgosterol

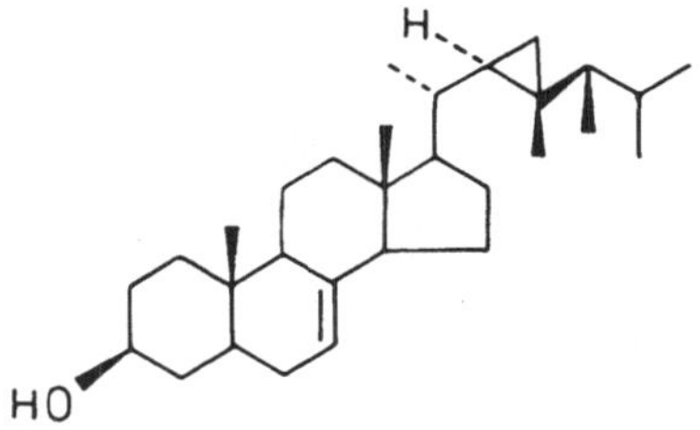

10 acanthasterol

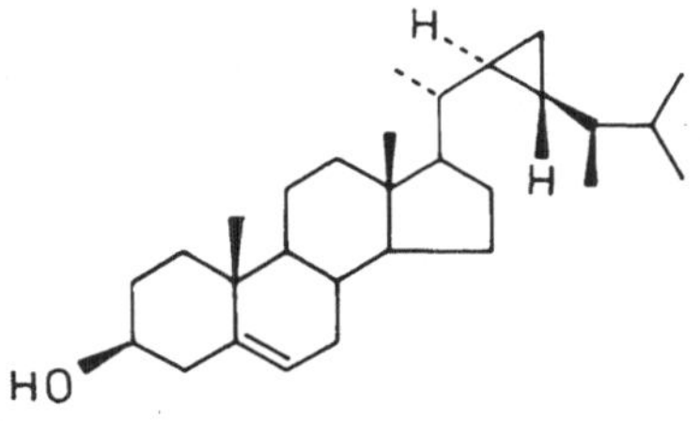

11 23-demethylgorgosterol

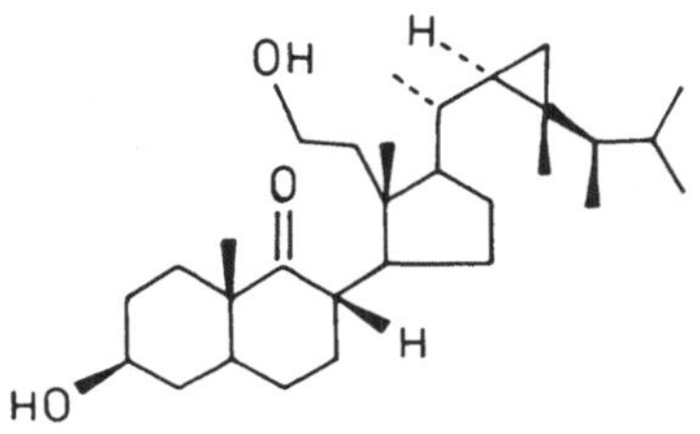

12

Figure 4. Cyclopropane-containing sterols.

Gorgosterol (9) (the history of which has been well recorded by Scheuer in his book, The Chemistry of Marine Natural Products[36]) was first discovered by Bergmann in 1943[37], but its structure was solved only in 1970, through a collaborative effort of three research groups[38], and later confirmed by X-ray diffraction analysis[39]. Additional cyclopropane-containing sterols are 23-demethylgorgosterol (11)[40] and the very unusual seco-compound (12)[41]. The structures of both compounds were also elucidated by X-ray diffraction analysis[41,42]. Some reports of attempts to synthesize cyclopropane-containing sterols, resulting in the synthesis of two stereoisomers of 23-demethylgorgosterol (11), have been published[43,44].

The discovery of the cyclopropane ring in gorgosterol stimulated a discussion of its origin and biological significance. The presence of such a ring was also considered[38] to be evidence in support of the hypothesis[21] that cyclopropanes may be intermediates in the introduction of methyl groups into the side chains of sterols.

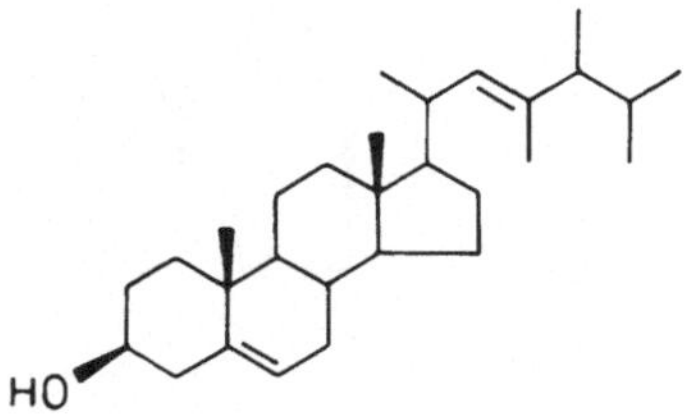

Figure 5.   23,24-Dimethyl sterols from marine sources.

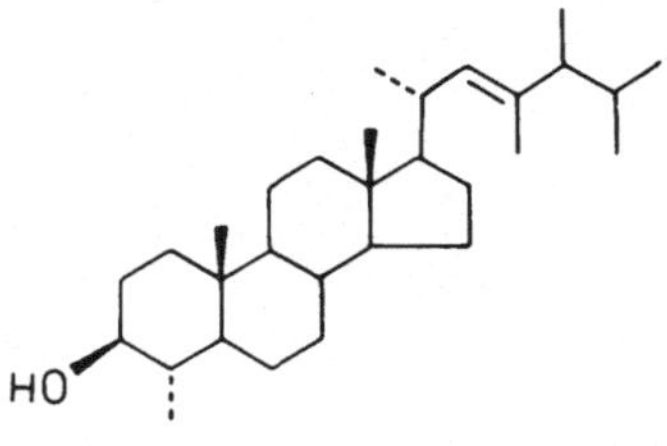

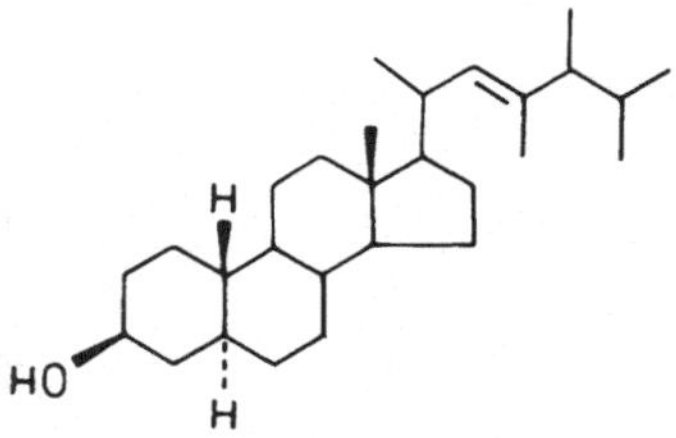

Although it was also emphasized[38] that cyclopropanation might be only a terminal step in the biosynthesis of gorgosterol, the conjecture quoted above was corroborated by the discovery in the soft coral Sarcophyta elegans of 23,24-dimethylcholesta-5,22-dien-3β-ol (13)[45], which, we can imagine, formed through an intermediate similar to 23-desmethylgorgosterol (11).

Very recently, a 4α-methylsterol (14) with the same side chain has been discovered in the toxic dinoflagellate Gonyaulax tamarensis[46]. It is remarkable that in this constituent of the phytoplankton the 4α-methylsterol (14) accounts for about 60% of the sterol mixture, the other component (40%) being cholesterol. The authors drew attention to the fact that 4α-methylsterols, intermediates in sterol biosynthesis, are accumulated under anaerobic conditions[47,48] and noted that the mass spectroscopic analyses of sterols from anaerobically-maintained, gorgonian-associated zooxanthellae were reported[49] to give a molecular ion m/e 428, assigned to "dihydrogorgosterol". Therefore, "dihydrogorgosterol" could be a sterol having a side chain in which the opening of the cyclopropane ring gives a substitution pattern similar to that of the 4α-methylsterol detected in Gonyaulax tamarensis.

We have also isolated[50] from the sponge Axinella polypoides (which contains a mixture of 19-norstanols discussed later) a 19-norstanol for which we propose the structure (15), on the basis of comparison of spectral data with those of the 23,24-dimethylated sterols.

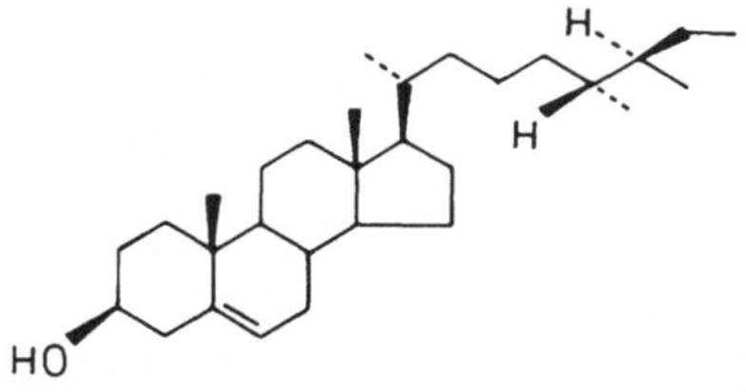

16 , aplysterol 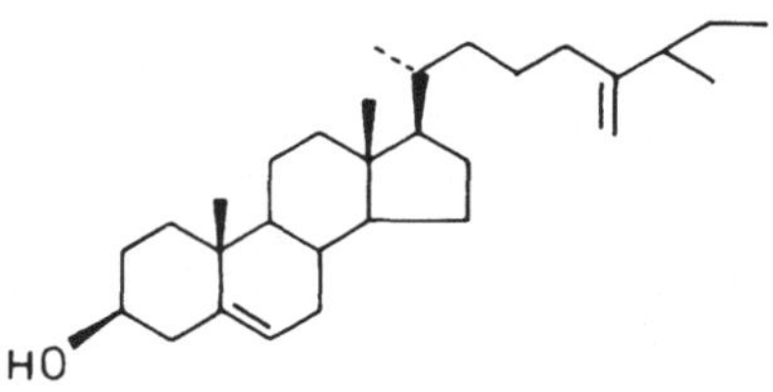   17 , didehydroaplysterol

Figure 6.   26-Methylsterols from sponges of the genus Verongia.

NON-CONVENTIONAL STEROLS FROM SPONGES

Since the extensive work of Bergmann on the sterols of invertebrates, it has been recognized that, in the animal kingdom, the
sponges contain the greatest variety of sterols, but only recently,
when sophisticated chromatographic techniques have been applied to
sterol analysis, have sterols of a completely new type been reported
from sponge sources.

The first biogenetically unprecedented sterols isolated from
a sponge were aplysterol (16) and 24,28-didehydroaplysterol (17),
which are the first examples of 26-alkylation in steroid biosynthesis (Figure 6). The structure of 26-methyl-24-methylenecholes-
terol was suggested for 24,28-didehydroaplysterol, on the basis of
spectroscopic data, along with degradative work[54], and later confirmed by synthesis[55].

The complete structure of 24,26-dimethylcholesterol, suggested
for aplysterol on spectral evidence and interrelation with didehydroaplysterol, was secured by single-crystal X-ray diffraction
studies of its p-iodobenzoate, which also revealed the

Table 4. Sponges containing aplysterol (16) and 24,28-didehydro-
aplysterol (17)[16].

Sponges	Source	% of the total sterol content
Order Dictioceratida		
Family Verongidae		
V. aerophoba	Naples	70%
V. archeri (hard)	Jamaican N. Shore	60%
V. archeri (soft)	British V. Islands	60%
V. fistularis	Bermuda	67%
V. thiona	La Jolla (California)	78%

stereochemistry of the side chain to be 24R, 25S[55]. These unique
sterols, first isolated from the orange sponge <u>Verongia aerophoba</u>,
appear to be confined to the family Verongidae. In an examination
of twenty-five sponges for their sterol composition, we have found
that all <u>Verongia</u> species examined, collected in different local-
ities and habitats, contained the two new sterols as 60-70% of the
total sterol content[16] (Table 4). These results provided useful
data for sponge systematics and added further evidence for segre-
gating Verongidae from Spongidae. On the basis of their distinct
amino acid patterns, they were considered by Bergquist and Hartman[56]
to be widely separated groups, even though they are classified as
belonging to the same order.

    At this time we cannot explain the biogenetic origin of these
sterols. Results from our laboratory have indicated that <u>Verongia
aerophoba</u> failed to incorporate either 1-^{14}C-acetate or 2-[14]C-
mevalonate into aplysterol and 24,28 -didehydroaplysterol. A
radiolabelling experiment using CH_3-^{14}C-methionine in the sponge
<u>V. aerophoba</u> also resulted in nonradioactive 26-methylsterols[57].

    In view of these results, and taking into account the limited
distribution of these sterols, which are confined to a single family, we can suppose that the sponge converts the exogenous sterols
taken from the food into aplysterol (16) and 24,28-didehydro-
aplysterol (17); such a conversion might include an internal transmethylation from a C_{29} precursor such as fucosterol. We do not
have any evidence to support this hypothesis, but in the case of

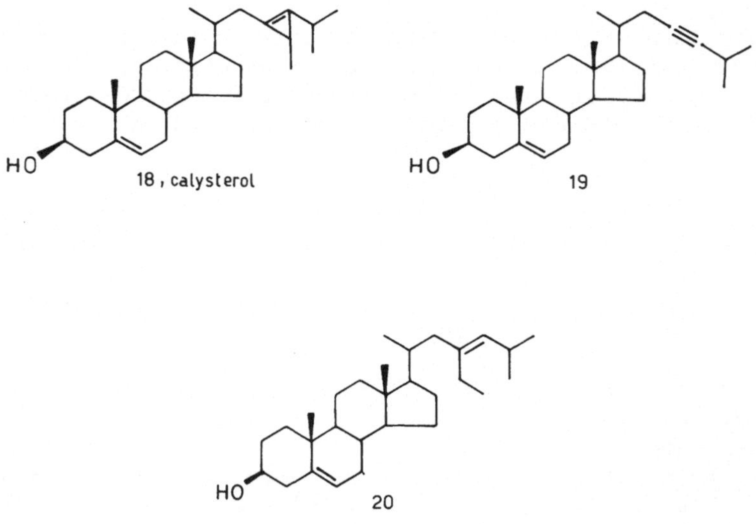

Figure 7. Non-conventional sterols from the sponge <u>Calyx nicaensis</u>.

the recently isolated calysterol (18), which includes the unique
feature of a cyclopropene ring in the side chain, we have obtained,
by tracer experiments, some data indicating that the sponge converts fucosterol into the latter (unpublished results).

Calysterol (18), the principal sterol component (90%) of the
sponge Calyx nicaensis, represents a further remarkable structural
variant in which there is attachment of "extra" carbon atoms to
the normal cholesterol skeleton at C-23 (Figure 7). The structure
suggested for this sterol was proposed,on the basis of spectral
data and chemical degradations, by Fattorusso and co-workers[58], who
obtained a 1,3-diketone derivative by ozonolysis of the corresponding stanol. The presence of a cyclopropene ring in the C_{10}
side chain was also supported by hydrogenation to a dihydroderivative having an nmr spectrum containing a three-proton, high field,
complex signal spread between 0.3-0.6 ppm which was assigned to
cyclopropyl hydrogens. Hydrogenation under more drastic conditions
gave β-sitostanol, confirming the structure and stereochemistry of
calysterol.

The same group has also discovered two further "unusual"
sterols, (19) and (20), from Calyx nicaensis[59]. The occurrence
of (19), the first example of an acetylenic functionality in a
steroid, was immediately regarded as a possible answer to the problem of the biochemical precursor of calysterol, itself, and also
the gorgonian sterols having a 22,23-cyclopropane. However, tracer
experiments have recently shown that Calyx nicaensis does not
incorporate ^{14}C-methyl methionine into calysterol; radiolabelling
experiments using labelled acetate in the sponge Calyx nicaensis
also resulted in no radioactive sterols.

On the basis of these results, we have proposed that, in the
sponge, calysterol might arise by modification of dietary sterols,
and with this in mind Calyx nicaensis was fed with a series of
labelled C_{29} sterols, such as fucosterol, stigmasterol, and
β-sitosterol. Table 5 shows the incorporation data, which indicate
that the sponge can metabolize injected $\left[7\text{-}^3\text{H}\right]$ fucosterol to produce labelled calysterol, while apparently stigmasterol and
β-sitosterol cannot act as precursors of this unique sterol.

In all the experiments in which labelled sterol precursors are
used, the main problem is often the separation of the precursor
from the metabolite. Calysterol and the conventional sterols used
as precursors have the same chromatographic behaviour, and this
caused serious problems of separation and purification. During
acetylation of calysterol at an elevated temperature, we observed
the formation of a slow-moving spot on tlc, along with the spot
corresponding to the calysterol acetate. When we allowed the
reaction to proceed for five hours, the slow-moving compound on
tlc became the major reaction product, which was shown by ms and

Table 5. Incorporation of $[7-^3H_2]$-C_{29} sterols into calysterol by the sponge <u>Calix nicaensis</u>

Administered precursors	Total fed (dpm)	Radioactivity recovered in sterols (%)	Radioactivity associated with calysterol (% of the recovered radioactivity)
$[7-^3H_2]$ fucosterol	4.8×10^6	53	2.7
$[7-^3H_2]$ stigmasterol	6.9×10^6	37	0.2
$[7-^3H_2]$ β-sitosterol	2.9×10^5	94	0.3

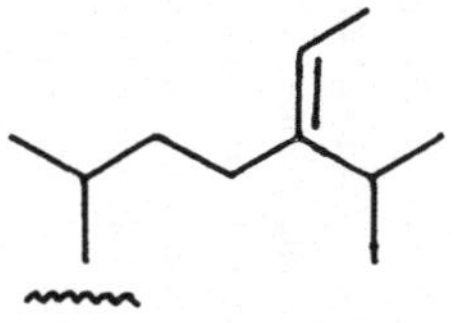

Fucosterol

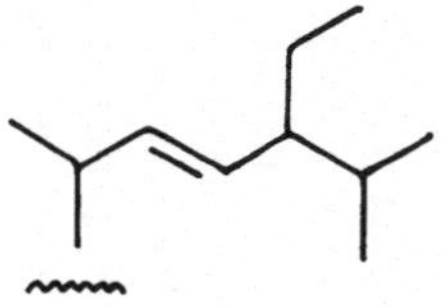

Stigmasterol

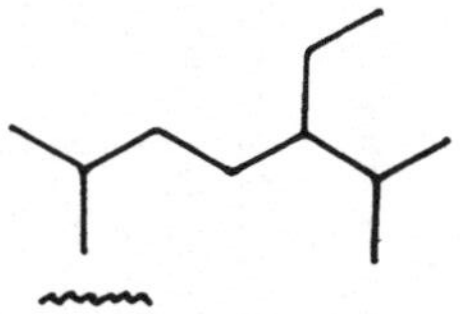

β-Sitosterol

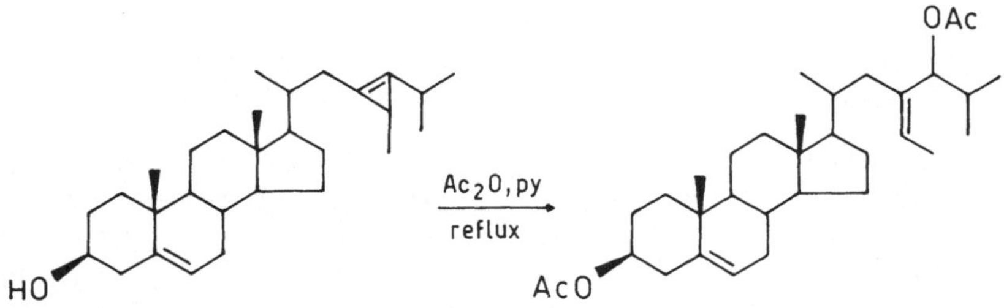

calysterol   21

Figure 8.   Acetylation of calysterol.

nmr to be a mixture of diacetates, the main component of which was identified as (21) (Figure 8). We took advantage of this observation and submitted the labelled mixture of sterols, after addition of the fucosterol, stigmasterol, or β-sitosterol carrier, to acetylation at reflux, and the diacetate fraction was purified by repeated chromatography of the parent diacetates and of the derived diols.

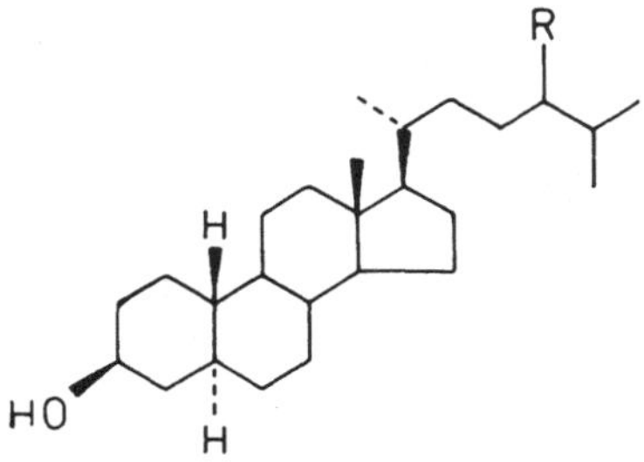

R=H , 24-nor	R=H , Δ^{22} - trans
R=H	R=Me , Δ^{22} - trans
R=Me	R=Me , $\Delta^{24(28)}$
R=Et	R=Et , Δ^{22} - trans

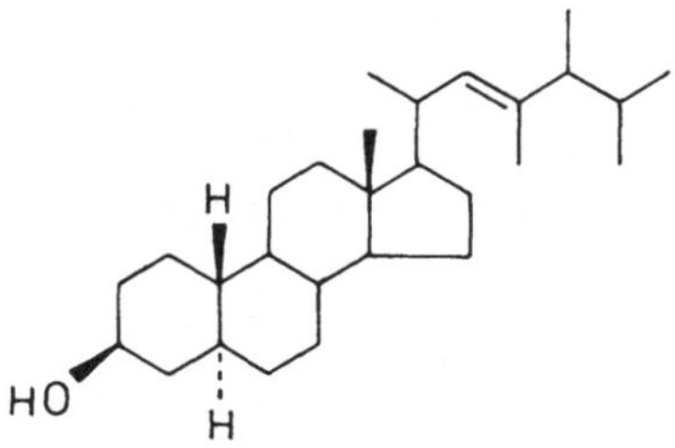

Figure 9.   19-Nor-stanols in the sponge Axinella polypoides.

In considering the origin of the minor sterols described from Calyx nicaensis, the acetylene (19) and the 24-ethyl-cholesta-5, 22-diene-3β-ol (20)could both be catabolites of calysterol.

Modifications of the sterol nucleus have also been found in sponges. The total sterol content of Axinella polypoides is a mixture of stanols having a 19-norcholestanol nucleus carrying conventional saturated and mono-unsaturated C_7 (24-nor), C_8, C_9 and C_{10} side chains[50] (Figure 9). As was briefly noted before, A. polypoides also contains a stanol having both the 19-norcholestanol nucleus and an unusual pattern of side chain alkylation which has already been encountered in a sterol from a soft coral and in a 4α-methyl sterol isolated from a dinoflagellate. The occurrence in A. polypoides of the latter, along with the 24-nor component, reflects the ability of the sponge to accept different sterol substrates and to remove the 19-methyl group from them. In fact, we have demonstrated by tracer experiments that the sponge very efficiently converts labelled cholesterol to 19-nor-cholestanol; of the total radioactivity associated with the sterol fraction after 290 hours' incubation, 78% was recovered in the 19-nor-stanol fraction and only 22% in the precursor[60]. Table 6 summarizes the incorporation data when A. polypoides was fed with 26-^{14}C-cholesterol.

Table 6. Incorporation of $\left[26\text{-}^{14}C\right]$-cholesterol into 19-nor-stanols by A. polypoides

Period of incubation	48 hours	290 hours
Total fed (dpm)	5.5×10^8	5.5×10^8
Total sterol recovered		
(mg)	220	305
(dpm)	2.03×10^6	11.3×10^6
Radioactivity recovered (%)	0.37	2.04
Radioactivity recovered (%)		
Precursor	80	22
19-nor-stanols	20	78

Further experiments designed to obtain more information on the biological conversion of cholesterol into 19-nor-cholestanol have shown that the presence of the Δ^5-double bond in the sterol nucleus must be regarded as a prerequisite for the removal of the 19-methyl group. In fact, when the sponge was fed with a mixture of $\left[^{14}C\right]$-cholesterol and $\left[7\text{-}\alpha\text{-}^3H\right]$-cholestanol, all the label incorporated into the 19-nor-stanols was due to ^{14}C (Table 7).

Table 7. ^{3}H:^{14}C ratios of the 19-nor-stanols isolated from
A. polypoides after administration of $[4-^{14}C]$-cholesterol
(3.2 x 10^8 dpm/mg) and $[7\alpha-^3H]$-cholestanol (6.65 x 10^8 dpm/mg)

	^{3}H (dpm)	^{14}C (dpm)	^{3}H/^{14}C
Administered substrates	7.2 x 10^7	2.8 x 10^7	2.5
Recovered 19-nor-stanols	--	4.8 x 10^6	0.0
Recovered substrates	5.6 x 10^6	1.1 x 10^6	5.1

Furthermore, radiolabelling experiments using $[1-^{14}C]$ acetate
resulted in non-radioactive sterols. We can therefore conclude
that the sponge is unable to synthesize its sterols de novo but
modifies the sterols taken up from the diet.

A further modification of the sterol nucleus has been found in
the stanols of Axinella verrucosa, which combine the unusual
3β-hydroxymethyl-A-nor-5α-cholestane nucleus with conventional
saturated and Δ^{22}-unsaturated C$_8$, C$_9$ and C$_{10}$ side chains[61].

Axinella verrucosa lacks the usual sterols. Feeding experiments
have shown that the sponge does not incorporate acetate into A-nor-
stanols but readily converts $[4-^{14}C]$-cholesterol into 3β-hydroxy-
methyl-A-nor-5α-cholestane[62]. Table 8 shows the incorporation data
when the sponge was fed with $[4-^{14}C]$-cholesterol; after 290 hours'
incubation, 20.4% of the total radioactivity administered was

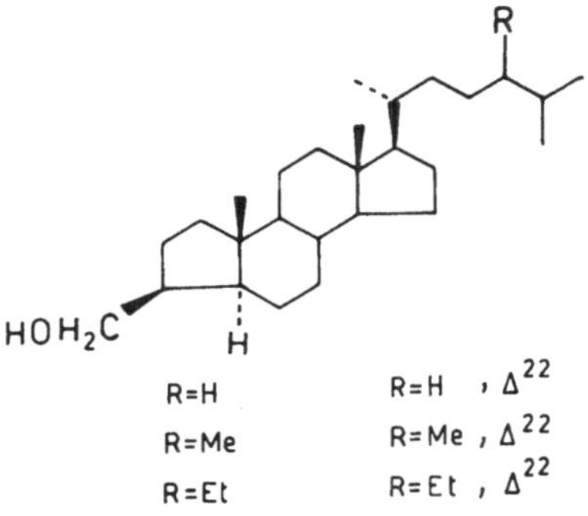

HOH$_2$C

R=H
R=Me
R=Et

R=H , Δ^{22}
R=Me , Δ^{22}
R=Et , Δ^{22}

Figure 10.   3β-Hydroxymethyl-A-nor-steranes in the sponge
Axinella verrucosa

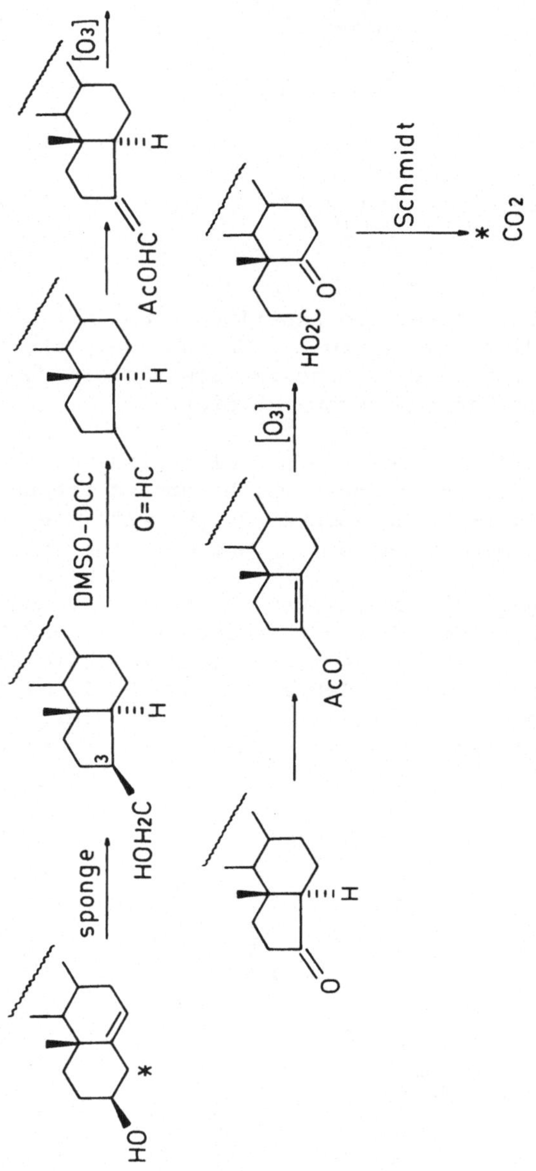

Figure 11. Carbon-3 of the A-nor-stanols is derived from carbon-4 of cholesterol.

Table 8.   Incorporation of [4-^{14}C]-cholesterol into 3β-hydroxymethyl-A-nor-5α-steranes by A. verrucosa

Period of incubation	48 hours	290 hours
Total fed (dpm)	5.55×10^8	5.55×10^8
Total sterol recovered		
(mg)	180	440
(dpm)	3.34×10^7	1.13×10^8
Radioactivity recovered (%)	6.0	20.4
Radioactivity in precursor(%)	63	34
In 3β-hydroxymethyl-A-nor-5α-steranes (%)	37	66

recovered in the sterol fraction, of which 66% was associated with the A-nor-stanols and only 34% with the precursor. These results suggest, as before, that in the sponge A. verrucosa these unique A-nor-stanols arise by modification (ring A-contraction) of dietary sterols.

Degradation experiments shown in Figure 11 indicated that all the label present in the A-nor-stanols biosynthesized by the sponge from [4-^{14}C]-cholesterol is located at position 3 of the A-nor-cholestane skeleton[63], so it is apparent that the ring contraction involves the formation of carbon-carbon linkage between C-4 and C-2 of cholesterol, while carbon-3 furnishes the hydroxymethyl carbon (Figure 12). Such a ring contraction is reminiscent of the ring B contraction occurring in the biosynthesis of the plant growth hormone gibberellic acid, which was shown to be derived from 7β-hydroxy-ent-kaurenoic acid[64], although the mechanism of this ring contraction is disputed[65].

Feeding experiments using cholesterol stereospecifically-labelled with tritium at positions 2, 3, and 4 are now in progress in our laboratory, in order to clarify details of the conversion of cholesterol to 3β-hydroxymethyl-A-nor-5α-cholestane.

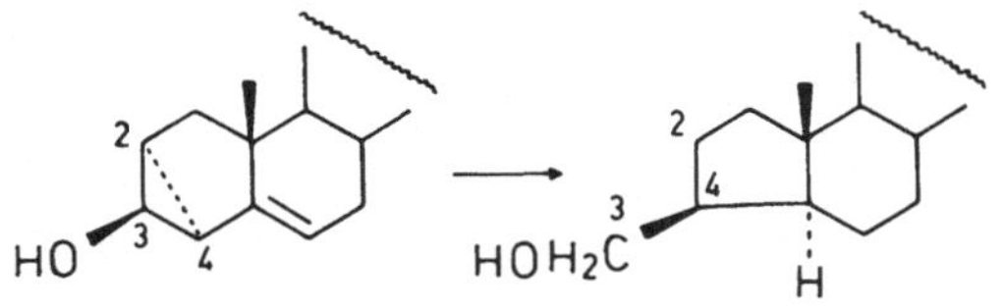

Figure 12. Biogenetic conversion of cholesterol into 3β-hydroxy-methyl-A-nor-cholestane

REFERENCES

1. E. Lederer, Quart. Rev. <u>23</u>, 453 (1969).

2. D. R. Idler, P. M. Wiseman, and L. M. Safe, Steroids <u>16</u>, 451 (1970).

3. D. R. Idler and P. Wiseman, Int. J. Biochem. <u>2</u>, 516 (1972).

4. M. Fryberg, A. C. Ochlschlanger, and A. M. Unrau, Chem. Comm. 1194 (1971).

5. A. Métayer, A. Quesneau-Thierry, and M. Barbier, Tetrahedron Lett., 595 (1974).

6. J. Viala, M. Devys, and M. Barbier, Bull. Soc. Chim. (France), 3626 (1972).

7. M. Kobayashi, R. Tsuru, K. Todo, and H. Mitsuhashi, Tetrahedron Lett., 2935 (1972).

8. M. Kobayashi, R. Tsuru, K. Todo, and H. Mitsuhashi, Tetrahedron <u>29</u>, 1193 (1973).

9. L. J. Goad, I. Rubinstein, and A. G. Smith, Proc. Roy. Soc. Ser B, <u>180</u>, 223 (1972).

10. A. G. Smith and L. J. Goad, Biochem. J. <u>146</u>, 35 (1975).

11. M. Kobayashi, K. Todo, and H. Mitsuhashi, Chem. Pharm. Bull. <u>22</u>, 236 (1974).

12. P. M. Boll, Acta Chem. Scand. <u>28</u>, 270 (1974).

13. T. R. Erdman and R. H. Thomson, Tetrahedron <u>28</u>, 5163 (1972).

14. A. Métayer and M. Barbier, C. R. Acad. Sci. Paris C <u>276</u>, 201 (1973); Chem. Comm. 424 (1973).

15. D. R. Idler, M. W. Khalil, J. D. Gilbert, and C. J. W. Brooks, Steroids <u>27</u>, 155 (1976).

16. M. De Rosa, L. Minale, and G. Sodano, Comp. Biochem. Physiol. <u>46 B</u>, 823 (1973).

17. A. G. Smith, I. Rubinstein, and L. J. Goad, Biochem. J. <u>135</u>, 443 (1973).

18. A. G. Smith and L. J. Goad, Biochem. J. <u>142</u>, 421 (1974).

19. P. A. Voogt, J. M. Van De Ruit, and J. W. A. Van Rheenem, Comp. Biochem. Physiol. 48 B, 47 (1974).

20. P. D. Voogt, Arch. Internat. Physiol. Biochim. LXXXI, 871 (1973); Neth. J. Zool. 24, 22 (1974); 24, 469 (1974).

21. M. Kobayashi, M. Nishizawa, K. Todo, and H. Mitsuhashi, Chem. Pharm. Bull. 21, 323 (1973).

22. J. L. Boutry, A. Alcaide, and M. Barbier, C. R. Acad. Sci. Paris D 272, 1022 (1971).

23. J. P. Ferezou, M. Devys, J. P. Allais, and M. Barbier, Phytochem. 13, 593 (1974).

24. M. Kobayashi and H. Mitsuhashi, Tetrahedron 30, 2147 (1974).

25. M. Kobayashi and H. Mitsuhashi, Steroids 24, 399 (1974).

26. M. Kobayashi and H. Mitsuhashi, Steroids 26, 605 (1975).

27. R. B. Clayton, J. Lipid. Res. 5, 3 (1964).

28. A. Kamazawa, N. Tanaka, S. Theshima, K. Kashiwada, Bull. Jap. Soc. Scient. Fish. 37, 211 (1971).

29. M. J. Walton and J. F. Pennock, Biochem. J. 127, 471 (1972).

30. J. P. Ferezou, M. Devys, and M. Barbier, Experientia 28, 408 (1972).

31. J. B. Guary and A. Kanazawa, Comp. Biochem. Physiol. 46A, 5 (1973).

32. D. Idler, L. M. Safe, and E. F. MacDonald, Steroids 18, 545 (1971).

33. Y. M. Sheikh and C. Djerassi, Tetrahedron 30, 4095 (1974).

34. K. C. Gupta and P. J. Scheuer, Tetrahedron 24, 5831 (1967).

35. Y. M. Sheikh, C. Djerassi, and B. M. Tursch, Chem. Comm. 217, 600 (1971).

36. P. J. Scheuer, Chemistry of Marine Natural Products, Academic Press, New York and London, 1973, pp. 78-80.

37. W. Bergmann and W. T. Pace, J. Amer. Chem. Soc. 65, 477 (1943).

38. R. L. Hale, J. Leclercq, B. Tursch, C. Djerassi, R. A. Gross,
     Jr., A. J. Weinheimer, K. Gupta, and P. J. Scheuer, J. Amer.
     Chem. Soc. 92, 2179 (1970).

39. N. C. Ling, R. L. Hale, and C. Djerassi, J. Amer. Chem. Soc.
     92, 5281 (1970).

40. F. J. Schmitz and J. Pattabhiraman, J. Amer. Chem. Soc. 92,
     6073 (1970).

41. E. L. Enwall, D. Van Der Helm, I Nan Su, T. Pattabhiraman,
     F. J. Schmitz, R. L. Spraggings, and A. J. Weinheimer,
     Chem. Comm. 215 (1972).

42. E. L. Enwall and D. Van Der Helm, Rec. Trav. Chim. 93, 53
     (1974).

43. Y. M. Sheikh, J. Leclercq and C. Djerassi, J. C. S. Perkin I,
     909 (1974).

44. G. D. Anderson, J. J. Powers, C. Djerassi, J. Fayos, and
     J. Clardy, J. Amer. Chem. Soc. 97, 388 (1975).

45. A. Kanazawa, S. Teshima, T. Ando, and S. Tomita, Bull. Jap.
     Soc. Scient. Fish. 40, 729 (1974).

46. Y. Shimizu, M. Alan, and A. Kobayashi, J. Amer. Chem. Soc.
     98, 1059 (1976).

47. C. Djerassi, G. W. Krakower, A. J. Lemin, L. H. Lin, J. S.
     Mills, and R. Villotti, J. Amer. Chem. Soc. 80, 6284 (1958).

48. A. D. Rahimtula and J. L. Gaylor, J. Biol. Chem. 247, 9 (1972).

49. L. S. Ciereszko, M. A. Johnson, R. W. Schmitz, and C. K. Kooms,
     Comp. Biochem. Physiol. 24, 899 (1968).

50. L. Minale and G. Sodano, J. C. S. Perkin I, 1888 (1974).

51. J. H. Block, Steroids 23, 421 (1974).

52. J. P. Engelbrecht, B. Tursch, and C. Djerassi, Steroids 20,
     121 (1972).

53. K. C. Gupta and P. J. Scheuer, Steroids 13, 343 (1969).

54. P. De Luca, M. De Rosa, L. Minale, and G. Sodano, J. C. S.
     Perkin I, 2132 (1972).

55. P. De Luca, M. De Rosa, L. Minale, R. Puliti, G. Sodano,
     F. Giordano, and L. Mazzarella, Chem. Comm. 825 (1973).

56. P. R. Bergquist and W. D. Hartman, Mar. Biol. 3, 247 (1969).

57. M. De Rosa, L. Minale, and G. Sodano, Comp. Biochem. Physiol.
     45B, 883 (1973).

58. E. Fattorusso, S. Magno, L. Mayol, C. Santacroce, and D. Sica,
     Tetrahedron 31, 1715 (1975).

59. E. Fattorusso, personal communication.

60. M. De Rosa, L. Minale, and G. Sodano, Experientia 31, 758
     (1975).

61. L. Minale and G. Sodano, J. C. S. Perkin I, 2380 (1974).

62. M. De Rosa, L. Minale, and G. Sodano, Experientia 31, 408
     (1975).

63. M. De Rosa, L. Minale, and G. Sodano, Experientia, in press.

64. J. R. Hauson and J. Hawker, Chem. Comm., 208 (1971).

65. J. E. Groebe, P. Medden, and J. MacMillan, Chem. Comm., 161
     (1975).

*
L. Minale, Laboratorio per la Chimica di Molecole di Interesse
Biologico del C.N.R., Via Toiano 2, Arco Felice, Napoli, Italy;
G. Sodano, Istituto di Chimica Organica dell'Universita,
Catania, Italy.

STEROLS FROM MARINE SOURCES

C. Djerassi, R.M.K. Carlson, S. Popov and T. H. Varkony

Department of Chemistry, Stanford University

Stanford, California U.S.A. 94305

The isolation of novel sterols from terrestrial (notably veg-
etable) sources has slowed down markedly during the past decade.
This is in striking contrast to the situation in the marine world,
which, during the past half dozen years, has yielded a plethora of
new sterols with unusual hydrocarbon side chains. Aside from
their intrinsic interest, these structures raise intriguing and
unanticipated questions about the course of sterol side-chain biogenesis. For the sterol cognoscenti and connoisseur, it is
sufficient to cite the structures of 24-nor-22-dehydrocholesterol
(1)[1] and gorgosterol (2)[2]. The former appears to have lost a
central carbon atom of the standard cholesterol side chain, while
the latter shows the hitherto unprecedented alkylation of every
methylene group of the cholesterol side chain. More new sterols
have been isolated from marine sources during the past decade than
during the entire past history of marine sterol chemistry--a complete reversal of the situation existing in the terrestrial sterol
field.

As in all current natural products chemistry[3], the chief
motivation for the search for new marine sterols is the question
of biological function or biosynthesis, rather than just chemical
structure elucidation for its own sake. As is pointed out elsewhere[3], a significant feature of the present status of marine
chemistry is that "we are just at the beginning of experimental
biosynthetic studies. Just as such work in plants followed earlier
work in animals by over a decade because of experimental difficulties, similar delays are being encountered in biosynthetic
studies with marine organisms. This is not only due to the compli-
cations associated with radioactive precursor incorporation but also
because frequently it is not even clear where and by whom the
natural product is biosynthesized."

111

An intermediate, albeit more speculative approach to the elucidation of biosynthetic pathways is the isolation of a variety of
closely related natural products, which frequently become "missing
links" in a hypothetical biogenetic scheme that eventually is
verified by biochemical experiments. One of the most striking
success stories in this regard is represented by the indole alkaloids[3], where the isolation of several hundred closely related
structures led to theories that eventually were modified and verified by tracer experiments.

We have decided to embark on precisely the same approach in
the marine sterol field. Taking 24-nor-22-dehydrocholesterol (1)
as an example, the apparent extrusion of a central carbon atom
from the standard cholesterol C_8 chain would suggest some bizarre
rearrangement or else a non-squalenoid biosynthetic precursor.
However, the recent isolation[4] of occlasterol (3)   suggests that
a much more likely path to the 24-nor-C_{26} marine sterols would be
one involving, first, standard C-24 methylation[5] and, second,
oxidative loss of the C-27 (see 3)   and C-26 methyl groups.

To date, over 100 different 3-hydroxy marine sterols have been
isolated, the majority of them being based on the seven nuclear
skeletons A-H collected in Figure 1. (The numbers in parentheses
refer to the number of naturally-occurring sterols known to contain
that skeleton.) In order to get a semi-quantitative indication of
the range and number of sterols that might be encountered among
marine sources, we developed a computer program that included

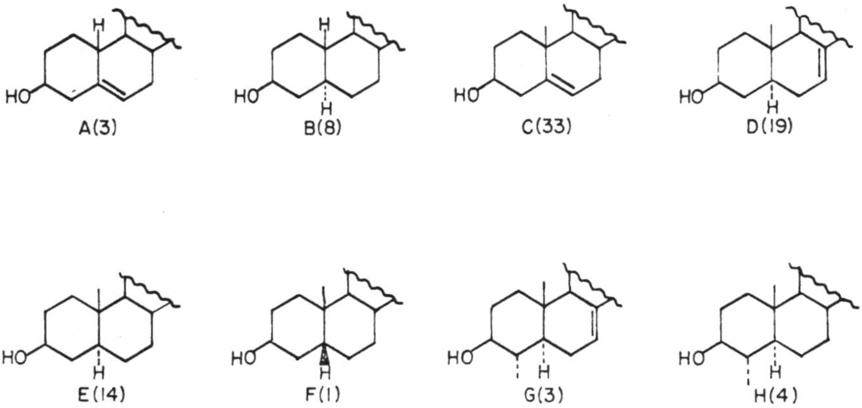

Figure 1.   Sterol nuclei of marine sterols.

1.   Starting material side chain:

2.   Methylation: $H-C\equiv C- \longrightarrow -\overset{\cdot}{\underset{+}{C}}\overset{\cdot}{-}\overset{\cdot}{C}\overset{\cdot}{-}C-$

   (no methylation occurs on tetra-substituted double bonds)

3.   Double bond generation: $-C\equiv C-C- \longleftarrow \overset{+}{\overset{\cdot\cdot\cdot}{-C}}\overset{\cdot\cdot}{-C}-C- \longrightarrow -C-C\equiv C-$

   (subsequent "migration" of double bond prohibited)

4.   Saturation: $-C\equiv C- \longrightarrow -C-C-$

5.   Degradation: $-C\equiv C-C\{CH_3 \longrightarrow -C\equiv C-C-$

6.   Cyclopropyl formation <u>via</u> carbonium ion intermediate:

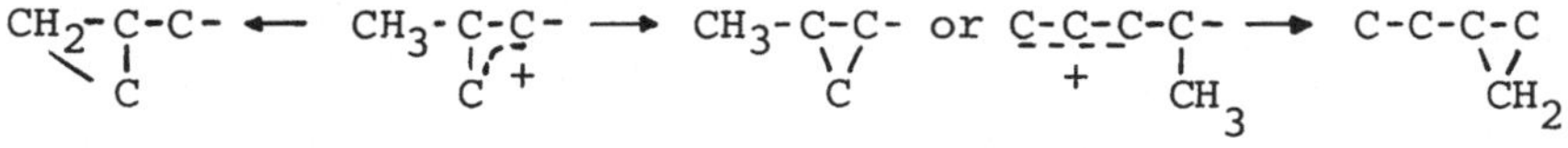

   $\underset{\diagdown C}{CH_2-C-C-} \longleftarrow \underset{\underset{+}{\diagup C}}{CH_3-C-C-} \longrightarrow \underset{\diagdown C \diagup}{CH_3-C-C-}$ or $\underset{+\ CH_3}{\overset{\cdot\cdot}{C}-C-C-C-} \longrightarrow \underset{\diagdown CH_2 \diagup}{C-C-C-C}$

7.   Cyclopropyl formation <u>via</u> olefins: $\underset{CH_3}{-C\equiv C-} \longrightarrow \underset{\diagdown CH_2 \diagup}{-C-C-}$

Figure 2.   Constraints for computer-assisted "biosynthesis" of
sterol side chains.

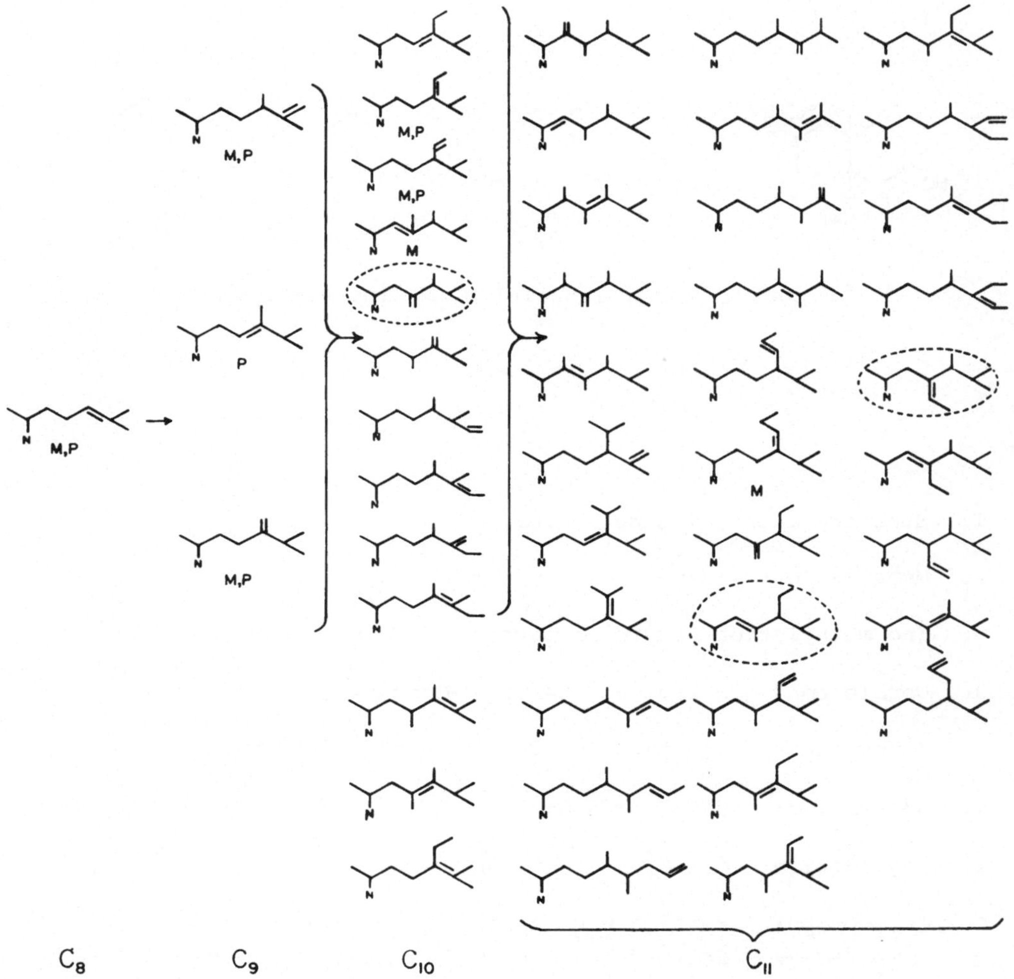

Figure 3. Computer-generated mono-olefinic side chains (methylation only).

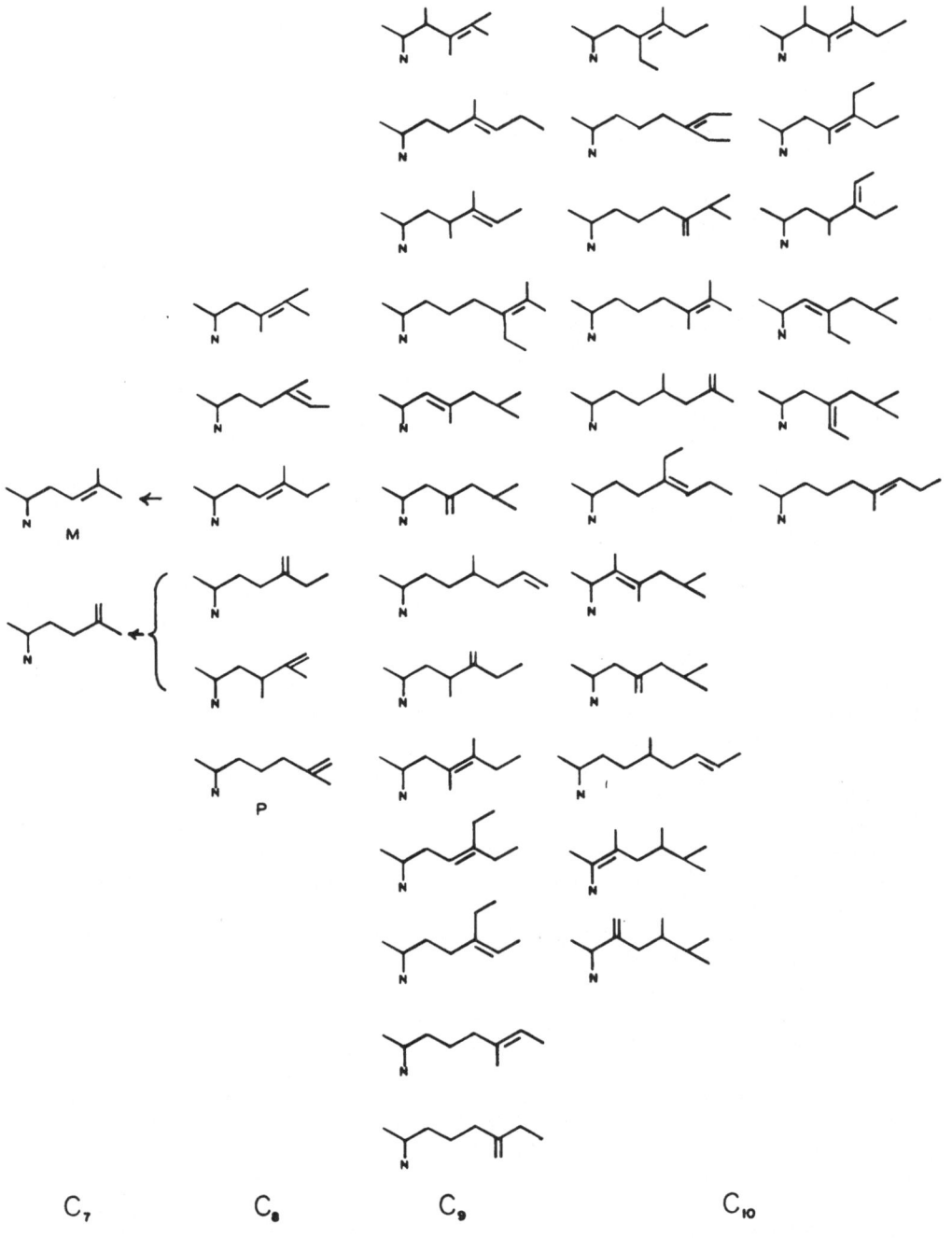

Figure 4. Computer-generated mono-olefinic side chains (degradation only).

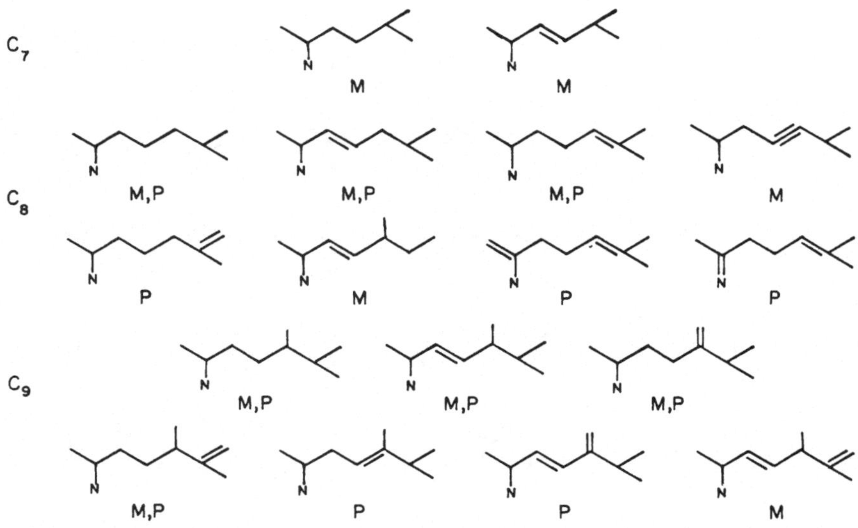

Figure 5. Naturally-occurring sterol side chains.

certain plausible though not all-inclusive[6] biosynthetic constraints
(see Figure 2). Typical structural examples of the resulting computer printout are shown in Figure 3 (C_8-C_{11} side chains generated
following constraints 2 and 3 of Figure 2) and Figure 4 (C_7-C_{10}
side chains generated according to the "degradation" constraint 5
of Figure 2), while those containing saturated or cyclopropyl-
containing side chains are not reproduced in the present paper.
The wide variety of hypothetical but plausible side chains predicted
by the computer program should be contrasted with the structures,
reproduced in Figures 5 and 6, of known C_7-C_{11} sterol side chains
of marine (M) and plant (P) origin.

Clearly, many exciting "missing links" might be anticipated,
and it almost depends on the reader's intuition and esthetic
preference to predict which might actually be found in nature. The
three circled structures in Figure 3 constitute examples which would
point toward interesting biosynthetic conclusions if such structures
were encountered in nature. The circled C_{10} side chain would indicate that methylation at position 23 proceeds in the same manner as
the well-documented C-24 methylation[5]; the two circled C_{11} side
chains require that C-23 methylation can occur even if a two-carbon
unit has been generated at C-24 or that a two-carbon unit can be
produced at C-23 after a single methylation at C-24.

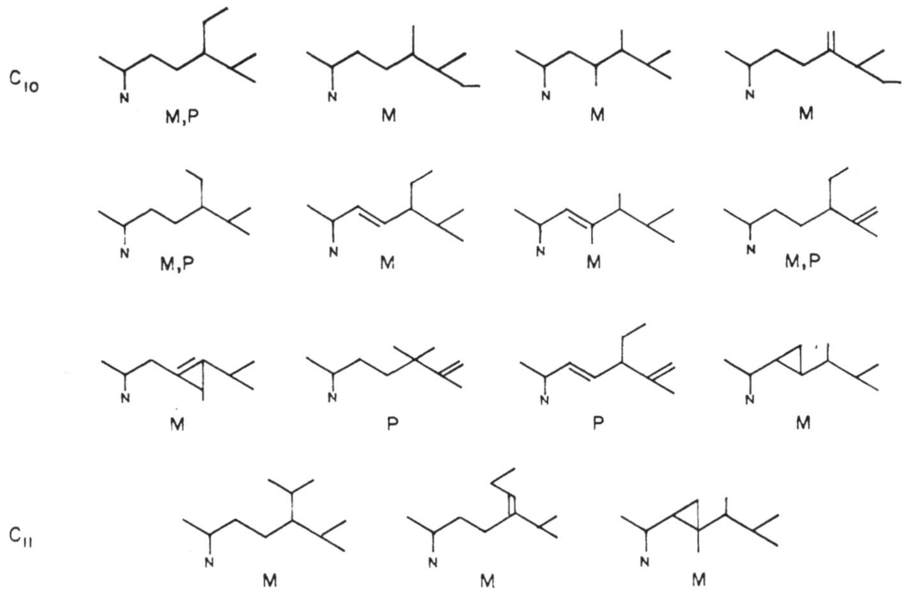

Figure 6.   Naturally-occurring sterol side chains.

        If we combine the possible saturated, mono-unsaturated and
cyclopropyl-containing side chains generated by the computer under
the constraints of Figure 2 with the seven most common sterol nuclei
summarized in Figure 1, we find (see Table 1) that only about 1100
complete sterol structures are produced. The purpose of this
exercise is to demonstrate that with likely biosynthetic constraints,
a perfectly manageable number of structures can be envisaged for
which other computer programs can be developed (work currently
underway in our laboratory) so as to facilitate and partially auto-
mate marine sterol identification by gas chromatographic-mass
spectrometric means. It is, of course, exceedingly unlikely that
even 50% of the structures enumerated in Table 1 will ever be
encountered. However, even a cursory inspection of Figures 3 and
4 suggests that if only 5-10% of these candidate structures is
detected in nature, a remarkable variety of exciting biosynthetic
leads will have been uncovered which have so far no precedent in
terrestrial sterol biosynthesis.

Table 1.   Total number of structural isomers based on seven nuclei and "biochemically" plausible C_7-C_{11} side chains.

Empirical Formula	Molecular Weight	Total No. Isomers	Olefins[*]	Isomers with Cyclopropyl-Substituted Side Chain
$C_{31}H_{56}O$	444	8	–	–
$C_{31}H_{54}O$	442	58	38	20
$C_{31}H_{52}O$	440	50	30	20
$C_{30}H_{54}O$	430	23	–	–
$C_{30}H_{52}O$	428	121	90	31
$C_{30}H_{50}O$	426	140	89	51
$C_{29}H_{52}O$	416	31	–	–
$C_{29}H_{50}O$	414	157	122	35
$C_{29}H_{48}O$	412	152	105	47
$C_{28}H_{50}O$	402	26	–	–
$C_{28}H_{48}O$	400	128	105	23
$C_{28}H_{46}O$	398	79	60	19
$C_{27}H_{48}O$	388	12	–	–
$C_{27}H_{46}O$	386	45	41	4
$C_{27}H_{44}O$	384	37	33	4
$C_{26}H_{46}O$	374	4	–	–
$C_{26}H_{44}O$	372	14	14	–
$C_{26}H_{42}O$	370	11	11	–
$C_{25}H_{44}O$	360	2	–	–
$C_{25}H_{42}O$	358	3	3	–
$C_{25}H_{49}O$	356	2	2	–
TOTAL		1103	743	254

[*]Nucleus and/or side chain

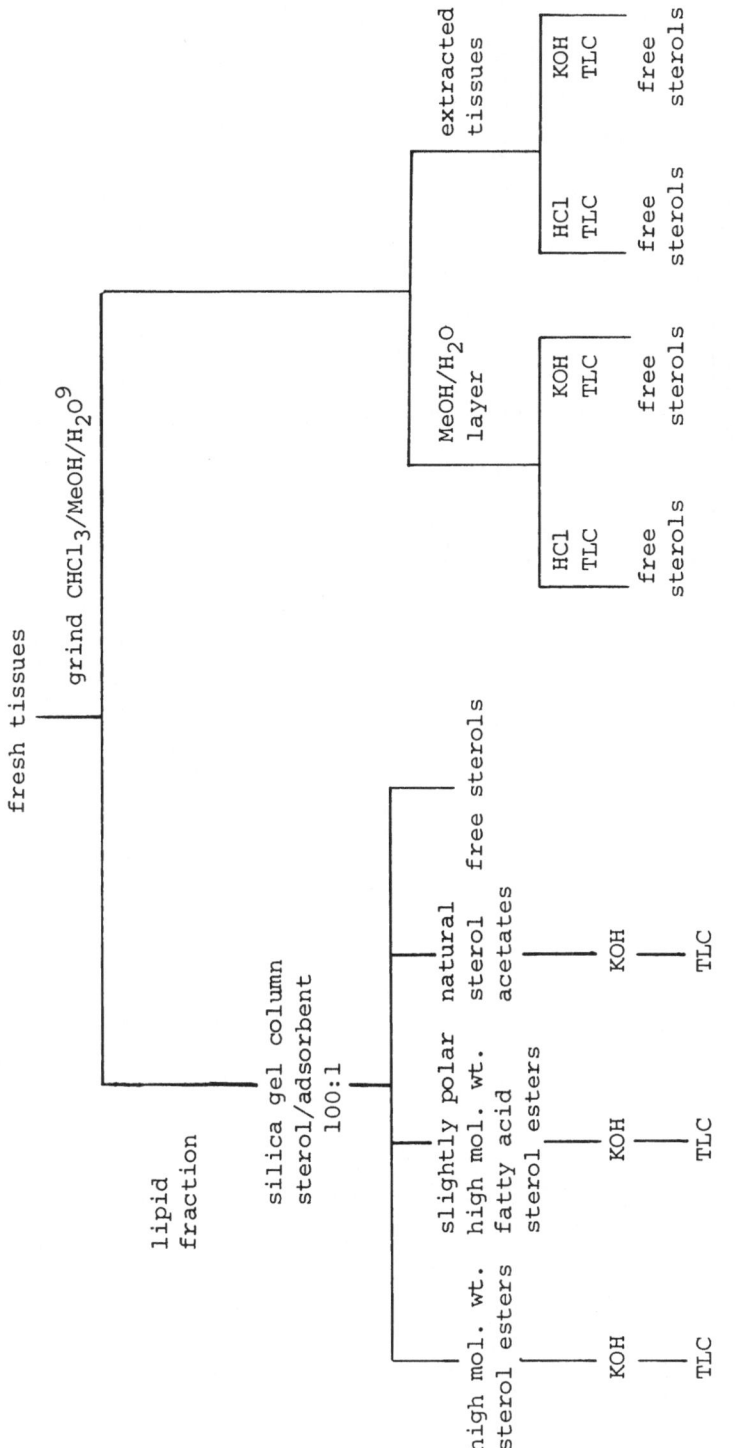

Scheme 1. Initial extraction and isolation scheme.

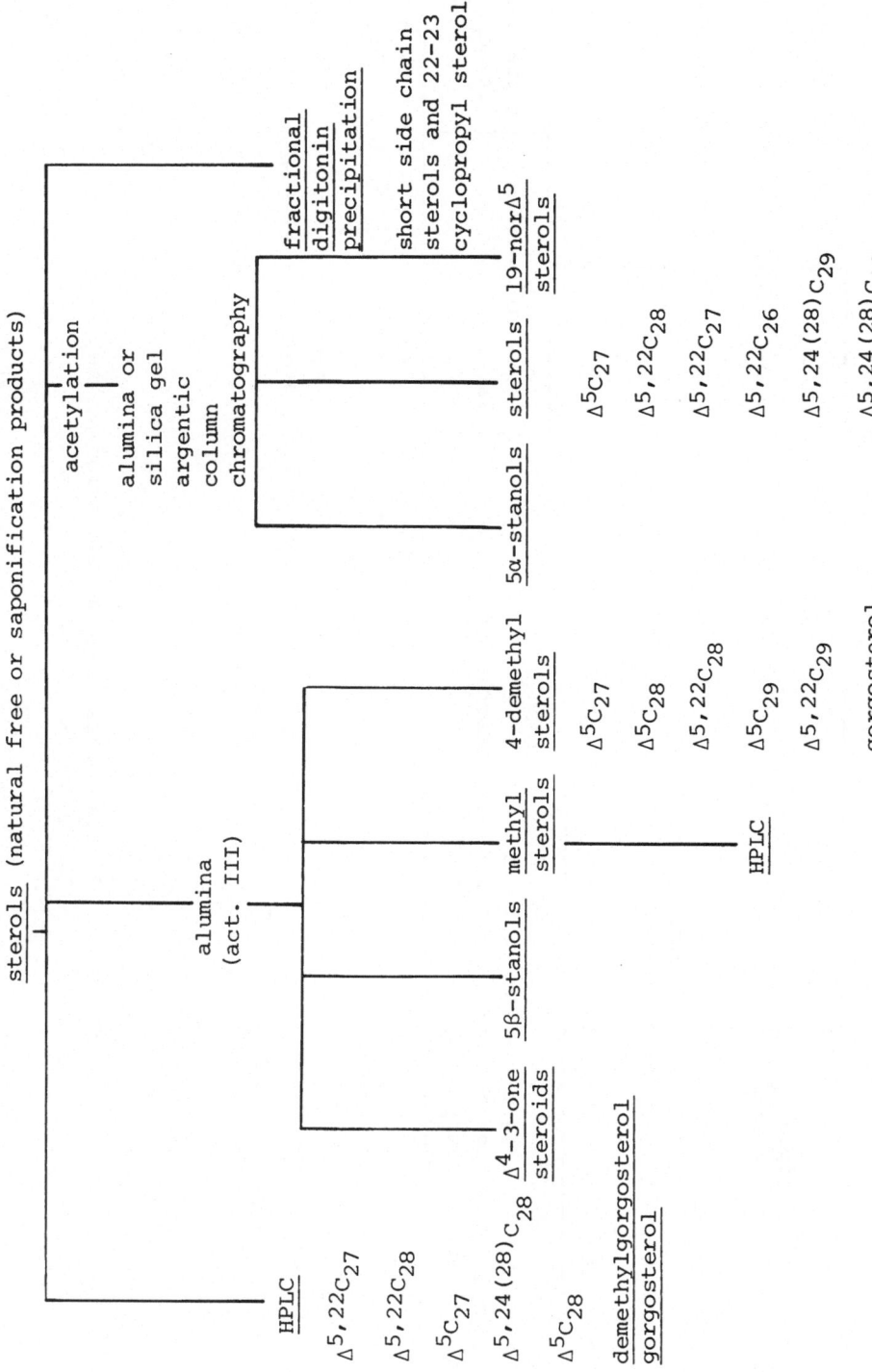

Scheme 2. Minor and trace sterol fractionation scheme.

Table 2. 4-Demethyl sterols present in Ps. Porosa and P. homomalla.

Sterol Reference Number	Structure (see ref. 7)	M^+	Number of Carbon Atoms	Relative Retention Time 3% OV-17 276 C	Percent of Sterol Mixture P. homomalla	Percent of Sterol Mixture Ps. porosa
1	*	274	19	0.13	0.3	0.4
2	*	288	20	0.25	<0.001	<0.001
3	*	290	20	0.25	<0.001	<0.001
4	*	300	21	0.53	--	0.05
5	*	302	21	0.23	0.3	0.4
6	*	312	21	0.57	--	0.01
7	*	314	22	0.57	0.01	0.06
8	*	314	22	0.31	--	<0.001
9	*	316	22	0.27	0.2	<0.001
10	*	316	21	0.57	0.06	0.7
11	*	318	22	0.27	--	0.001
12	*	318	21	0.57	<0.001	0.04
13	*	330	23	0.33	--	<0.001
14	*	344	24	0.6-0.8	--	<0.001
15	*	358	25	0.6-0.8	--	<0.001
16	*	360	25	0.6-0.8	--	<0.001
17	15	372	26	0.92	<0.001	0.08
18	*	384	27	1.17	<0.001	<0.001
19	1	384	27	0.93	1	0.3
20	2	386	27	1.00	50	6
21	16	386	27	1.15	<0.001	0.12
22	*	386	27	1.21	--	<0.001
23	10	388	27	1.00	0.1	0.08
24	3	398	28	1.13	10	11
25	7	398	28	1.29	9	3.5
26	17	398	28	1.35	--	0.04
27	4	400	28	1.27	9	12.5
28	11	402	28	1.27	0.13	0.12
29	5	412	29	1.36	3.7	1.5
30	8	412	29	1.65	4	9.5
31	6	414	29	1.54	0.5	1.1
32	12	414	29	1.65	0.005	0.01
33	*	414	29	1.62	0.02	--
34	14	416	29	1.33	0.03	--
35	*	416	29	1.53	--	0.002
36	*	424	30	2.25	<0.001	<0.001
37	9	426	30	2.20	13	51
38	*	426	30	3.38	0.05	0.03
39	*	426	30	1.53	--	0.05
40	*	426	30	2.15	--	<0.001
41	13	428	30	2.20	0.8	0.81
42	*	428	30	1.53	--	0.01

*Structure elucidation to be reported in a forthcoming publication.

With this hope in mind, we have developed[7] an isolation scheme (Schemes 1 and 2) which permits the detection of trace quantities of marine sterols and provides at least a useful mass spectrum starting with only a relatively small quantity (few hundred grams) of fresh marine animal or plant. Table 2 contains the results obtained by us in examining the sterols of Pseudoplexaura porosa and Plexaura homomalla. Of the forty-two 3-hydroxy-4-demethyl sterols encountered, more than half had never been noted before in nature, although several had been synthesized earlier (e.g., 19-norcholesterol (M^+ = 372); 17β-isopropyl-Δ^5-androsten-3β-ol (M^+ = 316), etc.). This approach is being applied by us to a variety of marine animals, notably from the California coastline, so as to facilitate subsequent feeding experiments with labeled precursors of selected candidate organisms.

The isolation of a given sterol, especially if it is present in only trace quantities, does not assure that it is of endogenous origin. If it arises by dietary incorporation, then it is crucial to follow the food chain in order to determine the prime candidate for biochemical experiments. A suitable example is the recent isolation and characterization of dinosterol (IV)[8] from the dinoflagellate Gonyaulax tamerensis. We have isolated the same sterol (6% of total sterol mixture) from P. homomalla, which is also known to contain gorgosterol (II). This could be first lead--admittedly a very tenuous one--to an exogenous biosynthetic precursor of gorgosterol (II) in P. homomalla, and it has the virtue of identifying possible candidates that should be fed as radioactive precursors to gorgosterol-containing animals. Of equal biosynthetic interest is the origin of dinosterol (IV) and other 23-methylated sterols in dinoflagellates.

ACKNOWLEDGMENT

Financial assistance by the National Institutes of Health (grants GM-06840 and RR-00612) is gratefully acknowledged.

REFERENCES

1. D. R. Idler, P. M. Wiseman, and L. M. Safe, Steroids 16, 451 (1970).

2. N. C. Ling, R. L. Hale, and C. Djerassi, J. Am. Chem. Soc. 92, 5281 (1970).

3. C. Djerassi, Pure Appl. Chem. 41, 113 (1975).

4. M. Kobayashi and H. Mitsuhashi, Steroids 24, 399 (1974); ibid. 26, 605 (1975).

5. E. Lederer, Quart. Rev. <u>23</u>, 466 (1969).

6. For instance, constraint No. 3 in Figure 2, prohibiting simple
    migration of a double bond, would formally eliminate the
    well-documented (F. F. Knapp, J. B. Greig, L. J. Goad and T. W.
    Goodwin, Chem. Comm., 707 (1971)) conversion of isofucosterol
    ($\Delta^{24(28)}$-double bond) into poriferasterol (Δ^{22}-double bond).

7. S. Popov, R. M. K. Carlson, A. Wegmann, and C. Djerassi,
    Steroids, in press.

8. Y. Shimizu, M. Alam, and A. Kobayashi, J. Am. Chem. Soc. <u>98</u>,
    1059 (1976).

9. E. G. Bligh, Can. J. Biochem. <u>37</u>, 911 (1959).

STEROLS AND OTHER METABOLITES OF SOME MARINE INVERTEBRATES

J. Stuart Grossert and Tiw Swee

Chemistry Department, Dalhousie University

Halifax, Nova Scotia, Canada B3H 4J3

The ocean quahaug, Arctica islandica Linne, is an ocean clam with a sturdy shell that is typically 8-10 cm in diameter. The shells are white, but are covered by a thin, leathery coat that varies in colour from pale greenish-brown to dark brown to almost black, depending on the location of the clam bed. The clam is a cool-water species that is abundant in the North Atlantic ocean, and it has been utilized commercially as a supplement to the regular clam fishery[1]. However, the quahaug has developed a reputation for having tough meat together with strong flavours and odours. As a result, its current commercial use is not great. Attempts have been made to diagnose the flavour problems[2] and Ackman has studied the lipids of the species in an attempt to cast light on the problem[3]. A major difficulty is that the poor taste is a sporadic occurrence and is not, at present, predictable unless the animal is eaten.

Disagreeable odours or flavours in an animal may be due to an intrinsic metabolic process of the animal or may, alternatively, reflect the environment of the particular collection of specimens. A knowledge of the metabolic constituents of the animal may give useful leads to the problem, although a complete analysis would be too massive an undertaking and in any case the relation between a given molecule and flavour or taste is not necessarily a simple one[4]. We have examined some of the products obtained from quahaug flesh by standard extraction techniques.

Frozen quahaugs[5] were shucked[6] and the flesh was manually separated into light-coloured meat and dark-coloured hepatopancreas (digestive gland). The extracts[7] of the two portions were centrifuged and separated into aqueous methanol and chloroform

125

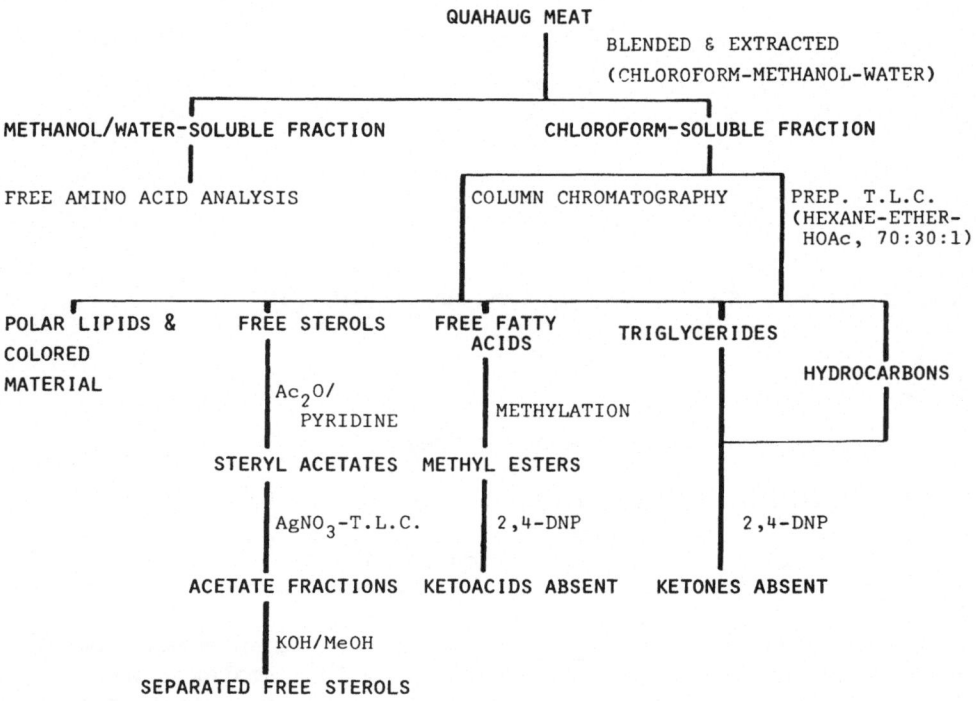

Scheme

layers, which were used for various analyses, as detailed in the Scheme.

Free amino acids were analyzed for by freeze-drying a portion of the aqueous methanol extract and introducing a citrate buffer solution of the residue into a Beckman Model 120A Automatic Amino Acid Analyzer[8]. Free amino acids were found to be present to the extent of 279.3 µmol/g of dried meat. A comparison between the amino acid composition of the quahaug and the hard clam Mercenaria mercenaria[9], in both a normal environment and a polluted environment is presented in Table 1.

The chloroform (lipid-containing) layers from the extraction process were separated by column or preparative-layer chromatography into fractions of different polarity. Results are detailed in Table 2. The methylated fatty acid fraction and the triglyceride fractions were tested with 2,4-dinitrophenylhydrazine reagent, and no isolable hydrazones could be detected. A detailed analysis of the fatty acids has been carried out by Ackman[3].

Table 1. Comparison of amino acid concentrations in the ocean Quahaug and in the hard clam M. mercenaria

Amino Acid	Ocean Quahaug		Hard Clam (mole/g dry wt.)	
	Wet Wt.	Dry Wt.	Normal	Stressed
Cysteic Acid	0.36	2.39	7.3	8.4
Aspartic Acid	2.06	13.74	27.0	28.9
Threonine	1.40	9.36	21.3	17.7
Serine	2.30	15.33	12.6	10.2
Glutamic Acid	2.98	10.91	54.9	43.4
Proline	0.78	5.18	--	--
Glycine	10.0	66.6	81.1	37.6
Alanine	9.47	63.13	76.3	52.2
Valine	1.24	8.26	16.2	12.0
Methionine	0.36	2.39	3.9	2.0
Isoleucine	1.18	7.87	14.2	9.6
Leucine	1.51	10.06	23.3	16.1
Tyrosine	0.81	5.38	11.6	7.9
Phenylalanine	0.70	4.68	12.4	8.0
Lysine	3.29	21.91	4.8	5.6
Histidine	0.99	6.57	6.5	4.0
Arginine	3.82	25.49	12.2	13.4

Table 2. Results from separation of the lipid-containing fraction from the ocean Quahaug extract

Source	Total Wt. of Meat (gm)	Total Wt. of Lipid (gm)	Separation Method	Total Wt. Sep. (mg)	Triglyceride mg/%	Fatty Acid mg/%	Sterol mg/%	Polar Lipid mg/%
Digest Gland (dark)	665	7.633	Column Chrom. #1	3029	547/ 19	808/ 27	438/ 14	1186/ 39
			Column Chrom. #2	4340	943/ 22	1348/ 31	332/ 8	1716/ 39
Light Meat	350	2.38	Prep. TLC	671	204/ 30	249/ 37	47/ 7	169/ 26
Total Flesh	---	---	Prep. TLC	1200	265/ 22	335/ 28	95/ 8	505/ 42
		Average Percentage			23	31	9	37

Table 3. Sterols of the ocean quahaug

TLC band fraction no.	R$_F$ of acetate bands	Weight of acetate (mg)	Acetate m.p. °C	Sterol m.p. °C	Sterol GLC peak no.	Relative amounts of sterols	Sterol structure no.
1	0.10	12.0	132–134	141–144	7	630	3d
2	0.22	5.9	129–131	130–132	9	222	3e
					10	8	3f
					1	4	3a
3	0.31	11.2	131–134	141–144	1	111	3a
					2	28	3g or 3h
					9	trace	3e
					10	trace	3f
4	0.40	5.7	129–131	127–130	3	333	3b
					6	1	3c
5	0.47	10.0	150–151	146–149	6	370	3c
6	0.62	35.8	116–120	136–149	4	1740	2a
					7	79	2b
					8	95	2c
					–	trace	2d
7	0.72	2.1	113–128	102–108	5	75	1a
					–	trace	1b
					–	trace	1c

    All attempts at a useful tlc separation of the free sterol
mixture[10] were unsuccessful, in contrast to results obtained with
Δ^7 sterols from a sea star[11]. However, the product from acetylation of the free sterol mixture could be separated by argentation
preparative-layer chromatography into seven bands[12]. Separations
of the sterol acetates were monitored by glc[13], mp, ir and mass
spectrometry, after which each fraction was hydrolyzed with metha-
nolic KOH. Each crystallized sterol was then characterized by mp,
glc, ir and ms; in some cases, ^{1}H nmr spectra were recorded also.
Interpretation of the results from these data permitted the conclusion to be drawn that six of the seven bands from argentation tlc
were mixtures, but, nevertheless, the data were adequate to permit
identification of fourteen sterols in the mixture, as detailed in
Table 3. The glc trace produced by the free sterol mixture is
reproduced in the Figure. Interpretation of this trace was made
possible by careful examination of the traces of the seven individual fractions.

    Fraction 7 consisted of saturated sterols. Although the C-28
(1b) and C-29 (1c) homologs of 5α-cholestan-3β-ol (1a) were present
in trace amounts, they were clearly visible in the mass spectrum of
the fraction. In a similar manner, 24ε-n-propylcholest-5-en-3β-ol
(2d) was observed as a discernible peak at m/e 428 in the ms of
fraction 6. Fraction 5 is presumed to be 24-epibrassicasterol
(3c), based on arguments made by Kobayashi and Mitsuhashi[14].

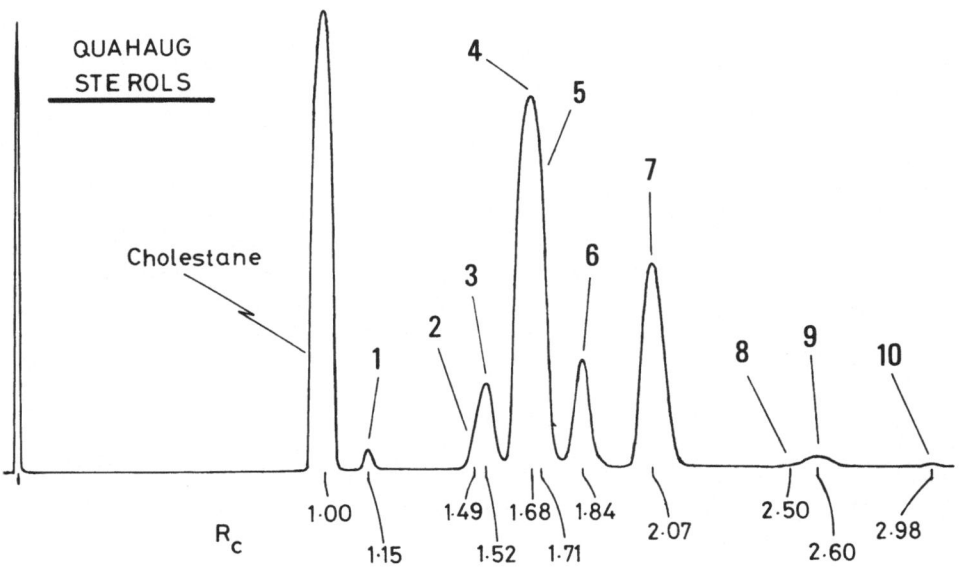

Figure 1.

Fraction 4 was almost pure cholesta-5,E-22-dien-3β-ol (3b), but fraction 3 was a mixture of at least four components. The major component was clearly the C-26 sterol (3a). However, the structure of the second component was not obvious. It may be either occelasterol (3g)[14] or cholesta-5,Z-22-dien-3β-ol (3h). Since its acetate is more polar than that of (3b), it is conceivable that (3h) is the more likely structure[14]. Sterols (3e) and (3f) were detectable in the fraction ms. Fraction 2 was primarily isofuco-sterol (3e), the minor components (3f and 3a) being removable by methanol recrystallization, and fraction 1 was pure 24-methylene-cholesterol (3d).

The sterol mixture of the quahaug is complex and, given fur-ther, more detailed, separations, would no doubt prove to be of comparable complexity to the sterol mixtures found in scallops[14,15]. These analyses have been achieved in large measure through argenta-tion tlc. We were intrigued by the fact that Δ^7 monoene free sterols could be separated from Δ^7+side-chain diene free sterols[11] but that the same was not true for Δ^5 free sterols. Examination of molecular models suggests that a silver cation could simultaneously complex with both double bonds in many Δ^7+side-chain dienes. This is not possible with Δ^5+side-chain dienes. However, a silver cation could easily complex simultaneously with the carbonyl oxygen and side-chain double bond in the Δ^5+side-chain diene acetates. This would occur most easily for $\Delta^5,24-28$ or $\Delta^5,25$ dienes, which are, in fact, the most polar on $AgNO_3$-tlc. As the double bond on the side chain becomes more hindered (cf. Idler's series[12]), so polarity decreases. It appears that the $AgNO_3$-tlc mobility of these compounds may be regarded as a sensitive probe for the conformational mobility of their side chains in the direction of the sterol nucleus.

ACKNOWLEDGEMENTS

We thank Drs. R. G. Ackman and J. C. Medcof (Environment Canada) for assistance with the ocean quahaugs; Dr. J. C. Verpoorte kindly provided the amino acid analyses; Dr. A. K. Lumb assisted with the early extractions, and Mr. J. K. LeDue provided services for glc and ms. We also thank Environment Canada, the National Research Council of Canada, and Dalhousie University for financial assistance.

REFERENCES AND NOTES

1. A. S. Merrill, J. L. Chamberlin, and J. W. Ropes, The Encyclo-
    pedia of Marine Resources, Ed. F. E. Firth, Van Nostrand
    Reinhold, New York, 1969, p.125; P. S. Parker and E. D. McRae,
    Jr., Fishery Industrial Research, 6(4), U. S. Fish and Wildlife
    Service (U. S. Department of Interior), 1970, p. 185.

2. J. C. Medcof, Report, Fisheries Research Board of Canada Bio-
    logical Research Station, St. Andrews, N.B., File No. 46-11-0.

3. R. G. Ackman, S. Epstein, and M. Kelleher, J. Fish. Res. Board
    Can. 31, 1803 (1974).

4. D. A. Forss, Prog. Chem. Fats Other Lipids 13, 220 (1972);
    I. Hornstein and P. F. Crowe, J. Gas Chromatog. 2, 128 (1964).

5. Collected from the Atlantic Ocean off Lunenburg County, Nova
    Scotia; stored for several months in National Harbours Board
    Cold Storage, Halifax, Nova Scotia, and obtained through the
    courtesy of Dr. R. G. Ackman.

6.   Carried out by heating the frozen animal in 80°C water for
     35 sec in order to open the shell.

7.   Obtained by blending the flesh in MeOH/H_2O/CHCl$_3$, according to
     the method of E. J. Bligh and W. J. Dyer, Can. J. Biochem.
     Physiol. 37, 911 (1959).

8.   Carried out through the courtesy of Dr. J. A. Verpoorte,
     Biochemistry Department, Dalhousie University.

9.   H. P. Jefferies, J. Invert. Pathol. 20, 242 (1972).

10. The crude, free sterol mixture was purified by recrystalliza-
     tion (3X) from MeOH-ether to give pure white needles, mp
     120-135°C.

11. J. S. Grossert, P. Mathiaparanam, G. D. Hebb, P. Price, and
     I. M. Campbell, Experientia 29, 258 (1973).

12. Plates for plc had 20% w/w AgNO$_3$-silica gel as adsorbent, were
     eluted five times with n-hexane-toluene (3:1 v/v) and were
     visualized with long wavelength uv light. This system appears
     to give better separations than does the hexane-benzene system,
     cf., D. R. Idler and L. M. Safe, Steroids 19, 315 (1972).

13. Glc analyses were carried out on an H-P 5750 fid gas chromato-
     graph (equipped with H-P electronic integrator) using a
     1829 x 3.2 mm stainless steel column (at 270°C), packed with
     10% UC W-98 on DMCS Chromosorb W, and an N_2 flow rate of
     100 ml/min.

14. M. Kobayashi and H. Mitsuhashi, Steroids 26, 605 (1975).

15. D. R. Idler, M. W. Khalil, J. D. Gilbert and C. J. W. Brooks,
     Steroids 27, 155 (1976), and references therein.

STUDIES ON STEROID METABOLISM IN THE ECHINODERM ASTERIAS RUBENS

S. I. Teshima, R. Fleming, J. Gaffney and L. J. Goad

Department of Biochemistry, The University

P. O. Box 147, Liverpool, L69 3BX, United Kingdom

Since the pioneering work of Bergmann[1] on the sterol composition of marine invertebrates, it has been apparent that the echinoderm classes Asteroidea (starfish) and Holothuroidea (sea cucumber) present unique problems in relation to steroid metabolism. It is now well established[2] that the echinoderm classes Crinoidea (feather stars), Ophiuroidea (brittle stars) and Echinodea (sea urchins) contain sterol mixtures which, in common with the majority of animals, are predominently Δ^5-compounds, and these animals have the capacity to synthesise cholesterol de novo from mevalonic acid via the intermediates squalene and lanosterol. The large quantities of C_{28} and C_{29} Δ^5-sterols found in these animals are presumed to be derived from the diet, since a C-24 methylation reaction has not been demonstrated. By contrast, the free sterols found in starfish and sea cucumbers are complex mixtures of Δ^7-sterols, with usually less than 5% of the mixture being comprised of stanols and Δ^5-sterols[2]. Both classes of animals have been shown to have a limited ability for de novo sterol biosynthesis, but the sequence apparently stops at the Δ^7-sterol stage[2,3], thus leading to the accumulation of 5α-cholest-7-en-3β-ol as the principal free sterol component found in many species. In starfish, the fate of dietary sterols has been examined, and it has been established that the usual dietary Δ^5-sterols can be metabolised to give Δ^7-sterols. Thus, dietary cholesterol is converted into 5α-cholest-7-en-3β-ol, while the various C_{26}, C_{28}, and C_{29} Δ^7-sterols found in starfish are derived by metabolism of the corresponding Δ^5-sterol taken in the diet[2,4,5]. The principal metabolic route from cholesterol proceeds with an oxidation catalysed by a 3β-hydroxysteroid dehydrogenase enzyme system to produce cholest-4-en-3-one, which is converted by two NADPH-requiring reductions into 5α-cholestan-3β-ol. Introduction of a Δ^7-bond into the latter compound then produces cholest-7-en-3β-ol[4].

The belief that starfish contain only Δ^7-sterols as major components has now required revision, with the discovery that starfish, and indeed all five classes of echinoderms, contain high concentrations of sterol sulphate, of which cholesterol sulphate is the main constituent[6,7,8]. It is striking that while the free sterols may contain high proportions of C_{28} and C_{29}-sterols, the sterol sulphates are largely C_{27}-sterols.

The sterol sulphates are derived in the main by preferential sulphation of the cholesterol component of the dietary sterol or the 5α-cholestanol and cholest-7-en-3β-ol derived from it by the metabolic route outlined above[7,8,9]. None of the sulphated cholesterol appears to arise by de novo synthesis in the starfish from mevalonate, although a proportion of the cholest-7-en-3β-yl sulphate may arise by the de novo route[8,9]. The biochemical role of steryl sulphates in starfish and other echinoderms is at present obscure. Cholesterol sulphate has been identified in small amounts in mammalian and insect issues, and it has variously been suggested as (a) a precursor of pregnenolene sulphate and other steroid hormones[10], (b) a membrane constituent[11], (c) an excretory product[12,13] and (d) a vector in cation transport at plasma membranes[6].

The steroid hormone biochemistry of echinoderms has received relatively little attention, although other aspects of the hormonal control of gonad development and oocyte maturation, particularly in starfish, has been more extensively studied[14,15]. The only reports of the occurrence of steroid hormones in echinoderms are the tentative identifications of oestradiol-17β and progesterone in the ovaries of Pisaster ochraceus[16] and Strongylocentrotus franciscanus[17] and the isolation of progesterone from the ovaries of Asterias amurensis[18]. The interconversion of oestrone and oestradiol-17β was observed with the eggs of the sea urchin Dendraster excentricus[19], while gut tissue preparations from the urchin Strongylocentrotus franciscanus efficiently converted oestradiol-17β into oestradiol-3-sulphate[20], which was suggested to be an excretory product. Radioactive progesterone injected into the starfish Asterias rubens and Marthasterias glacialis was metabolised to yield labelled 3β-hydroxy-5α-pregnan-20-one and polar metabolites, one of which was identified as 3β,6α-dihydroxy-5α-pregnan-20-one[21]. Steroid conjugates, soluble in aqueous methanol, were also produced in these incubations.

The steroid biochemistry of starfish and sea cucumbers is particularly interesting on account of their content of toxic steroidal saponins. Similar saponins are apparently absent from the other three classes of echinoderm[22]. The steroidal saponins of the sea cucumbers, termed holothurins, have a steroidal aglycone based upon the lanostane skeleton, while the sugar components have been identified as D-glucose, D-xylose, quinovose and 3-O-methyl glucose[2,23-25].

Several steroidal aglycones have been obtained by hydrolysis of asterosaponins, the toxic compounds found in starfish[2]. These are based upon either the cholestane or the pregnane skeleton[2]. The first aglycone structure to be elucidated was that of marthasterone ($3\beta,6\alpha$-dihydroxy-5α-cholesta-9(11),24-dien-23-one), isolated from Marthasterias glacialis[26-28]. Subsequently, the main aglycone of the asterosaponins of Acanthaster planci, Asterias amurensis, and Asterias forbesi was identified [29-32] as $3\beta,6\alpha$-dihydroxy-5α-pregnan-9(11)-en-20-one. In the intact saponins, the 3β-hydroxyl group is esterified to a sulphate moiety, while a tetra-saccharide is attached by a glycosidic linkage to the 6α-hydroxyl group. In asterosaponin A, from Asterias amurensis, the sugars are deoxy compounds, two molecules of fucose and two molecules of quinovose[33]. In the saponin of Marthasterias glacialis, one molecule of quinovose is replaced by glucose[34]. The steroidal aglycone was recovered in earlier work after acid hydrolysis of the asterosaponin. However, the isolation of the 20-hydroxy steroid aglycones, thornasterol A and thornasterol B, after more careful and controlled hydrolysis of the saponins from Acanthaster planci, has led to the suggestion that 20-hydroxylated C_{27} (and perhaps C_{28} and C_{29}) steroids are the natural products and that the C_{21}-pregnane aglycones are perhaps artifacts produced during acid hydrolysis of the saponins[35].

The holothurins and asterosaponins are biologically-active compounds. They are toxic to fish; they cause developmental abnormalities in fertilised sea urchin eggs; they are haemolytic agents; they are active against some fungi; they have antiviral properties; and they produce irreversible destruction of excitability at a cholinergic neuromuscular junction (reviewed in reference 2). It is suggested that in starfish asterosaponins may play a role as a spawning inhibitor. In starfish, the gonads undergo a seasonal development, culminating in the release of gametes[14]. The gonad-stimulating substance (GSS), a polypeptide produced by the radial nerve, passes across the coelemic cavity to the ovaries, which are stimulated to produce 1-methyladenine, a meiosis-inducing hormone which promotes oocyte maturation[14,15] prior to their discharge from the gonads. It has been suggested that, to prevent premature spawning, the action of GSS is counteracted by the asterosaponins[36,37].

An antimeiotic factor has been reported in starfish ovaries[38], and the active principle which inhibits oocyte germinal vesicle breakdown has been partially characterised as a steroid glycoside of a different type than that of the asterosaponins[39]. In this context it is worth noting that progesterone and 3β-hydroxy-5α-pregnan-20-one will induce germinal vesicle breakdown in frog oocytes[40-42], although the former is apparently inactive with starfish oocytes[18].

With the above information in mind, we have embarked upon a research programme to investigate steroid metabolism in starfish. The aims are to determine whether starfish have the enzyme systems required for conversion of cholesterol or other C_{27}-sterols into C_{21}-steroids such as progesterone and if these compounds can be further metabolised by pathways similar to those found in higher animals[43] and other phylla of invertebrates[2] to give C_{19} and C_{18} hormones. It is further hoped that these studies will give an insight into asterosaponin biosynthesis and the role of steroidal compounds in gonad development and oocyte maturation in starfish.

METABOLISM OF CHOLESTEROL BY ASTERIAS RUBENS

We have previously described the conversion by Asterias rubens of cholesterol into 5α-cholestan-3β-ol and 5α-cholest-7-en-ol[4] and into cholesterol sulphate[7,8]. In these studies there was evidence suggesting the production of other labelled metabolites, and this section describes the isolation and identification of some of these more minor metabolites of cholesterol.

Cholesterol-$\left[4\text{-}^{14}C\right]$ (50μCi) was emulsified in 0.2 ml of 0.5% Tween 80 and injected into the body cavity of a male specimen of Asterias rubens, which was then maintained for twelve days in an aquarium held at 10°C. The animal was then lyophilised and extracted by refluxing with chloroform-methanol (2:1), and the lipid was separated into various fractions by column and thin-layer chromatography (TLC) as outlined in Scheme 1.

Fraction FP was analysed by TLC (Si gel; $CHCl_3$-MeOH, 97:3), which revealed four radioactive bands with R_f values corresponding to 3β-hydroxy-5α-pregnan-20-one (3%), 5α-pregnane-3β,20ξ-diol (18%), 5α-cholestane-3β,6α-diol (79%), and an unknown (2%). No radioactivity was detected in zones corresponding to progesterone, 17α-hydroxyprogesterone, or androstendione. This initial TLC system could not separate testosterone, 5α-pregnane-3β,20α-diol, and 5α-pregnane-3β,20β-diol. TLC of this labelled material on two other systems revealed that no radioactivity was associated with testosterone and only a small proportion of label ran with 5α-pregnane-3β,20α-diol. The major labelled compound co-chromatographed with 5α-pregnane-3β,20β-diol. After addition of carrier and purification by preparative TLC, a portion was acetylated and the radioactivity was shown to chromatograph on TLC with 5α-pregnane-3β,20β-diol diacetate. A further portion of the suspected labelled 5α-pregnane-3β,20β-diol was oxidised with Jones reagent and the radioactive product was found to have an R_f corresponding to 5α-pregnane-3,20-dione. Moreover, after addition of carrier 5α-pregnane-3,20-dione, the material crystallised to constant specific activity after an initial fall (112, 79, 69, 67, 72 dpm/mg for successive crystallisations).

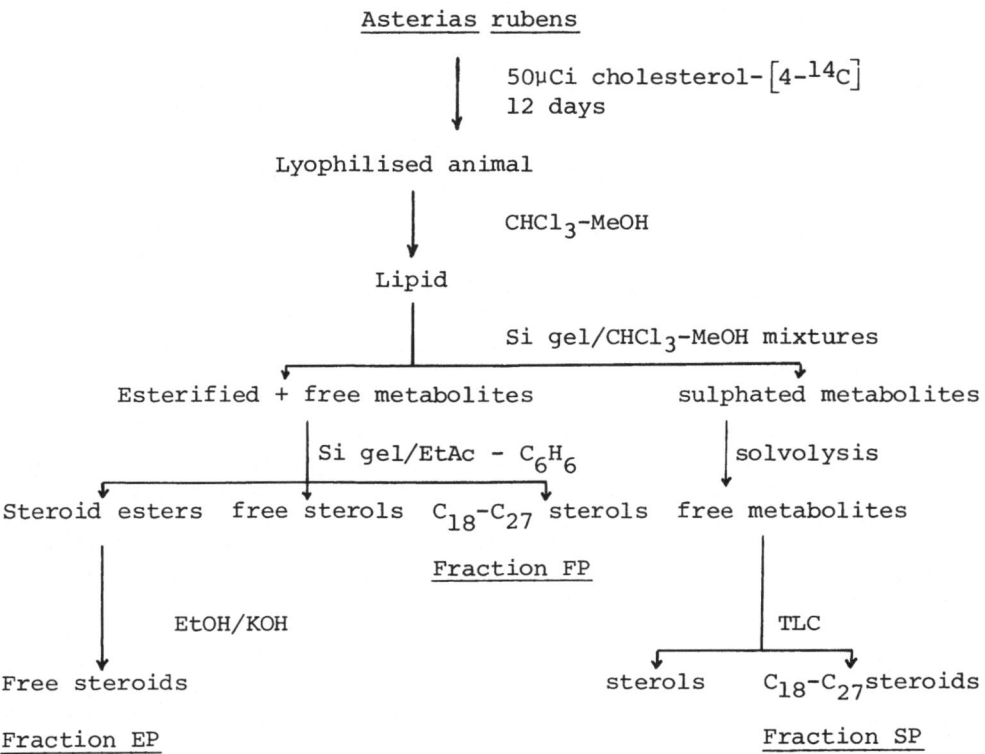

Scheme 1. Fractionation of the lipids obtained from Asterias
rubens after twelve days' incubation with cholesterol-$[4-^{14}C]$.

     The radioactive material recovered by preparative TLC of
Fraction FP, which had an R_f corresponding to 3β-hydroxy-5α-pregnan-
20-one, may have contained labelled dehydroepiandrosterone or
pregnenolone, since these three steroids were not resolved on this
first TLC system. Accordingly, the radioactive material was examined on other TLC systems, but only 3β-hydroxy-5α-pregnan-20-one
was found to be labelled. Acetylation and Jones oxidation gave
radioactive products which co-chromatographed with authentic
3β-acetoxy-5α-pregnan-20-one and 5α-pregnane-3,20-dione,
respectively.

     The labelled steroid co-chromatographing with 5α-cholestane-
3β,6α-diol was rechromatographed on a second TLC system and then
crystallised with carrier 5α-cholestan-3β,6α-diol. This resulted
in a large decrease in the specific radioactivity before a constant
value was attained (1870, 1150, 900, 540, 560, 540 dpm/mg for
successive crystallisations). Portions of the material from the

final crystallisation were either acetylated or treated with Jones reagent. The radioactive diacetate (specific activity 470 dpm/mg) co-chromatographed with authentic 5α-cholestane-3β,6α-diol diacetate on TLC, as did the labelled diketone (550 dpm/mg) with marker 5α-cholestane-3,6-dione. Examination of the labelled 5α-cholestane-3β,6α-diol diacetate on silver nitrate-silica gel TLC showed a major radioactive peak chromatographing with the carrier diacetate, but there was evidence of two other very minor labelled compounds. These may have represented the remaining small amounts of the apparently highly labelled compounds which were lost in the initial crystallisation of the 5α-cholestane-3β,6α-diol. The identity of this other labelled material is at present unknown.

After alkaline hydrolysis, the esterified steroids of Fraction EP were analysed by TLC to reveal radioactive compounds co-chromatographing with 3β-hydroxy-5α-pregnan-20-one (16%), 5α-pregnane-3β,20β-diol (38%), cholestan-3β,6α-diol (29%), 3β,6α-dihydroxy-5α-pregnan-20-one (5%, identification only tentative), and an unknown steroid (12%). The first three labelled compounds were characterised by essentially the same methods as those employed for Fraction FP.

In a similar manner, the steroids obtained by solvolysis from the sulphated steroids (Fraction SP) were identified as 3β-hydroxy-5α-pregnan-20-one (16%), 5α-pregnane-3β,20β-diol (42%), and cholestane-3β,6α-diol (23%); an unidentified metabolite accounted for 19% of the radioactivity.

On the basis of the metabolites identified, a partial route for cholesterol metabolism may be represented by Scheme 2. The introduction of a 6α-hydroxyl group to give 5α-cholestane-3β,6α-diol, and the occurrence of this compound in the free form and, particularly, the 3β-sulphated form is of significance to a consideration of asterosaponin biosynthesis. Conceivably the radioactive companion compound which was lost during crystallisation could be cholesta-9(11)-en-3β,6α-diol, which might be anticipated as an asterosaponin precursor; certainly a re-examination of the labelled metabolites seems warranted.

The formation of the C_{21}-steroids reveals that the starfish has the necessary enzyme system for sterol side-chain cleavage. In Scheme 2, the route is shown, proceeding from cholesterol to give, first, progesterone, although this compound was not detected in the present work. 3β-Hydroxy-5α-pregnan-20-one could then be produced by reduction of the progesterone, as demonstrated previously[21]. However, it is conceivable that 5α-cholestan-3β-ol might act as a substrate for side-chain cleavage to give 3β-hydroxy-5α-pregnan-20-one as the product.

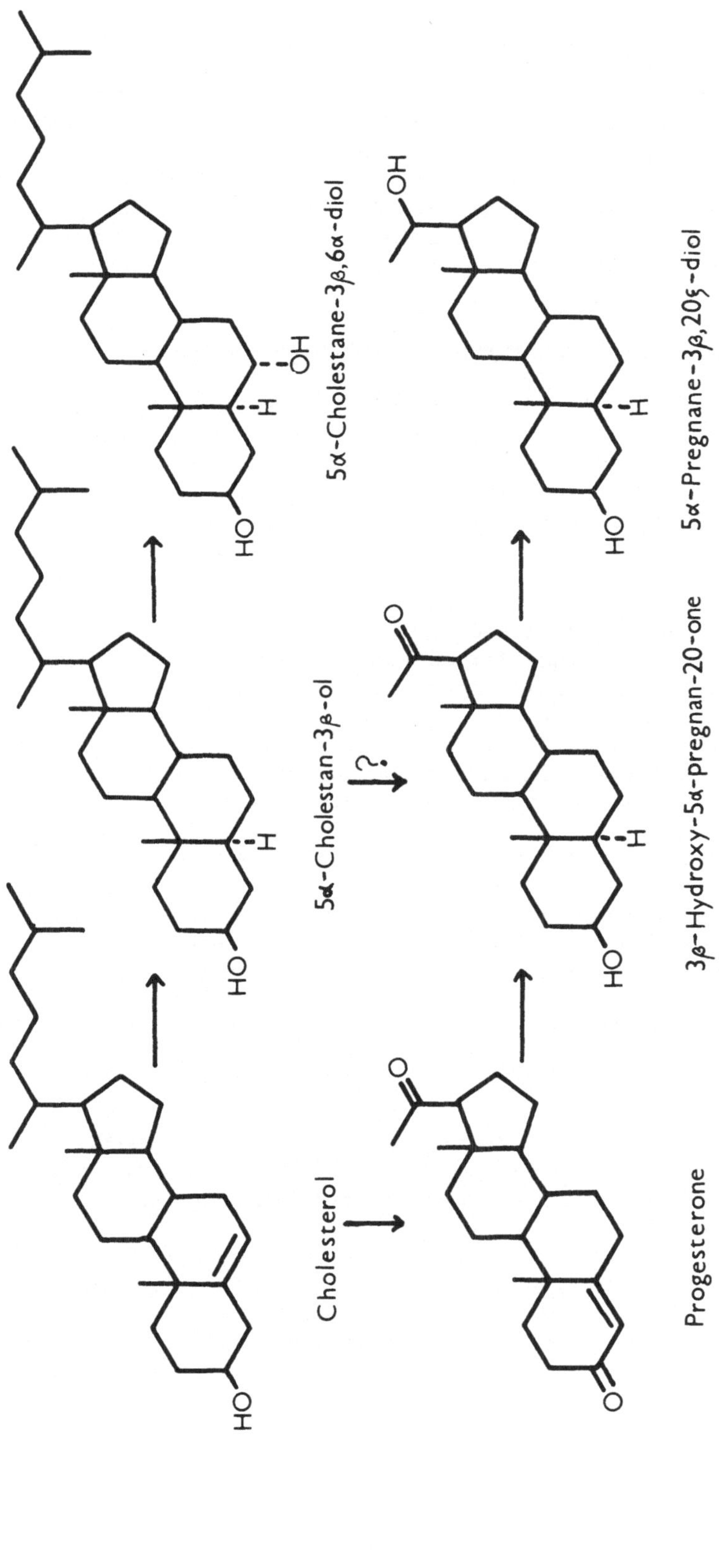

5α-Cholestane-3β,6α-diol

5α-Cholestan-3β-ol

Cholesterol

5α-Pregnane-3β,20ζ-diol

3β-Hydroxy-5α-pregnan-20-one

Progesterone

Scheme 2. Some products of cholesterol metabolism in the starfish Asterias rubens.

In the removal of the side chain of cholesterol in mammals, hydroxylation at C-20[43] occurs as a preclude to loss of the C_8 side chain as isocaproic acid. It is therefore noteworthy that thornasterol A has a 20-hydroxyl group[35]. The present demonstration that starfish have at least a limited capacity for the production of C_{21}-steroids suggests that, although pregnane aglycones can be produced as artifacts during acid treatment of thornasterol[35], some of the C_{21}-steroid aglycone may nevertheless be a genuine natural product.

METABOLISM OF PROGESTERONE BY ASTERIAS RUBENS

In a previous study[21], 5α-pregnane-3,20-dione, 3β-hydroxy-5α-pregnan-20-one, and 3β,6α-dihydroxy-5α-pregnan-20-one were identified as products of the in vivo incubation of progesterone-$[4-^{14}C]$ with A. rubens. In view of the above identification of 5α-pregnane-3β,20β-diol as a metabolite of cholesterol, reinvestigation of the metabolites of progesterone was undertaken.

Specimens of A. rubens were injected with progesterone-$[4-^{14}C]$ (5μCi) in a Tween 80 emulsion and the animals maintained in aquaria for four days. The technique for extraction was designed to separate water-soluble steroidal conjugates from more lipophilic metabolites. The starfish were homogenised with chloroform-methanol-water (5:2:5) and filtered, and the upper aqueous methanol layer of the extract was acidified to pH 4.0 and extracted twice with n-butanol to give the "steroid sulphate" fraction.

Analysis of the chloroform extract by TLC revealed a number of metabolites, including the previously identified[21] 3β-hydroxy-5α-pregnan-20-one as a major constituent, and a small proportion of the radioactivity co-chromatographed with 3β,6α-dihydroxy-5α-pregnan-20-one. About 30% of the recovered radioactivity had a low polarity consistent with the R_f of steroidal fatty acyl esters, while another metabolite had an R_f corresponding to 5α-pregnane-3β,20ξ-diol. These various metabolites were recovered by column chromatography followed by preparative TLC.

The radioactive 3β-hydroxy-5α-pregnan-20-one co-chromatographed with authentic material on TLC and crystallised to constant specific radioactivity after addition of carrier steroid.

The pregnane-3β,20ξ-diol fraction was rechromatographed with carriers on a TLC system[44], which just separated the 20α- and 20β-epimers, and it appeared that both compounds were labelled. Oxidation with Jones reagent gave the common product 5α-pregnane-3,20-dione, which was crystallised with carrier to constant specific activity.

The relatively non-polar metabolites which had an R_f on TLC similar to synthetic 3β-palmitoxy-5α-pregnan-20-one was subjected to alkaline hydrolysis to give a radioactive steroid which co-chromatographed and co-crystallised with carrier 3β-hydroxy-5α-pregnan-20-one. Acetylation gave a monoacetate which, again, co-chromatographed and co-crystallised to constant specific activity with the authentic material 3β-acetoxy-5α-pregnan-20-one.

TLC of the "steroid sulphate" material from the aqueous extract revealed four main metabolites. The most minor metabolite had an R_f similar to the marker asterosaponin, but the low level of radio-activity did not permit its further investigation. A second component co-chromatographed with 3β-sulphoxy-5α-pregnan-20-one, while the two most heavily labelled bands were barely separated and were slightly more polar than 3β-sulphoxy-5α-pregnan-20-one. Solvolysis, followed by isolation of the steroids, permitted their identifications as 3β-hydroxy-5α-pregnan-20-one and 5α-pregnane-3β-20ξ-diol, with the 20β-epimer of the latter apparently predominating. At this stage it is not clear if the diol was sul-phated at the 3β- or 20-position or, indeed, if any 3,20-disulphate was present.

To further investigate progesterone metabolism in A. rubens, portions of individual tissues were incubated with progesterone-$[4-^{14}C]$ for twenty-four hours. When incubations were performed with gonad tissues, the same products that were found in the in vitro incubations were observed. However, striking differences were apparent when the proportions of the various products produced by male and female gonad tissues were compared. This is illustrated by the thin-layer radio-scans shown in Figure 1. With the male incubation, all the starting progesterone had been metabolised and the two major peaks chromatographed with 3β-hydroxy-5α-pregnan-10-one and the fatty acyl ester of this steroid; only a small radioactive peak ran in the region expected for steroid sulphate. By contrast, in the female incubation a small amount of unchanged progesterone remained; the major metabolite was again 3β-hydroxy-5α-pregnan-20-one, but very little of this compound was esterified to fatty acid, whereas now the steroid sulphate fraction was heavily labelled. Solvolysis of the sulphate fraction and TLC showed that the major sulphated steroid was 3β-hydroxy-5α-pregnan-20-one, with only about 10-15% of 5α-pregnan-3β,20ξ-diol present in this conjugated form.

The difference in the relative amounts of esterified and sul-phated metabolites of progesterone produced by male and female gonad tissue was found to be reproducible using tissue samples from a number of animals of varying stages of sexual maturity. How-ever, it was notable that with preparations from animals in which the gonads were approaching full development prior to spawning, the production of steroid sulphate declined markedly, with a concomi-tant accumulation of free 3β-hydroxy-5α-pregnane-20-one (Table 1).

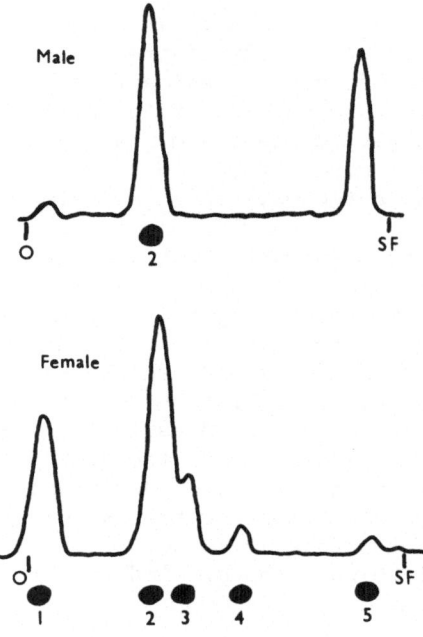

Figure 1. Thin-layer radioscans of the total lipid extracted from
male and female gonads of A. rubens after incubation with proges-
terone-[4^{14}C]. Marker spots were: 1, 3β-sulphoxy-5α-pregnan-20-
one; 2, 3β-hydroxy-5α-pregnan-20-one; 3, progesterone; 4, 5α-
pregnan-3,20-dione; 3β-palmitoxy-5α-pregnan-20-one.

    Incubations of progesterone-[4-^{14}C] with body-wall tissue
resulted in reduction to 3β-hydroxy-5α-pregnan-20-one and 5α-
pregnane-3β,20ξ-diol, with an appreciable proportion (30-60%) of
these compounds being sulphated. Only a very small proportion
(<5%) of the product from either male or female tissue chromatographed as a fatty acyl ester. With incubations employing pyloric
caecae tissue, reduction of progesterone appeared to be virtually
complete and some 80-95% of the labelled product was of a very
polar nature, which suggests it was a conjugate. It was established that this material was not steroidal sulphate, but its
nature is at present obscure.

    Reduction of progesterone-[4-^{14}C] to 5α-pregnane-3,20-dione
and 3β-hydroxy-5α-pregnan-20-one was achieved using cell-free
homogenates prepared from both gonad and pyloric caecae tissue of
A. rubens. Δ^4-3-Keto steroid reductase activity was found in a
105,000g "microsomal" pellet, which suggests that the reductase
enzymes are membrane-bound. There was an absolute requirement for
NADPH as co-factor. Using stereospecifically-tritiated samples
of NADPH, it was established that the 4-pro-S hydride of NADPH

Table 1. Production of sulphated, free and esterified metabolites of progesterone-$[4\text{-}^{14}C]$ produced by gonad tissue from specimens of <u>Asterias</u> <u>rubens</u> at different stages of sexual maturity.

	Specimen No.	Gonad Index	Metabolites[*]		
			Sulphate	Free	Ester
Male	1a[†]	0.08	12	41	47
	1b	0.08	15	40	45
	2	0.10	4	70	26
	3	0.13	14	59	25
	4	0.19	--	66	33
	5	0.25	--	60	39
Female	1a[†]	0.07	25	70	5
	1b	0.07	22	72	6
	2	0.12	25	73	2
	3	0.14	20	70	9
	4	0.26	7	83	9
	5	0.27	6	86	6

[*]Relative amounts of metabolites are expressed as percentages of the total peak area after thin-layer radioassay.

[†]Incubations 1a and 1b in each case are duplicates performed with samples of tissue from the same animal.

is transferred to the 3α- and 5α-positions in the reduction of the Δ^4-3-keto system of both cholest-4-en-3-one and progesterone. This is the same stereospecificity as was observed previously for the Δ^4-3-keto steroid reductase system[45,46].

CONCLUSIONS

    In none of the studies outlined above was evidence obtained for any extensive metabolism of cholesterol or progesterone to yield C_{19}- or C_{18}-steroids. It is, however, quite possible that conversion to yield only small amounts of such sterols would not have been detected by the methods employed. The ability of starfish to produce C_{18}-steroids such as oestradiol-17β, previously tentatively reported in starfish gonads[16], therefore remains unresolved.

    Steroid sulphates are well known as products of steroid metabolism in mammalian tissues[45], but, to our knowledge, fatty acyl esters of 3β-hydroxy C_{21}-steroids have not been reported previously. The production of such steroid esters by <u>A</u>. <u>rubens</u>, and in

particular their facile synthesis by male gonad tissue, is thus of
considerable interest. Further studies seem warranted to establish
the nature of the fatty acyl moieties, to establish the esterification mechanism involved in their elaboration, and, perhaps most
important, to elucidate a biochemical role for these compounds.

The formation by A. rubens of 5α-cholestane-3β,6α-diol from
cholesterol and the fact that this compound also occurs in a
sulphated form suggest that this diol may be an early precursor in
the pathway to the asterosaponins. This perhaps suggests that the
asterosaponins are derived by appropriate metabolism of dietary
sterol. This is also suggested by the C-24 methyl group of thornasterol B, since C-24-alkylated sterols are generally regarded as
arising solely from biosynthesis in plants and, therefore, as
being of dietary origin when found in animals[2].

ACKNOWLEDGEMENT

We thank the Science Research Council for financial support.

REFERENCES

1. W. Bergmann, in Comparative Biochemistry, Vol. 3, p. 103,
    eds. M. Florkin and H. S. Mason, Academic Press, London and
    New York, 1962.

2. L. J. Goad, in Biochemical and Biochemical Perspectives in
    Marine Biology, Vol. 3, p. 213, eds. D. C. Malins and J. R.
    Sargent, Academic Press, London and New York, 1976.

3. A. G. Smith and L. J. Goad, Biochem. J. 146, 25 (1975).

4. A. G. Smith and L. J. Goad, Biochem. J. 146, 35 (1975).

5. P. A. Vogt and J. W. A. van Rheenen, Comp. Biochem. Physiol.
    54B, 473 (1976).

6. L. R. Björkman, K.-A. Karlson, I. Pascher, and B. E.
    Samuelsson, Biochem. Biophys. Acta 270, 260 (1972).

7. R. M. Goodfellow and L. J. Goad, Biochem. Soc. Trans. 1, 759
    (1973).

8. R. M. Goodfellow, Ph.D. Thesis, Univ. of Liverpool, 1974.

9. S. I. Teshima and L. J. Goad, paper in preparation.

10. R. B. Hochberg, S. Ladany, M. Welch, and S. Lieberman, Biochemistry 13, 1938 (1974).

11. G. Bleau, F. H. Bodley, J. Longpre, A. Chapdelaine, and K. D. Roberts, Biochem. Biophys. Acta 352, 1 (1974).

12. J. S. D. Winter and A. M. Bongiovanni, J. Clin. Endocrinol. Metab. 28, 927 (1968).

13. R. F. N. Hutchins and J. N. Kaplanis, Steroids 13, 605 (1969).

14. H. Kanatani, Int. Rev. Cytol. 35, 253 (1973).

15. A. W. Scheutz, Biol. of Reproduction 10, 150 (1974).

16. C. R. Boticelli, F. C. Hisaio, and H. H. Wotiz, Proc. Soc. Exp. Biol. Med. 103, 875 (1960).

17. C. R. Boticelli, F. C. Hisaio, and H. H. Wotiz, Proc. Soc. Exp. Biol. Med. 106, 887 (1961).

18. S. Ikegami, H. Shirai, and H. Kanatani, Zool. Mag. 80, 26 (1971).

19. R. R. Hathaway and R. E. Black, Gen. Comp. Endocrinol. 12, 1 (1969).

20. J. E. Creange and C. M. Szego, Biochem. J. 102, 898 (1967).

21. J. Gaffney and L. J. Goad, Biochem. J. 138, 309 (1973).

22. T. Yasumoto, M. Tanaka, and Y. Hashimoto, Bull. Jap. Soc. Sci. Fish. 32, 673 (1966).

23. P. J. Scheuer, Fortschr. Chem. Org. Naturstoffe 27, 322 (1969).

24. E. Premuzic, Fortschr. Chem. Org. Naturstoffe 29, 417 (1971).

25. J. S. Grossert, Chem. Soc. Rev. 1, 1 (1972).

26. A. M. Mackie and A. B. Turner, Biochem. J. 117, 543 (1970).

27. A. B. Turner, D. S. H. Smith, and A. M. Mackie, Nature 233, 209 (1971).

28. D. S. H. Smith, A. B. Turner, and A. M. Mackie, J. Chem. Soc. Perkin 1, 1745 (1973).

29. S. Ikegami, Y. Kamiya, and S. Tamura, Agr. Biol. Chem. 36, 1777 (1972).

30.   Y. M. Sheikh, B. M. Tursch, and C. Djerassi, J. Amer. Chem.
      Soc. 94, 3278 (1972).

31.   Y. Shimizu, J. Amer. Chem. Soc. 94, 4051 (1972).

32.   J. W. ApSimon, J. A. Buccini, and S. Badripersaud, Can. J.
      Chem. 51, 850 (1973).

33.   S. Ikegami, Y. Hirose, Y. Kamija, and S. Tamura, Agr. Biol.
      Chem. 36, 2453 (1972).

34.   S. H. Nicholson and A. B. Turner, J. Chem. Soc. Perkin 1,
      1357 (1976).

35.   I. Kitigawa, M. Kobayashi, T. Sugawara, and I.Yasioka,
      Tetrahedron Lett., 967 (1975).

36.   S. Ikegami, S. Tamura, and H. Kanatani, Science 158, 1052
      (1967).

37.   S. Ikegami, Y. Kamiya, and S. Tamura, Agr. Biol. Chem. 36,
      1087 (1972).

38.   L. V. Heilbrunn, A. B. Chaet, A. Dunn, and W. L. Wilson,
      Biol. Bull. 106, 158 (1954).

39.   S. Ikegami and S. Tamura, Experientia 29, 325 (1973).

40.   A. W. Scheutz, Proc. Soc. Expl. Biol. Med. 124, 1307 (1967).

41.   S. Subtelney, L. D. Smith, and R. E. Ecker, J. Expl. Zool.
      1968, 39 (1968).

42.   M. K. Sanyal and E. R. Sibre, Proc. Soc. Exp. Biol. Med. 144,
      483 (1973).

43.   H. L. J. Makin, The Biochemistry of Steroid Hormones, Blackwell,
      London, 1975.

44.   B. P. Lisboa, Steroids 6, 605 (1965).

45.   I. Björkhem, H. Daniellson, and K. Wikrall, Eur. J. Biochem.
      36, 8 (1973).

46.   I. Björkhem, J. A. Gustaffson, and O. Wrange, Eur. J. Biochem.
      37, 143 (1973).

47.   S. Bernstein and S. Solomon, in Chemical and Biological Aspects
      of Steroid Conjugation, Springer Verlag, Heidelberg, 1970.

X-RAY DIFFRACTION AND THE STRUCTURE OF MARINE NATURAL PRODUCTS

J. Finer, K. Hirotsu, and J. Clardy

Ames Laboratory, USERDA, and Department of Chemistry

Iowa State University, Ames, Iowa 50011 USA

While the success of X-ray diffraction in the elucidation of molecular structuré is well known to natural products chemists, the physical principles responsible for this success appear to be only dimly perceived. Various idiosyncratic views range from disgust at an approach that smacks of cheating to contentment with an expeditious way to secure structures. In spite of this wide range of relatively uninformed views, a single-crystal X-ray analysis is taken as the ultimate measure of molecular structure. Why should this be the case?

X-rays are monochromatic ($\sim$1Å) electromagnetic radiation that scatters from the electrons in the diffracting medium. There is an analogy between X-ray diffraction and light diffraction. When the wavelength of the illuminating light is roughly the same size as the object illuminated, the scattering pattern is not a simple picture of the object[1]. The unfamiliar resulting pattern is called the transform of the illuminated object. While this represents a technical complexity, it has redeeming features. Since there is no one-to-one correspondence between object and transform, part of the transform can be sacrificed and a recognizable object recovered. Unfortunately, X-ray sources are not intense enough to give a measurable pattern from a single molecule. The purpose of a crystal is simply to hold large numbers of molecules ($\sim 10^{+17}$) in an orderly fashion for the diffraction experiment.

The analogy between light and X-rays is sadly deficient in one respect: while light waves can be bent and recombined with lenses, there are at present no lenses for X-rays. The crystallographer must collect the scattered waves and use the computer as his lens

to recombine them. Unfortunately, to combine the waves properly
one must know their phases, and this information is irretrievably
lost. If this problem can be overcome, an X-ray analysis will
reveal the entire molecular structure.

The advantages of an X-ray structure determination are:
(1) It is performed nondestructively on sub-milligram quantities.
(2) The technique is "prejudice free." (3) There is a lavish amount
of experimental data for each unknown.

SOME SIMPLE ILLUSTRATIVE EXAMPLES: DINOSTEROL AND DICTYOXEPIN

Dinosterol (1) is an unusual C_{30} sterol from the toxic dinoflagellate Gonyaulax tamarensis. Most of the structure was secured
by Professor Shimizu and his collaborators[2]. The substitution at
C(23) was unusual and had analogy only in gorgosterol (2)[3], acanthasterol[4], and demethylgorgosterol[5]. Unfortunately, the chemical and
spectral work did not define the crucial stereochemistry of the
C(22) double bond or the configuration at C(24).

While crystalline, dinosterol (1) did not form adequate crystals for X-ray diffraction purposes, but the p-iodobenzoate did.
This had the added benefit of introducing an atom that was very
easily located. The phase problem could be overcome in this fashion.

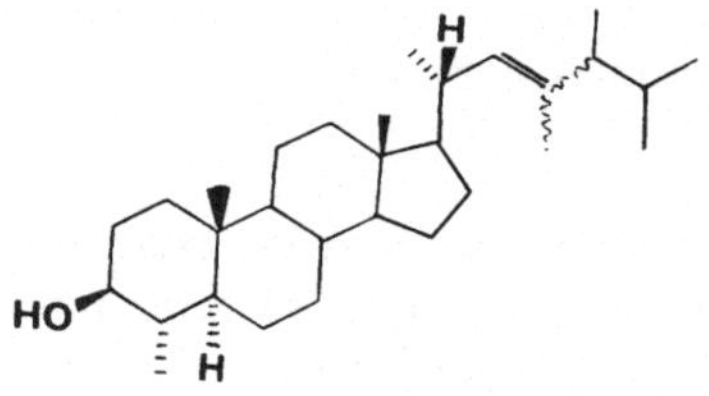

1

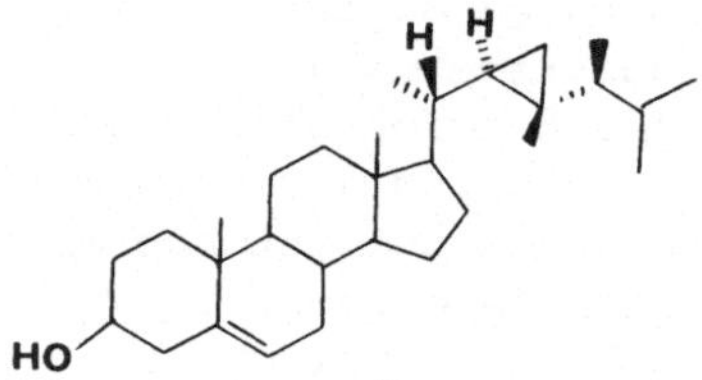

2

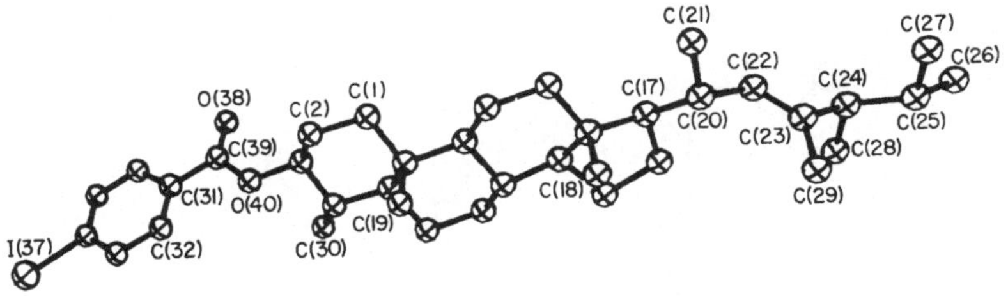

Figure 1. A computer-generated perspective drawing of the final
X-ray model of dinosterol p-iodobenzoate. Hydrogen atoms are
omitted for clarity. The absolute configuration was deduced from
anomalous scattering from the iodine.

The crystals belonged to the orthorhombic crystal class, with
a = 8.081, b = 10.658(3), and c = 40.212(7)Å. All unique data
with $\theta \leq 114$ were collected and the structure routinely solved
by the heavy atom method. After correction for Lorentz, polarization, and background effects, a total of 1889 (69%) reflections
were judged observed. Assuming that the position (three variables)
and type (one variable) of each atom would satisfactorily define
the structure, this problem is overdetermined by a factor of 1889
(obs.)/160 (variables) ∿12. Introduction of a heavy atom such as
iodine, in addition to simplifying the phase determination procedure,
has the added benefit of allowing the absolute configuration to be
determined. In this case, the average discrepancy between observed
and calculated data was 4.9% for the structure shown and 8.0% for
the enantiomer.

The configuration of the double bond is E, and C(24) has the
R absolute configuration; a final drawing of the X-ray model is
shown in Figure 1.

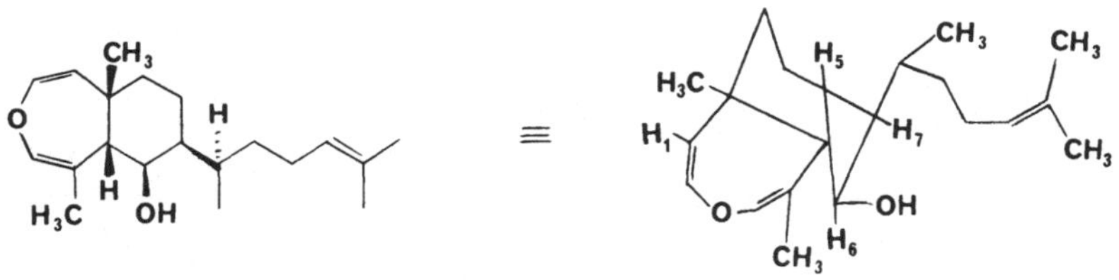

3

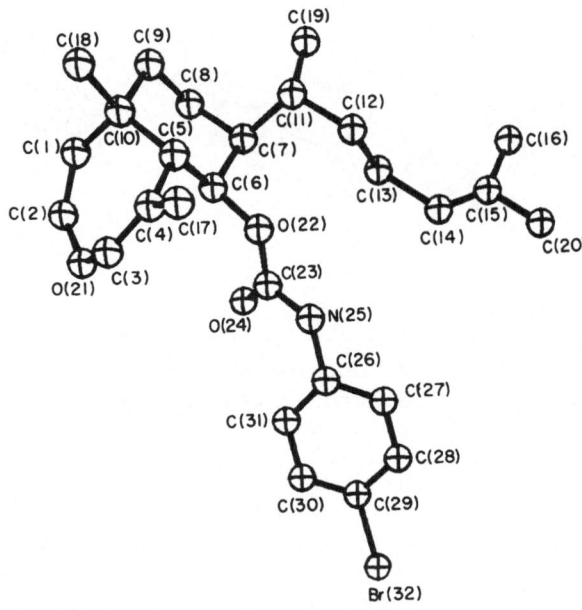

Figure 2.   A computer-generated drawing of the final X-ray model
of the p-bromophenylurethane of dictyoxepin.

Dictyoxepin (3) was isolated by Professor K. L. Erickson from
the brown alga Dictyota acutiloba[6].   In this case, the chemical
and spectral work did not clearly indicate even a two-dimensional
structure.   The mass spectrum indicated $C_{20}H_{32}O_2$, and the heteroatoms seemed to belong to a vinyl ether and a secondary alcohol.
The carbon skeleton contained three double bonds and two rings.

Since the parent material was not crystalline, the p-bromophenylurethane was prepared.   The crystals formed in the relatively
rare hexagonal crystal class and the systematic absences conformed
to $P6_1$.   Cell constants were $\underline{a} = \underline{b} = 14.238(3)$ and $\underline{c} = 23.306(3)$Å.
After correction for Lorentz, polarization, and background effects,
1369 (74%) of the unique reflections with $2\theta \leq 114^\circ$ were judged
observed.   Solution by the heavy atom method and least-squares
refinement proceeded routinely to reveal structure (3) for dictyoxepin.   In this case, the difference between the structure and
its mirror image is only a modest 5.3% vs. 5.4%.   This shows the
marked advantage of using I rather than Br for determining absolute
configurations with CuK_α X-rays and the care which must be exercised
in using only Br for absolute configurations.

As can be seen from Figure 2, the ring junction is cis with
the bridgehead methyl, C(18) being equatorial and the bridgehead
hydrogen being axial.   This conformation is apparently the same

conformation that the free alcohol adopts in solution. The pmr
spectrum of (3) indicates that H(6) must be axial, as it displays
axial-axial (11 Hz) and axial-equatorial (4 Hz) coupling to
neighboring protons. The 4,5-dihydrooxepin system of (3) is quite
rare in the terpenoids but has been reported in the sesquiterpenoids occidenol[7] and miscandenin[8]. The structure of the related
metabolite dictyolene (4) was deduced from chemical and spectral
information[6].

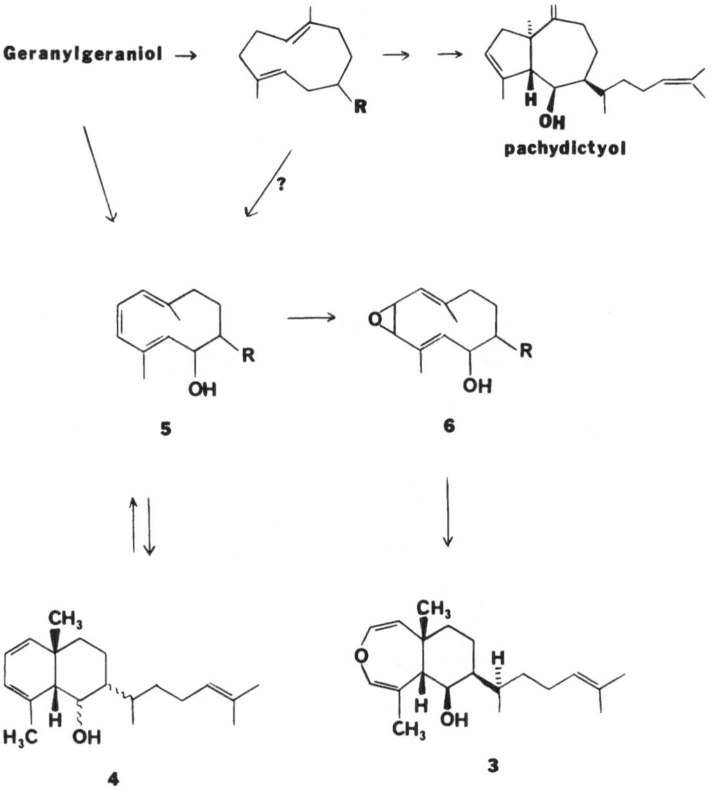

Scheme 1

A plausible biogenesis of dictyoxepin is given in Scheme 1[9].
Cyclization and oxidation of geranylgeraniol gives <u>trans</u>,<u>cis</u>,<u>trans</u>-
cyclodecatriene (5) and its corresponding epoxide (6). A thermally
allowed disrotatory ring closure of (4) affords dictyolene (4),
while a [3,3] sigmatropic shift in the epoxide gives dictyoxepin (3).

MORE COMPLEX ILLUSTRATIVE EXAMPLES: DOLABELLADIENE AND ILIMAQUINONE

It is still possible to solve the phase problem in the absence
of heavy atoms. A few initial phases are chosen arbitrarily or
assigned symbolic values, and the other phases are derived by various mathematical relations[10]. Since these relations make no
specific chemical assumptions other than the fact that electron
density is non-negative everywhere in a crystal, they are called
"direct methods." Currently there exist excellent, almost routine

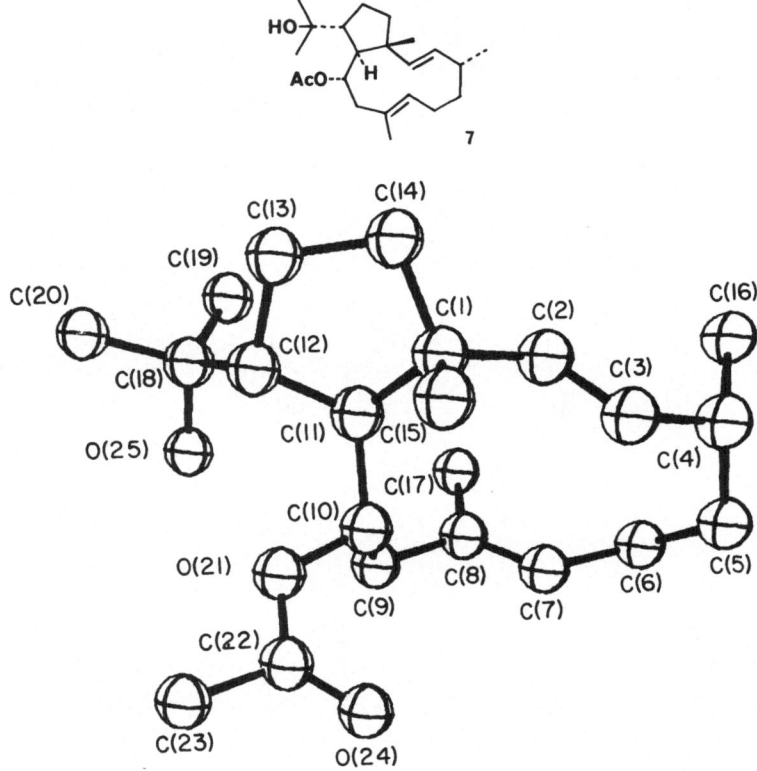

Figure 3. A computer-generated perspective drawing of (7). Hydrogens are not shown, and no absolute stereochemistry is implied.

programs for the application of direct methods to the interpretation of diffraction data[11]. There are pitfalls for the unwary and
pathologies for the connoisseur, but the majority of moderate-sized
($\leq$ 40 nonhydrogen atoms) organic structures can be solved routinely.

Dolabelladiene (7) is a crystalline member of a series of
diterpenes isolated from the digestive gland of <u>Dolabella</u> <u>cali-</u>
<u>fornica</u> by Professor D. J. Faulkner and his colleagues at Scripps
Institution of Oceanography[12]. The molecular formula was $C_{22}H_{36}O_3$,
with the heteroatoms present as a secondary acetate and a tertiary
hydroxyl. The carbon skeleton had two rings and two double bonds
but resembled no known diterpene skeleton.

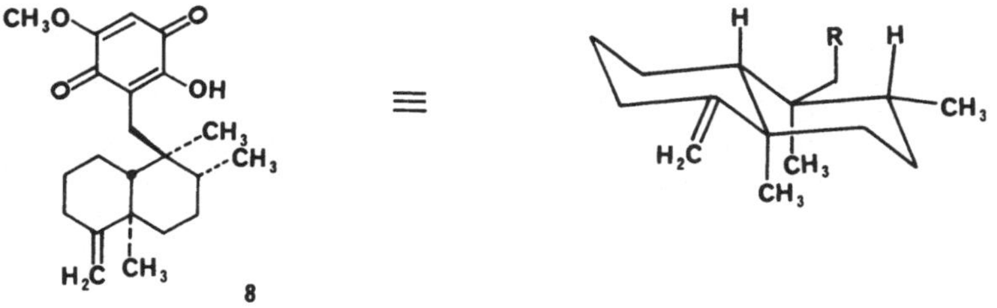

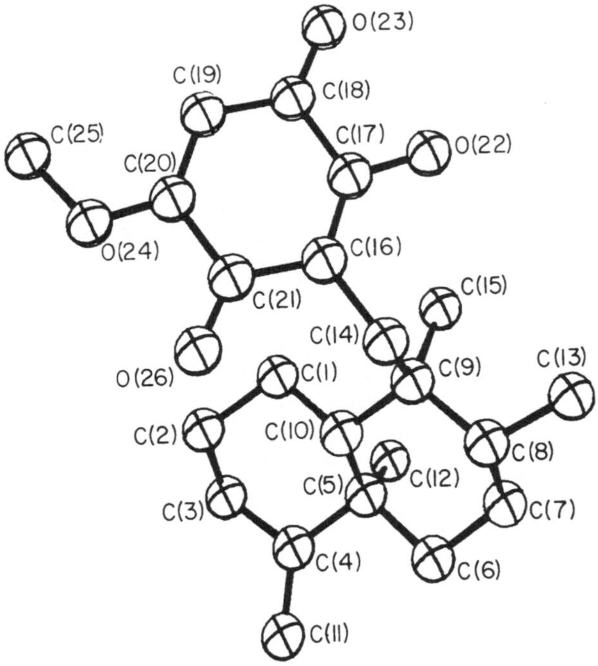

Figure 4. A perspective drawing of the final X-ray model of
ilimaquinone. No absolute configuration is implied.

The structure was solved by the direct method of X-ray analysis
and is shown in Figure 3.   The crystals belonged to the common
chiral space group $P2_12_12_1$, with a = 8.778(1), b = 9.470(1), and
c = 25.785(2)Å.   After standard data collection and workup, 1540
(90.4%) reflections were judged observed.   The structure is shown
in Figure 3, and it can be seen that the double bonds at C(2) and
C(7) both have the E configuration.   The rings are joined in a
trans fashion.   The chemistry and biosynthetic speculations will
be presented elsewhere by Professor Faulkner.

Without the presence of significant anomalous scatterers, the
X-ray technique cannot define the absolute stereochemistry.

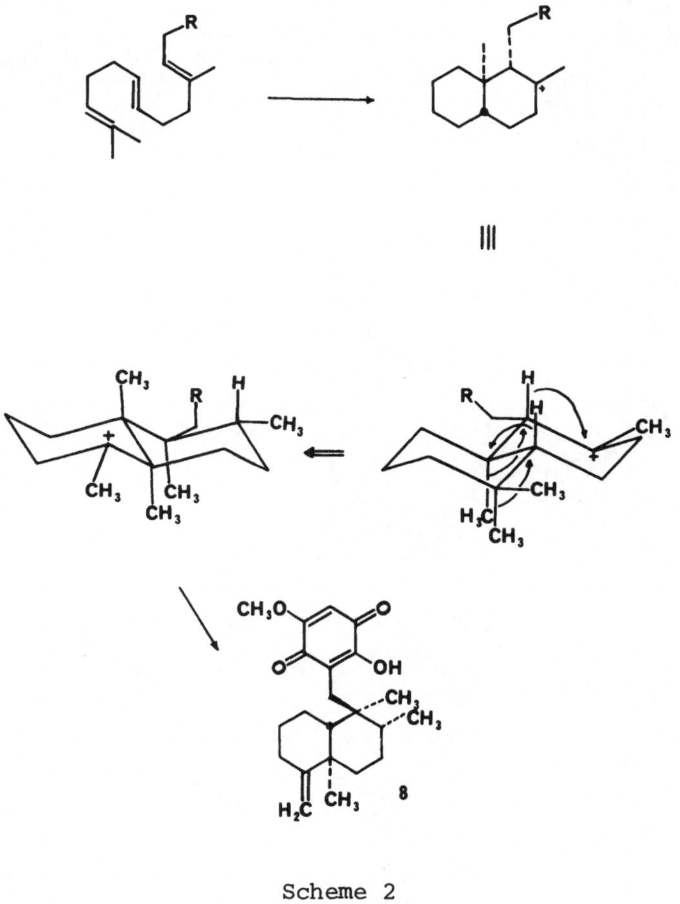

III

Scheme 2

Ilimaquinone (8) was isolated from a sponge by Professor P. J. Scheuer and his colleagues at the University of Hawaii. While the material was always crystalline, it was some years before a suitable X-ray crystal was obtained! Additional difficulties became apparent. Crystal symmetry is beneficial for the application of direct methods; ilimaquinone had none. Small molecular size is usually helpful; ilimaquinone crystallized in the triclinic space group P_1, with a = 7.013(7), b = 7.733(5), c = 22.648(23)Å, α = 93.26(7), β = 125.44(6), γ = 100.35(6)° and Z = 2. The crystal scattered poorly, with only 1313 (54%) observable reflections.

While the structure of ilimaquinone was more difficult than that of dolabelladiene, it did eventually yield to standard techniques. The final residual was .057.

The structure of the bicyclic sesquiterpene is shown in Figure 4. It is clearly related to the recently described compound avarol[13]. A plausible biosynthesis is given in Scheme 2. While the stereochemistry is what would be expected from axial-axial rearrangements, it is comforting to receive ocular confirmation.

ISOCONCINNDIOL HYDROPEROXIDE - A CAUTIONARY TALE

Recently Professor W. Fenical isolated a diterpenoid material from Laurencia snyderae. The crystals formed in the space group $P2_1$, with a = 6.797(2), b = 13.550(4), c = 10.677(3), and β = 96.81(3)°. The 1143 (82%) observed reflections did not yield to direct methods in any routine fashion. After substantial effort, twenty-three atoms were located (we believed the molecular formula $C_{20}H_{32}O_3$) which made some sense. As is customary, we had identified each atom as carbon, leaving the matter of correct atom attribution for later. A drawing of the model is presented in Figure 5.

Oxygen atoms are typically located by their small size and short bond distances on such a model. By these criteria we had four oxygen atoms (*'s), two of them as a hydroperoxide moiety! There was also a very poorly behaved carbon atom (+) with a C-C bond distance of 1.70 Å! After consideration of biogenetic and spectral data, the structure shown as (9) was arrived at. The new diterpene skeleton can be imagined as arising from concinndiol(10)[14] by solvolysis, rearrangement, and elimination to (11), followed by singlet oxygen oxidation (cf. Scheme 3). The relative stereochemistry at the five centers in common with concinndiol are identical. This last example is one in which X-ray diffraction, by itself, was not sufficient to elucidate the structure.

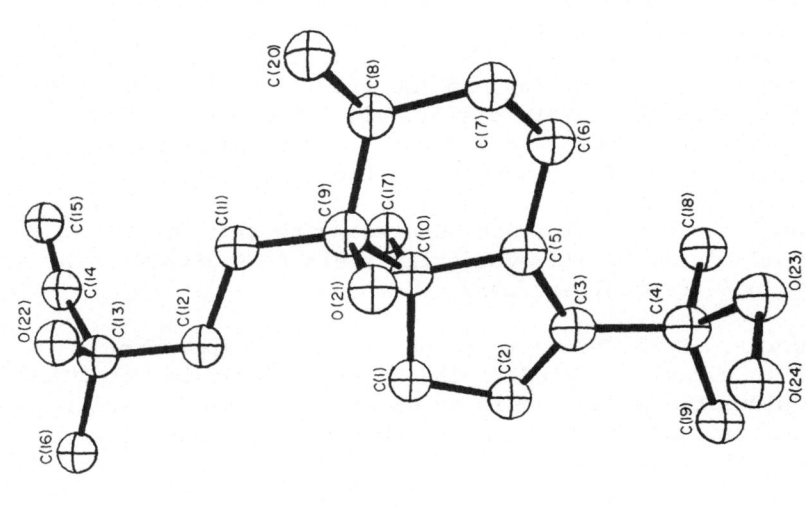

Figure 6. A drawing of the final X-ray model of isoconcinndiol hydroperoxide.

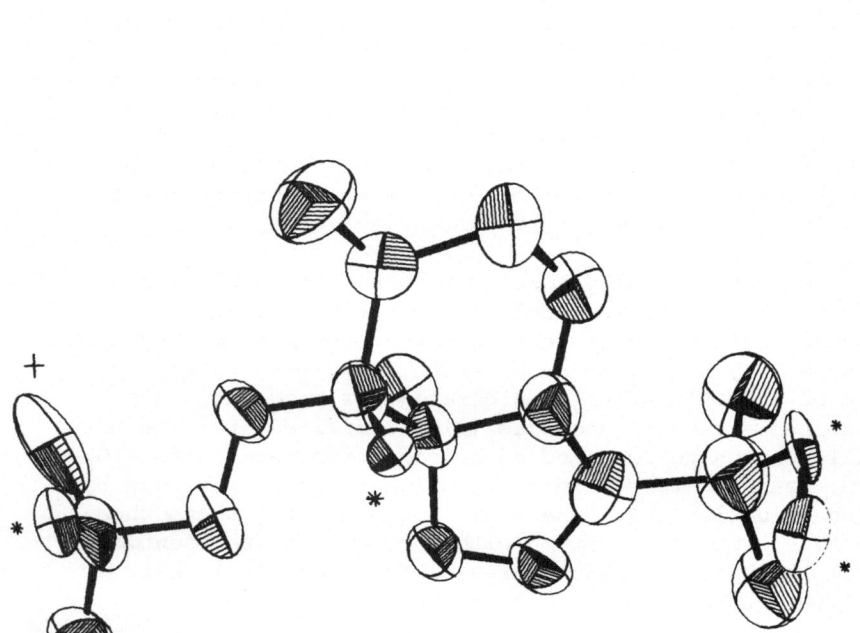

Figure 5. A drawing of an early X-ray model of isoconcinnidol hydroperoxide. Atoms thought to be oxygens are indicated with *'s and a poorly behaved carbon with a +.

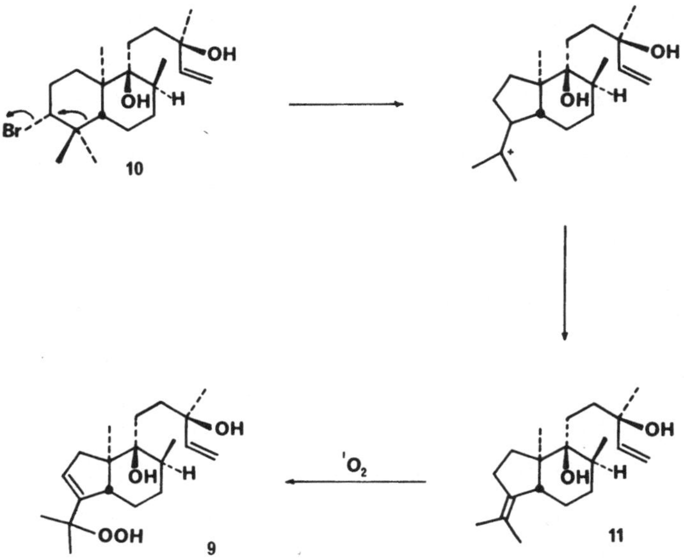

Scheme 3

REFERENCES

1. G. Harburn, C. A. Taylor, and T. R. Welberry, Atlas of Optical
    Transforms, Cornell University Press, Ithaca, New York, 1975.

2. Y. Shimizu, M. Alam, and A. Kobayashi, J. Amer. Chem. Soc.
    98, 1059 (1976).

3. N. C. Ling, R. L. Hale, and C. Djerassi, J. Amer. Chem. Soc.
    92, 5281 (1970) and references therein.

4. K. C. Gupta and P. J. Scheuer, Tetrahedron 24, 5831 (1968);
    Y. M. Sheikh, C. Djerassi, and B. M. Tursch, Chem. Commun.,
    217 (1971).

5. F. J. Schmitz and P. Pattabhiraman, J. Amer. Chem. Soc. 92,
    6073 (1970).

6. H. H. Sun, S. M. Waraszkiewicz, K. L. Erickson, J. Finer, and J. Clardy, in press.

7. B. Tomita and Y. Hirose, Tetrahedron Lett., 235 (1970).

8. W. Herz, P. S. Subramanian, P. S. Santhanam, K. Aota, and A. L. Hall, J. Org. Chem. $\underline{35}$, 1453 (1970); P. J. Cox and G. A. Sim, Chem. Commun., 428 (1973).

9. A. G. Hortmann, D. S. Daniel, and J. E. Martinelli, J. Org. Chem. $\underline{38}$, 728 (1973); A. G. Hortmann, J. Org. Chem. $\underline{33}$, 5785 (1968).

10. M. M. Woolfson, An Introduction to X-ray Crystallography, Cambridge University Press, London, 1970.

11. G. Germain, P. Main, and M. M. Woolfson, Acta Crystal. $\underline{B24}$, 274 (1970).

12. C. Ireland, D. J. Faulkner, J. Finer, and J. Clardy, J. Amer. Chem. Soc. $\underline{98}$, 4664 (1976).

13. S. Rosa, L. Minale, R. Riccio, and G. Sodano, J. Chem. Soc. Perkin I, 1408 (1976).

14. J. L. Sims, G. H. Y. Lin, R. Wing, and W. Fenical, Chem. Commun.,470 (1973).

CAULERPIN

Bhim C. Maiti and Ronald H. Thomson

Department of Chemistry, University of Aberdeen

Meston, Walk, Old Aberdeen, Scotland

Caulerpin is an unusual orange-red pigment first isolated by Santos[1] from the green algae <u>Caulerpa</u> <u>serrulata</u>, <u>C</u>. <u>sertularioides</u> and <u>C</u>. <u>racemosa</u> var. <u>clavifera</u> collected in the Philippines. We have also found it in the latter species and in <u>C</u>. <u>taxifolia</u>, both obtained from Sri Lanka.

The proposed structure (1) for caulerpin, $C_{24}H_{18}N_2O_4$, was based on the following evidence. Its aromatic nature was deduced from the low-field nmr resonances[*] at δ 8.21 (2H) and 7.6-7.0 (8H), the uv absorption at λ_{max} (EtOH) 222, 270, 292, and 317 nm, and other spectroscopic data. The presence of the two methyl ester functions was indicated by the nmr singlet (6H) at δ 3.83, ν_{CO} at 1685 cm^{-1}, and the mass spectrum, and the NH groups were denoted by peaks at 3380 cm^{-1} and δ 11.36, suggesting that they were hydrogen-bonded to the ester carbonyls. On alkaline hydrolysis caulerpin gave a di-acid which was decarboxylated by heating with copper bronze in quinoline, but the product was poorly characterised. While this evidence is consistent with the proposed structure (1), it falls short of proof, and biogenetically it looks improbable. Some thirty phenazines, e.g. (2), comprise a well-known group of bacterial pigments[2] which are known[3] to be derived from shikimic acid, but there is no precedent for a dinaphthopyrazine. Another difficulty, namely the colour of caulerpin, is more apparent than real. Although it is orange-red in the solid state, its solutions are yellow and show only end-absorption in the visible region, which might be compatible with structure (1).

---

[*]Solvent not reported.

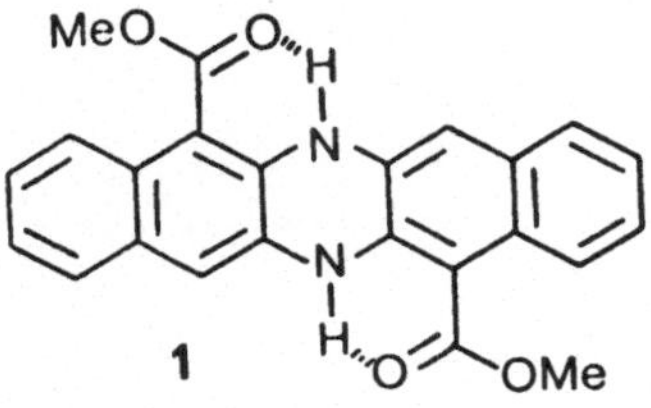

1

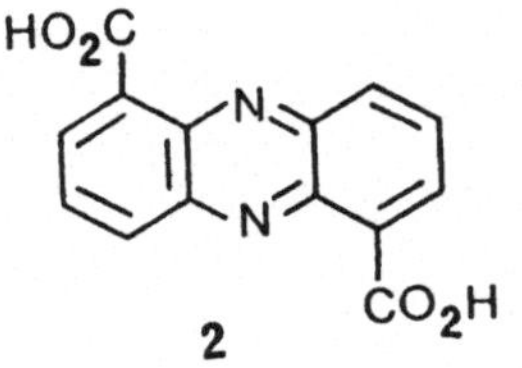

2

After reinvestigation of this pigment, we now regard caulerpin as the di-indolo pentacyclic compound (3). We have confirmed the spectroscopic evidence and now assign the low-field nmr signal at δ 8.05 (in CDCl$_3$) to the vinyl protons adjacent to the pyrrole rings. On reduction with lithium aluminium hydride, caulerpin formed the diol (4), accompanied by loss of colour (λ_{max} 299 nm and a shift of the low-field singlet to δ 6.80 (DMSO). On hydrogenation over platinum, caulerpin formed a tetrahydro derivative (5) showing indolic uv absorption at 286 and 294 nm, while the nmr spectrum was characterised by the absence of a 2H singlet at low-field and the presence of coupled signals at δ 3.4 and 3.8 for methylene and methine protons. It is now clear that structure (1) cannot be correct.

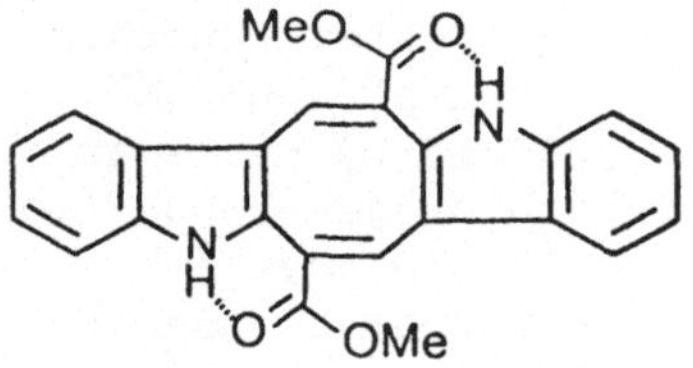

3

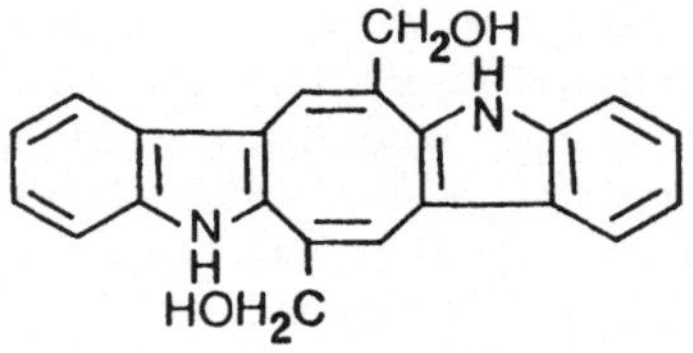

4

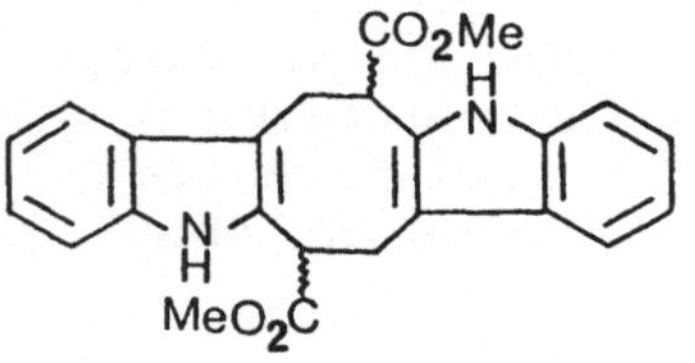

5

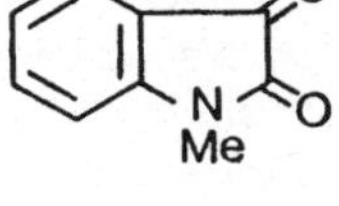

6

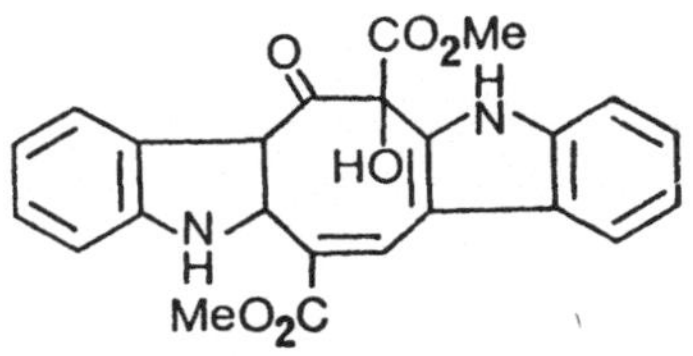

7

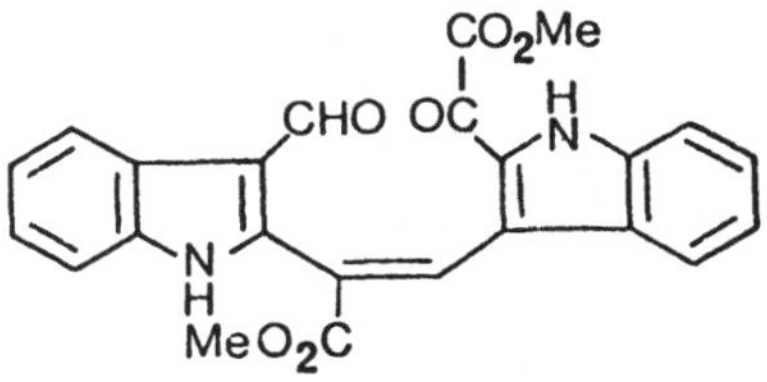

8

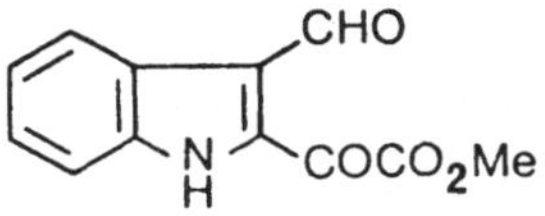

9

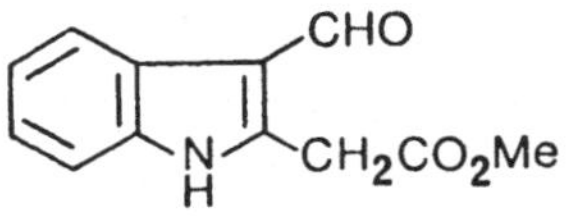

10

Direct evidence for the presence of an indole nucleus was obtained by oxidation of the di-N-methyl derivative of caulperpin with sodium periodate and a catalytic amount of ruthenium tetroxide, which gave N-methylisation (6) as the main product (10%). A similar oxidation of the parent compound gave no useful result, but, using periodate-osmium tetroxide, partial oxidation occurred, yielding (7)(8%), (8)(9%), and (9)(1%). The aldehyde (8) showed complex carbonyl absorption in the infrared from 1730 to 1635 cm^{-1}, while the nmr spectrum comprised singlets at δ (CDCl$_3$), 9.68 (CHO), 9.4 br and 8.1 br (NH), 8.65 (-CH=), 3.99 and 3.95 (MeO), and aromatic resonances.

The combined evidence now indicates that caulerpin is indeed the di-indolo compound (3), but as the yields from degradative reactions were low, a synthesis was desirable. This was achieved by converting 2-indolylacetic ester[4] into the aldehyde (10) by a Vilsmeier reaction, and this was then condensed with itself by refluxing in methanol containing catalytic amounts of piperidine and diethylamine. This gave synthetic caulerpin identical in all respects to the natural material. As expected, the yield was low. A recent study[5] of this condensation, under various conditions similar to ours, has shown that usually both trans- and cis-isomers arise, the former predominating. The formation of (3) requires a double cis self-condensation of (10). Both cis-trans and

SCHEME

trans-trans condensations lead to polymeric products, and the bulk of the crude product did not move from the base line on thin-layer chromatography

Caulerpin is not planar, but models show that it can adopt a conformation in which about half the molecule is very roughly in one plane, and it is noteworthy that the uv spectrum of the indolylacrylic ester (11)[6] (λ_{max} (MeOH) 224, 276, 325 nm) is very similar to that of caulerpin.

It is now evident that caulerpin is probably derived in vivo from tryptophan, possibly by way of the acrylic ester (see 11).

ACKNOWLEDGEMENTS

We thank Dr. S. Balasubramanian and Dr. S. Sotheeswaran for collecting algal material and Dr. M. Mahendran for the original isolation of caulerpin from C. taxifolia.

REFERENCES

1. G. Aguilar-Santos, J. Chem. Soc. (C), 842 (1970).

2. N. N. Gerber, in C. R. C. Handbook of Microbiology, Vol. III, 329, C. R. C. Press, Cleveland, 1973.

3. R. B. Herbert, F. G. Holliman, and J. B. Sheridan, Tetrahedron Lett., 639 (1976); U. Hollstein, G. E. Krisov, and D. L. Mock, Tetrahedron Lett., 3267 (1976).

4. V. Snieckus and K. S. Bhandari, Tetrahedron Lett., 3375 (1969).

5. G. R. Newkome and J. M. Robinson, J. Org. Chem. 41, 2536 (1976).

6. H. H. Inhoffen, K.-H. Nordsiek, and H. Schäfer, Annalen 668, 104 (1963).

RECENT RESULTS IN THE CHEMISTRY OF MEDITERRANEAN ALGAE

E. Fattorusso

Istituto di Chimica Organica, Università di Napoli

Naples, Italy

About two years ago our research group in Naples, in collaboration with that of Professor Piatelli in Catania, commenced a
general survey of the chemical constituents of Mediterranean algae.
Initially we studied algae belonging to the division Rhodophyta,
class Florideophyceae; we examined about fifty species harvested
along the east coast of Sicily and representative of all the six
orders in which the class Florideophyceae is generally divided.
In the extracts of these seaweeds we investigated some groups of
metabolites: free amino acids, low-molecular-weight carbohydrates
and sterols. I reported the results of this study at the meeting
on the Chemistry of Marine Products which was held last year in
Aberdeen and, therefore, I will not dwell upon them here.

During our investigation we observed a quite simple chromatographic pattern of the ether-soluble fraction of almost all the
species examined. This was indicative of the absence, at least in
substantial quantities, of unusual metabolites, which often are of
great importance for chemotaxonomic purposes. However, the lipid
fraction of an alga belonging to the order Gigartinales, Sphaerococcus coronopifolius, was more promising from this point of view.
In fact, we observed the presence of remarkable amounts of two
compounds, which we isolated by the usual chromatographic techniques.

The first, bromosphaerol, has the molecular formula $C_{20}H_{32}Br_2O$
(1). Catalytic hydrogenation of bromosphaerol yielded a dihydro
derivative, and therefore the remaining three degrees of unsaturation implied by the molecular formula must be due to rings.
Spectral properties (ν_{max} 3600-3380 cm^{-1}) indicated the presence
in bromosphaerol of a hydroxyl group which must be tertiary, as it

was unaffected by Jones reagent and acetic anhydride-pyridine.
The presence of two secondary methyl groups at δ 0.97 (d, J = 6 Hz,
H-19) and 0.90 (d, J = 6 Hz, H-20) and two tertiary methyl groups
at δ 1.38 (s, H-16) and 1.31 (s, H-15) was deduced from the NMR
spectrum. Decoupling experiments indicated that the secondary
methyls belong to an isopropyl group; irradiation at δ 1.93 (H-18)
caused both doublets at δ 0.90 and 0.97 to collapse into singlets.
The tertiary methyl, which resonates at δ 1.38, must be α to the
hydroxyl group, since it was shifted downfield in the spectrum of
bromosphaerol acetate, obtained from bromosphaerol after treatment
with acetyl chloride in hot xylene. The somewhat deshielded position of the signal at δ 1.31 may be rationalized by its location
β to a bromine atom.

   In the NMR spectrum the following signals are also present:
a doublet of doublets at δ 4.0 (bromomethine), an AB-type quartet
at δ 3.81 (bromomethylene), and a further coupled AB system at
δ 5.76 (vinyl hydrogens).

   Oxidation of bromosphaerol acetate with chromic anhydride in
pyridine furnished the two isomeric ketones (2) and (3), both
showing in the IR spectrum six-membered or larger ring ketone
absorption (2, ν_{max} 1683 cm^{-1}; 3, ν_{max} 1677 cm^{-1}).

   The NMR spectra of these compounds provided information on
the part structure C(1)-C(5), C(9)-C(10) and C(18)-C(20). The NMR
spectrum of (2) displayed the following signals: δ 6.4 (1H, A part
of an ABX system, J_{AB} 10 Hz, J_{AX} 4.5 Hz, H-3), 6.03 (1H, B part of
an ABX system, J_{BX}, very small, H-2), 3.83 (1H, d, J 11 Hz, H-10),
2.42 (1H, m, H-4) and 1.98 (1H, d, J 11 Hz, H-9). By irradiation
at δ 3.83, the doublet at δ 1.98 was simplified to a singlet and,
conversely, saturation of the signal at δ 1.98 caused the doublet
at δ 3.83 to collapse into a singlet. Irradiation at δ 6.40
simplified the multiplet at δ 2.42, while irradiation at δ 2.42
converted the signal at δ 6.40 to an A part of an AB system.

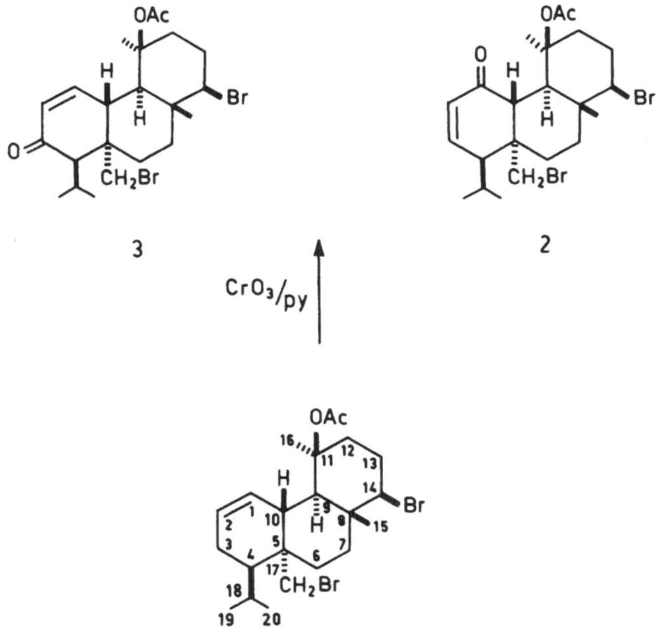

3

CrO$_3$/py

2

The NMR spectrum of the ketone (3) showed the following diag-
nostically important signals: δ 2.48 (1H, d, J 5.5 Hz, H-4), 1.07
(3H, d, J 7 Hz, H-19) and 0.88 (3H, d, J 7 Hz, H-20). The assignments were confirmed by decoupling; irradiation at δ 2.03,
tentatively the frequency of H-18, simplified the doublet at δ 2.48
and at the same time collapsed the two doublets of the isopropyl
methyls into singlets.

Information of the part structures C(11)-C(14) and C(8)-C(9)
was obtained from the spectral analysis of the two derivatives of
bromosphaerol (4) and (5). The NMR spectrum of (4) showed that
only two olefinic protons are present in the molecule, thus demonstrating that dehydration yielded a tetrasubstituted double bond.
Furthermore, the methyl group at C-11, which in the NMR spectrum of
bromosphaerol appeared at δ 1.38, resonates at δ 1.68, thus confirming the location of the hydroxyl group in bromosphaerol.
Finally, a multiplet at δ 3.46 was assigned to the proton of the
diallylic methine.

The unconjugated triene (5) was obtained from (4) by treatment
with diazabicyclononene. The most significant features of the
NMR spectrum of (5), when compared with that of (4), were as
follows. In the olefinic region, signals of four protons are
present, thus indicating that dehydrobromination generated two new
olefinic protons. Moreover, a 2H multiplet at δ 2.58, assigned to
the diallylic methylene protons at C-12, is present. Finally, the
upfield shift of the methyl group at C-8 from δ 1.30 in (4) to
δ 1.20 in (5) confirms its position relative to bromomethine group.

All these data led us to propose structure (1) as the most
favourable for bromosphaerol. This structure is also in agreement
with ^{13}C NMR. An X-ray crystallographic analysis performed by
Professor C. Pedone and Doctor B. Di Blasio of Istituto Chimico of
the University of Naples on a single crystal of (5) confirmed the
structure of bromosphaerol and showed the relative stereochemistry
represented by the formula (1), in which the configurations at
C-4, C-5, C-8 and C-10 were deduced from those of (5), while the
others were assigned as follows. The bromine atom at C-14 must be
equatorial as indicated by the coupling constants (J = 9 Hz, ax-ax,
and J = 4 Hz, ax-eq) of the bromomethine proton to the adjacent
methylene protons. The coupling constant (11 Hz) between H-9 and
H-10 in the NMR spectrum of the ketone (2) indicated a _trans_
relationship between these two protons. Finally, the stereochemistry at C-11 was deduced from the dehydration reaction of
bromosphaerol, taking into account that under the experimental conditions used (POCl_3-pyridine) a _trans_ elimination must take place[1].

As for the second compound present in Sphaerococcus coronopifolius, we attempted to correlate it with bromosphaerol.

6

However, when this had been accomplished, Fenical et al.[2] reported the X-ray structure determination of this compound, which they named sphaerococcenol A (6). Unfortunately, in the study of the chemical constituents of Sphaerococcus coronopifolius, an abso- lutely unintentional duplication of work occurred, and the research group of Dr. Fenical established the structure of bromosphaerol shortly after our paper had been submitted for publication. They determined the absolute stereochemistry, showing that bromosphaerol possesses the antipodal configuration of (1)[3].

As previously pointed out, we also examined some water-soluble metabolites present in red seaweeds. In our communication at Aberdeen we reported the structure determination of a new amino acid, pyrrolidine-2,5-dicarboxylic acid, isolated from Schyzimenia dubyi[4].

In pursuing the investigation of the free amino acids of Rhodophyta, we have now isolated from Ceramium rubrum two further pyrrolidine derivatives, namely pyrrolidine-2,4-dicarboxylic acid and 4-hydroxypyrrolidine-2,4-dicarboxylic acid. Assignment of the structure of pyrrolidine-2,4-dicarboxylic acid, $C_6H_9NO_4$, was based on the spectral data [m/e 159 (M^+), 114 (M^+-COOH), 68 (M^+-COOH-HCOOH); ν_{max} 1725 and 1590 cm^{-1}; δ 4.35 (1H, m, H-2), 3.6 (3H, m, H-4 and H-5), 2.5 (2H, m, H-3)]. Confirmation was obtained from aromatization of its dimethyl ester with selenium, to afford pyrrole-2,4-dicarboxylic acid dimethyl ester. The structure of 4-hydroxypyrrolidine-2,4-dicarboxylic acid, $C_6H_9NO_5$, followed from the spectral data [ν_{max} 1720, 1580 cm^{-1}; δ CF_3COOH 4.3 (1H, t, J 6 Hz, H-2), 4.02 (2H, s, H-5), 3.08 (2H, d, J 6 Hz, H-3); mass spectrum of its dimethyl ester: 185 (M^+-H_2O), 144 (M^+-$COOCH_3$), 126 (M^+-H_2O-$COOCH_3$)]. Moreover, on aromatization with selenium, its dimethyl ester gave pyrrole-2,4-dicarboxylic acid dimethyl ester.

The stereochemistry of these amino acids remains to be assigned. However, the molecular rotation of pyrrolidine-2,4-dicarboxylic acid $[M_D$ -73.20^0 (H$_2$O) and -47.21^0 (HCl)] suggests a 2S,4S configuration.

Recently our chemical survey of the Mediterranean marine algae was extended to the brown seaweeds in order to obtain information of potential chemotaxonomic value. At present we have examined and quantitatively determined the free amino acids, amino sulfonic acids, sugars and sterols of the following eleven species.

Sphacelariales
    Stypocaulaceae
        Halopteris filicina
        H. scoparia
    Cladostephaceae
        Cladostephus verticillatus
Dictyotales
    Dictyotaceae
        Dictyopteris membranacea
        Dictyota dichotoma
        Dilophus ligulatus
Fucales
    Cystoseiraceae
        Cystoseira adriatica
        C. crinita
        C. fimbriata
        C. stricta
    Sargassaceae
        Sargassum vulgare

The results obtained from this very limited number of species were not very interesting, and only two facts seem worthy of mention here. N-methyl taurines, previously found only in Rhodophyta and considered to be characteristic of this phylum, have been detected in several brown algae. Furthermore, sterol analysis confirmed the general occurrence of fucosterol in Phaeophyta but also showed that cholesterol and 24-methylenecholesterol, previously reported as trace components[5], are often present in substantial quantities.

In contrast to those of the red algae, the lipid fractions of brown seaweeds showed a quite complex chromatographic pattern for several species. This led us to devote more attention to the investigation of the uncommon lipophilic compounds.

We have examined the extracts of three species: Cystoseira crinita, Dictyota dichotoma and Dilophus ligulatus. From Cystoseira crinita, after saponification and silica gel chromatography, we isolated two new linear terpenoids, oxocrinol and crinitol.

7

Oxocrinol (7), which is present in the alga as the acetate, has the
molecular formula $C_{14}H_{24}O_2$, deduced from precision mass measurement.
Part structures a, b, and c were deduced from the NMR spectrum and
decoupling experiments.

Further information on the structure of (7) was provided by
ozonolysis of oxocrinol acetate, which afforded, after decomposition with diazomethane, laevulinic acid methyl ester, 2,6-hepta-
nedione and acetylglycolic acid methyl ester.

All these data established the structure of oxocrinol, except
for the location of a double bond, which could be at C-6 or C-7.
Double irradiation experiments performed on oxocrinol acetate
eliminated the second possibility. Irradiation at δ 5.28 (1H, bt,
J = 7 Hz) simplified the triplet at δ 4.48 (2H, d, J = 7 Hz, H-1)
into a singlet and, upon irradiation at δ 1.59 (H-9), the triplet
at δ 2.29 (2H, t, J = 7 Hz, H-10) collapsed to a singlet, while the
vinyl signal at δ 5.04 (1H, bt, J = 6 Hz, H-6) was unaffected.

8

The stereochemistry of the double bonds was assigned on the basis of the chemical shifts of both vinyl methyl signals, which are consistent with these groups, being _trans_ to the olefinic protons[6]. Crinitol (8), which is present in the alga in the free state, has the molecular formula $C_{20}H_{34}O_2$. In its NMR spectrum, the five vinyl methyls resonate between δ 1.56 and 1.70, while a complex signal spread between δ 4.15 and 5.65 was associated with the four olefinic hydrogens.

The formation of a diacetate indicated the presence of two hydroxyl groups, one of which must be primary and the other secondary on the basis of the NMR spectrum [signals at δ 3.98 (2H, d) and 4.24 (1H, q)] . The basic skeleton of crinitol and the location of the double bonds were established by its reduction with sodium in liquid ammonia, which gave 2,6,10,14-tetramethylhexadeca-2,6,10,14-tetraene. This result also indicated that the primary allylic hydroxyl must be at C-1. The location of the secondary hydroxyl group was established as follows. Positions 4, 8, and 12 could be excluded on the basis of multiplicity of the NMR signal of the carbinol methine proton and decoupling experiments, which indicated its position α to an olefinic hydrogen. Of the remaining three possible positions, only position 9 was consistent with the fragmentation pattern of the mass spectrum [m/e 288 (M^+ - H_2O), 189 (a - H_2O), 123 (b), 121 (c, - H_2O), 85 (d) and 69 (e)] .

Concerning the biogenesis, farnesol and geranylgeraniol appear to be the precursors of oxocrinol and crinitol, respectively. However, for oxocrinol, other possibilities cannot be excluded; for instance, it could be derived from a monoterpene by addition of two acetate units or from geranylgeraniol by oxidative elimination of a C-6 fragment.

9

About three years ago, Hirschfeld et al.[7] reported the isolation of a new diterpene alcohol, pachydictyol A (9) from a Pacific
brown alga, Pachydictyon coriaceum. This compound possessed a
hydroazulene skeleton which was unique among diterpenes.

Recently we examined three species belonging to this family,
namely Dictyopteris membranacea, Dictyota dichotoma and Dilophus
ligulatus. From the last two algae we isolated five new compounds,
dictyol A (10), B (11), C (12), D (13) and E (14), strictly related
to pachydictyol A and a new diterpene dilophol (15), with a
germacrane-type skeleton, possibly related biogenetically to them.
Compounds (1), (11), (12), and (13) were present in D. dichotoma,
while (14) and (15) were found in D. ligulatus. It should be noted
that dictyol A, B, C, and D were also isolated by Professor Minale
and co-workers from the digestive gland of the mollusk Aplysia
depilans, which is known to feed on D. dichotoma. The structure
of dictyol A and B were independently determined by our group and
that of Professor Minale, and the pertinent papers were published
separately.

The structure of dictyol C, D and E were established in collaboration with Professor Minale and his co-workers.

The structure of dictyol A (10) was assigned on the basis of
NMR spectral data (Table 1) and chemical evidence. On reduction
with sodium in liquid ammonia, it yielded pachydictyol A and compounds (16) and (17). The stereochemistry at C-2 was deduced by
considering that the closure of the tetrahydrofuran ring requires
a cis relationship between H-1 and H-2. The NMR spectrum of
dictyol B (11) (Table 1) is very similar to that of pachydictyol A.
The most important differences are the presence of an additional
carbinol methine signal and the splitting of the exomethylene
signal into two broad singlets.

On reduction with sodium in liquid ammonia, dictyol B furnished
compound (16), which was also obtained from dictyol A; moreover,
Jones oxidation yielded the corresponding diketone identified on
the basis of its spectral properties.

Table 1. H^1-chemical shifts of (10), (11), (12), (13) and (14) (CDCl$_3$). Multiplicities and coupling constants are shown (Hz) in parentheses.

Proton Position	(10)	(11)	(12)	(13)	(14)
1	3.13 (b)	--	2.21 (b)	2.62 (b)	--
2	4.62 (d,7)	--	--	4.50 (b)	--
3	5.46 (b)	5.21 (m)	5.26 (b)	5.56 (b)	5.32 (b)
5	2.85 (b)	--	2.74 (b)	2.94 (t,10)	--
6	4.06 (m)	3.65-4.03 (b)	3.87 (dd;9,3)	3.86 (d,9)	4.18 (d,9)
9	5.46 (b)	3.65-4.03 (b)	--	--	--
14	5.09 (t,7)	5.05 (t,7)	5.14 (t,6)	5.11	5.14 (t,6)
16	1.61 (s)	1.59 (s)	1.62 (s)	1.61 (s)	1.62 (s)
17	1.86 (s)	1.75 (s)	1.85 (s)	1.92 (s)	1.82 (s)
18	4.36 (s)	4.81 and 5.08 (bs)	1.22 (s)	5.11, 4.86	4.75 (s)
19	0.99 (d,6)	1.00 (d,6)	1.00 (d,6)	0.98 (d,6)	1.23 (s)
20	1.68 (s)	1.67 (s)	1.70 (s)	1.70 (s)	1.70 (s)

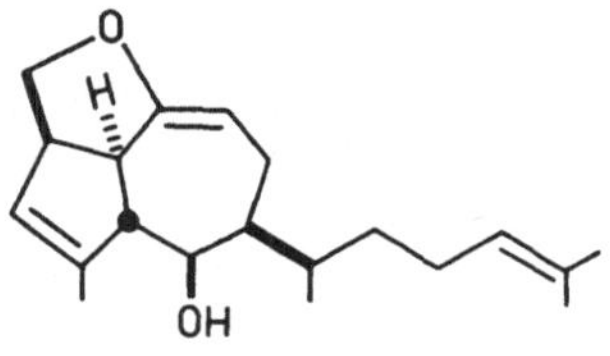

10

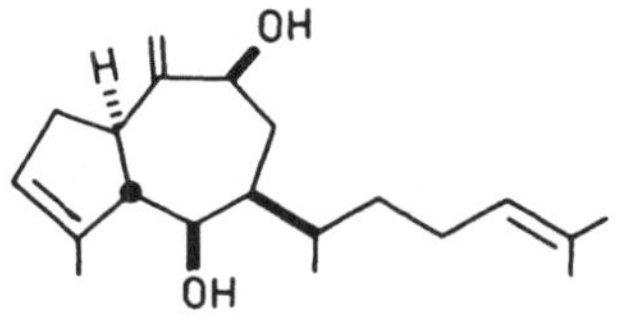

11

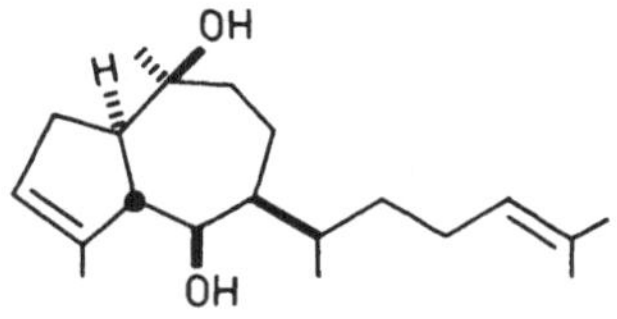

12

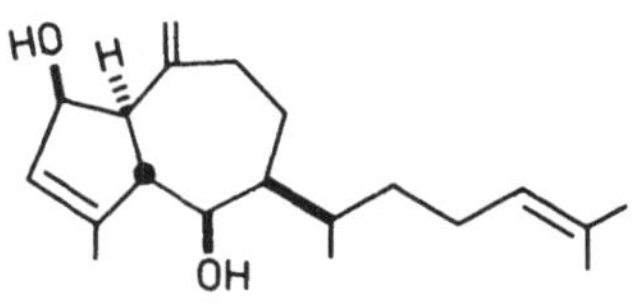

13

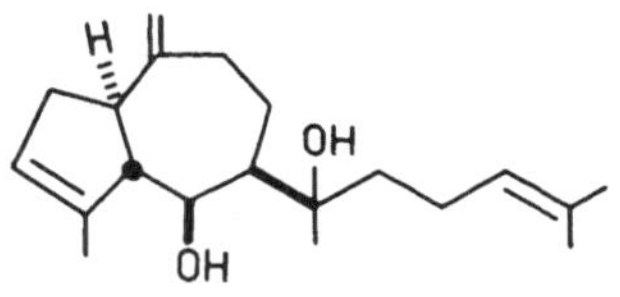

14

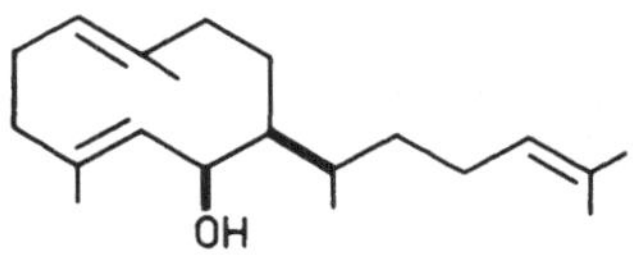

15

The stereochemistry of C-9 was tentatively assigned from the NMR spectrum after gradual addition of Eu(fod)₃. The NMR spectrum of dictyol C (12) (Table 1) is strongly reminiscent of that of pachydictyol A, apart from the replacement of the exomethylene signal by a 3H signal at δ 1.22, characteristic of a methyl group linked to an oxygen-bearing carbon atom.

All the spectral data led us to propose structure (12) for dictyol C, which was confirmed by ^{13}C NMR data and dehydration, which afforded pachydictyol A (9). The stereochemistry at C-10 was deduced from further examination of the NMR spectrum in the presence of variable amounts of Eu(fod)₃.

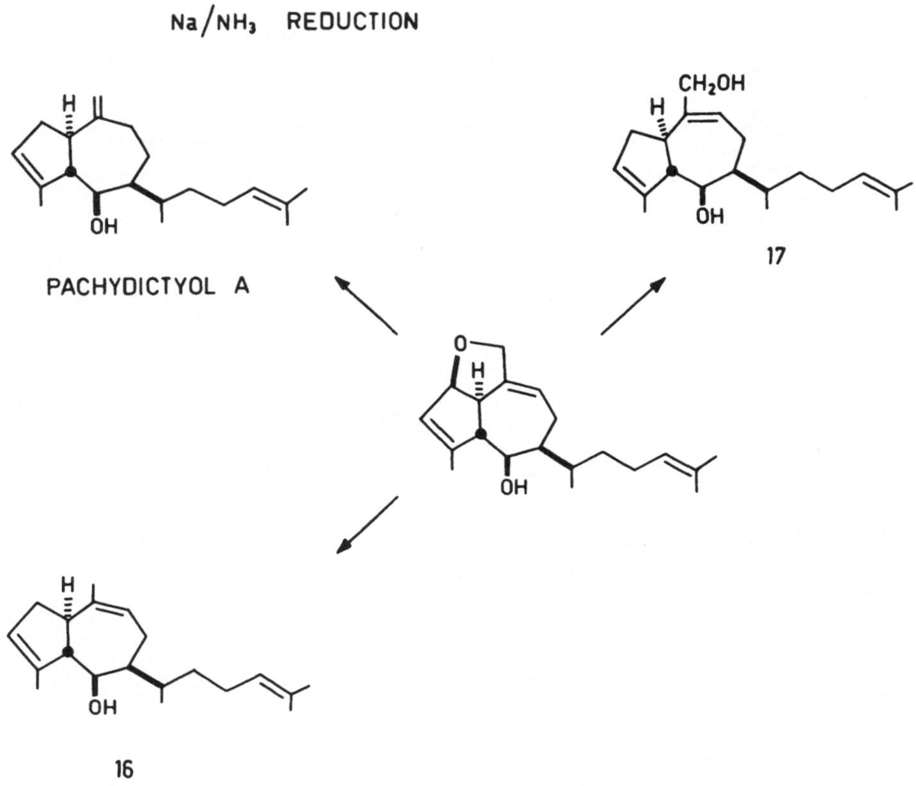

Na/NH₃   REDUCTION

PACHYDICTYOL A

17

16

The most important features in the NMR spectrum of dictyol D
(13) (Table 1), when compared with that of pachydictyol A, are the
appearance of an allylic carbinol methine proton as a multiplet at
δ 4.5, which, on irradiation at δ 5.56, was converted to a sharp
doublet, the splitting of the <u>exo</u>methylene signal into two broad
singlets, and the downfield shift of the H-3 and 4-Me signals.
These data are fully explained by structure (13), which was confirmed by reduction of dictyol D with lithium in ethylamine to a
dihydroderivative which was shown to be identical to 10,18-dihydro-
pachydictyol A obtained from pachydictyol A by the same treatment.
The stereochemistry at C-2 was deduced from the NMR spectrum after
gradual addition of Eu(fod)₃; the hydroxyl group must be nearer
H-5 than H-1, since the former shows a larger paramagnetic shift.

The most important feature in the NMR spectrum of dictyol E
(Table 1) is the absence of the 11-Me doublet, present at ca. δ 1.00
in the spectra of pachydictyol A and dictyol A, B, C and D, which
is replaced by a 3-H singlet at δ 1.23. This is indicative of

the location of the hydroxyl group in the side chain. Also, the ^{13}C NMR data and the Eu(fod)$_3$-induced shifts fully agree with the proposed structure.

The NMR spectrum of dilophol $\big[\delta$ (CCl$_4$) 0.98 (3H, d, J = 6 Hz, 11-Me), 1.46 (3H, s, 10-Me), 1.60 (6H, s, 4-Me and trans-15-Me), 1.70 (3H, s, cis-15-Me) and δ (C$_6$D$_6$) 4.55 (1H, m. H-6), 4.87 (1H, b, H-1), 5.02 (1H, A part of an AB system, J = 9 Hz, H-5) and 5.25 (1H, bt, J = 7 Hz, H-14)$\big]$ indicated the presence of an allylic hydroxyl group and at least three double bonds in the molecule.

Aromatization with selenium gave 1,4-dimethyl-7-(1',5'-dimethylhex-5'-enyl) azulene, identified by comparison with a sample obtained from pachydictyol A by the same treatment. This result indicated that the additional degree of unsaturation implied by the molecular formula of dilophol must be due to a ten-membered ring and revealed the identity of the side chain and the position of the methyl groups on the nucleus. Lastly, oxidation of dilophol with potassium permanganate and sodium periodate, which yielded laevulinic acid, permitted location of the nuclear double bonds.

The stereochemistry of dilophol was tentatively assigned on the basis of its NMR spectrum, NOE experiments, Eu(fod)$_3$ resonance data, and biogenetic considerations.

The results of the present communication (considering that, especially for brown algae, they were obtained by examining a limited number of species) do not allow chemotaxonomic and phylogenetic speculations which might be made when more abundant data on the chemistry of the algae has been accumulated. However, our results appear to confirm that, among the various classes of compounds so far examined, terpenoids are the best candidates as taxonomic markers.

To the restricted occurrence of halogenated monoterpenes in Bonnemaisoniales and Rhizophyllidaceae and halogenated sesquiterpenes in Rhodomelaceae[8] previously described, we can now add that of the hydroazulene diterpenes in algae belonging to the family Dictyotaceae.

All the results that I have presented in this communication were obtained from collaborative research performed at the Institute of Organic Chemistry of the University of Catania by Professor M. Piattelli and Drs. V. Amico, G. Impellizzeri, S. Mangiafico, G. Oriente, S. Sciuto and C. Tringali and at the Institute of Organic Chemistry of the University of Naples by Drs. S. Magno, L. Mayol, C. Santacroce and D. Sica.

REFERENCES

1.   D. H. R. Barton, A. S. Campus-Neves, and R. C. Cookson, J.
     Chem. Soc., 3500 (1956).

2.   W. Fenical, J. Finer, and J. Clardy, Tetrahedron Lett., 731
     (1976).

3.   W. Fenical, private communication.

4.   G. Impellizzeri, S. Mangiafico, G. Oriente, M. Piattelli,
     S. Sciuto, E. Fattorusso, S. Magno, C. Santacroce, and D. Sica,
     Phytochem. 14, 1549 (1975).

5.   G. W. Patterson, Lipids 6, 120 (1971).

6.   R. B. Bates and C. M. Gale, J. Am. Chem. Soc. 82, 5749 (1960);
     R. B. Bates, D. M. Gale, and B. J. Gruner, J. Org. Chem. 28,
     1087 (1973).

7.   D. R. Hirschfeld, W. Fenical, G. H. V. Lin, R. M. Wing, P.
     Radlick, and J. J. Sims, J. Am. Chem. Soc. 95, 4049 (1973).

8.   For reviews, see: P. J. Scheuer, Chemistry of Marine Natural
     Products, Academic Press, New York, 1973; J. T. Baker and
     V. Murphy-Steinmann, Compounds from Marine Organisms, Roche
     Research Institute of Marine Pharmacology, Dee Why, Australia,
     1975.

DITERPENOID SYNTHESIS IN BROWN SEAWEEDS OF THE FAMILY DICTYOTACEAE

Frank J. McEnroe, Kenneth J. Robertson and William Fenical

Institute of Marine Resources, Scripps Institution of

Oceanography, La Jolla, California, USA 92093

Brown seaweeds (Phaeophyta), taxonomically placed within the family Dictyotaceae, are common inhabitants of the shallow water and intertidal community, particularly in subtropical and tropical ecosystems. No less than thirteen genera have been assigned to this distinct family (Figure 1), all of which are characterized by their flattened thalli, which emanate from single apical cell division. Despite extreme herbivore activity in most subtropical and tropical areas, seaweeds of the Dictyotaceae are observed in abundance and are apparently avoided by many fishes and crustacea. Bioassay experiments with extracts from various species in this family have shown that substances toxic to fish[1], bacteria[2-4], and viruses[5] are produced by these algae. Subsequent studies of the natural products chemistry of a few species, particularly those of the genus Dictyopteris, have shown the production of a diverse group of substances including sesquiterpenes[6], C_{11} hydrocarbons and sulphur-containing compounds[7], and C_{21} hydroquinones of mixed biogenesis[8]. C_{21} chromanols, also of mixed biogenetic origin, have been reported from the closely related alga Taonia[9] (Figure 2). Presumably these unique substances are involved in providing selective advantage against predation.

Within the Dictyotaceae, a smaller group exists which is highly comparable both morphologically and chemically. These plants are considered separable into three genera, Pachydictyon, Dictyota and Dilophus, based upon the number of cell layers and their arrangement along the outer fronds of the plant. It is exceedingly difficult to separate these genera based upon their gross morphology, and, based upon current studies, Pachydictyon and Dictyota species are also inseparable chemically.

<u>Division</u>
Phaeophyta

<u>Order</u>
Dictyotales

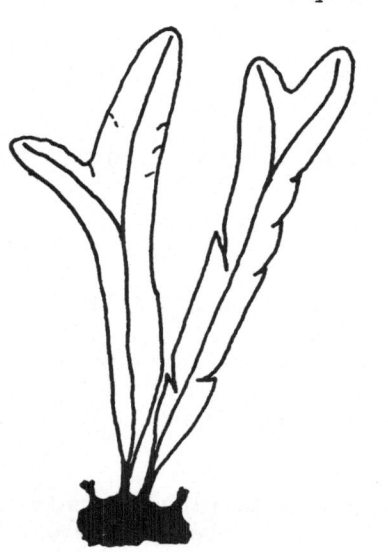

<u>Family</u>
Dictyotaceae

<u>Genera</u>

<u>Chlanidophora</u>
<u>Dictyopteris</u>
<u>Dictyota</u>
<u>Dilophus</u>
<u>Glossophora</u>
<u>Lobophora</u>
<u>Pachydictyon</u>
<u>Padina</u>
<u>Spathoglossum</u>
<u>Stypopodium</u>
<u>Syringoderma</u>
<u>Taonia</u>
<u>Zonaria</u>

Figure 1.   The family Dictyotaceae[13].

        In our early work on the Pachydictyon-Dictyota-Dilophus group,
we isolated and characterized perhaps the most frequent component
of this algal group, pachydictyol A, from <u>Pachydictyon coriaceum</u>[10]
(Figure 3).   Pachydictyol A was the first example of a diterpenoid substance which contained the perhydroazulene ring system.   In
more recent studies, two derivatives of this skeleton, dictyol A
and B, have been described as constituents of <u>Dictyota dichotoma</u>
from the Mediterranean[11,12] (Figure 3).   In this paper we wish to
report our continuing studies of the diterpenoid components from
<u>Pachydictyon</u> and <u>Dictyota</u> species from the Gulf of California,
Mexico.   Our studies indicate that some of the diterpenoids from
this group have a common biogenetic precursor and that the taxonomic separation of these genera may have been based upon weak
morphological arguments.

        Two species of the genus <u>Dictyota</u> collected near Puerto Peñasco,
in the northern Gulf of California, have been subsequently studied.
To our surprise, and considering taxonomic separation, <u>D</u>. <u>hespera</u>
contained large amounts of pachydictyol A.   <u>Dictyota flabellata</u>,
on the other hand, contained a complex mixture of new compounds;
we have assigned the major one as the corresponding methylene

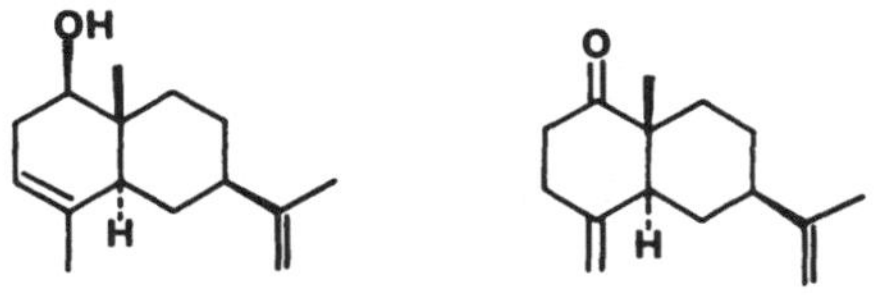

Dictyopteris divaricata (Japan)

Dictyopteris australis/plagiogramma (Hawaii)

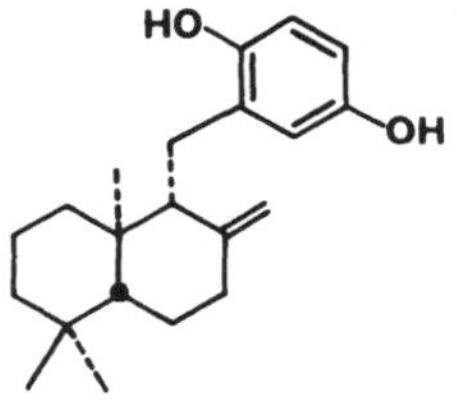

Dictyopteris undulata
(California)

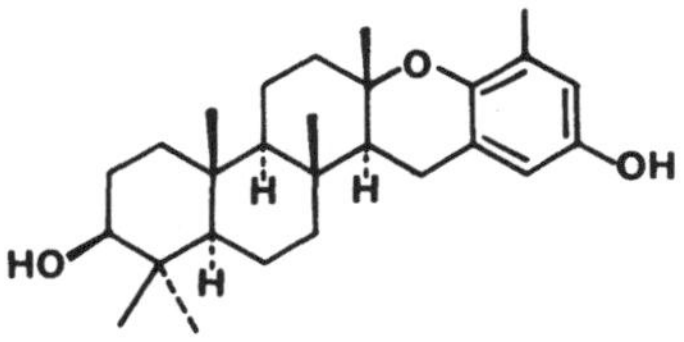

Taonia atomaria (Canary Is.)

Figure 2

epoxide of pachydictyol A. Pachydictyol A epoxide (1) was isolated
as a colourless oil, and the formula was confirmed as $C_{20}H_{32}O_2$ by
high resolution mass spectrometry. Compound (1) was recognized as
an alcohol by intense absorptions at 3400 cm^{-1} in its infrared
spectrum. Like pachydictyol A, (1) reacted slowly with acetic
anhydride/pyridine (50°C) to yield the corresponding monoacetate.
The second oxygen function was clearly assigned as an epoxide, since
(1) reacted smoothly with potassium hydroxide in methanol to
incorporate solvent (Figure 4). The incorporation product (2) was
assigned as the tertiary alcohol isomer, since neither hydroxyl
would acetylate at room temperature. Lithium aluminum hydride

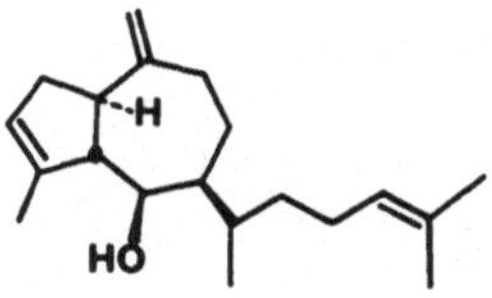

pachydictyol A
Pachydictyon coriaceum[10]
Dictyota hespera

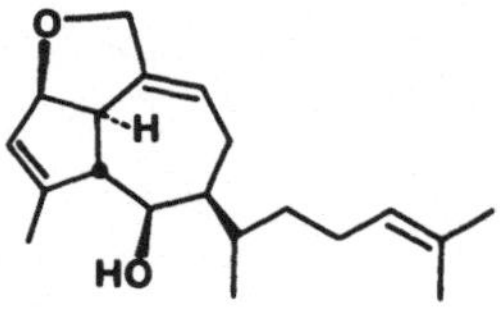

dictyol A 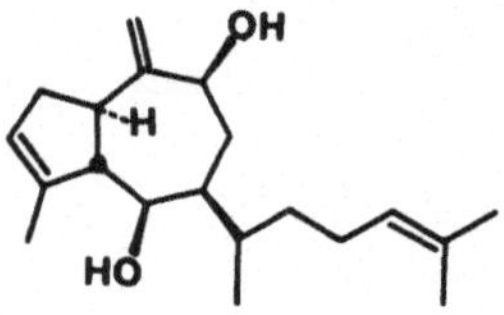   dictyol B

Dictyota dichotoma var. implexa[11,12]

Figure 3

reduction provided the diol (3) in high yield. It was clear that the reduction had stereoselectively generated a new methyl group, since a three-proton singlet was now present at δ 1.18 in the nmr spectrum of purified (3). Dehydration of (3) with phosphorous oxychloride/pyridine at 0°C went smoothly, to give a complex mixture of olefins. The major products were the alcohols (4) and (5) (pachydictyol A), obtained in a 7:3 ratio. We have assigned the stereochemistry of (1) and (3) at the tertiary oxygen-bearing carbon, based upon the products obtained in the dehydration of (3). Since no products were obtained via elimination of the lone bridge-head proton, and assuming a trans-diaxial transition state for elimination, the tertiary hydroxyl in (3) must be cis-oriented to the bridgehead proton. The molecular model of (3) suggests that clean trans-diaxial eliminations should be favorable to yield (4) and (5).

Spectral data for the epoxide (1) were in close harmony with those from pachydictyol A: [1]H nmr (CDCl$_3$, 220 MHz) δ 0.98 (d, 3H, J = 7 Hz), 1.36-1.91 (multiple bands, 9H), 1.20 (2H, m), 1.57 (3H,s),

1.66 (3H, s), 1.74 (3H, s), 2.34 (1H, d, J = 5 Hz), 2.48 (2H, m), 2.62 (1H, d, J = 5 Hz), 3.84 (1H, dd, J = 8, 3 Hz), 5.07 (1H, dd, J = 7, 7 Hz), 5.18 (1H, bs). ^{13}C Nmr (CDCl$_3$) δ 15.7, 17.5, 17.7, 20.6, 25.9, 25.9, 31.2, 34.9, 39.7, 43.9, 48.8, 50.6, 58.0, 62.2, 74.6, 124.2, 125.0, 131.3, 141.2.

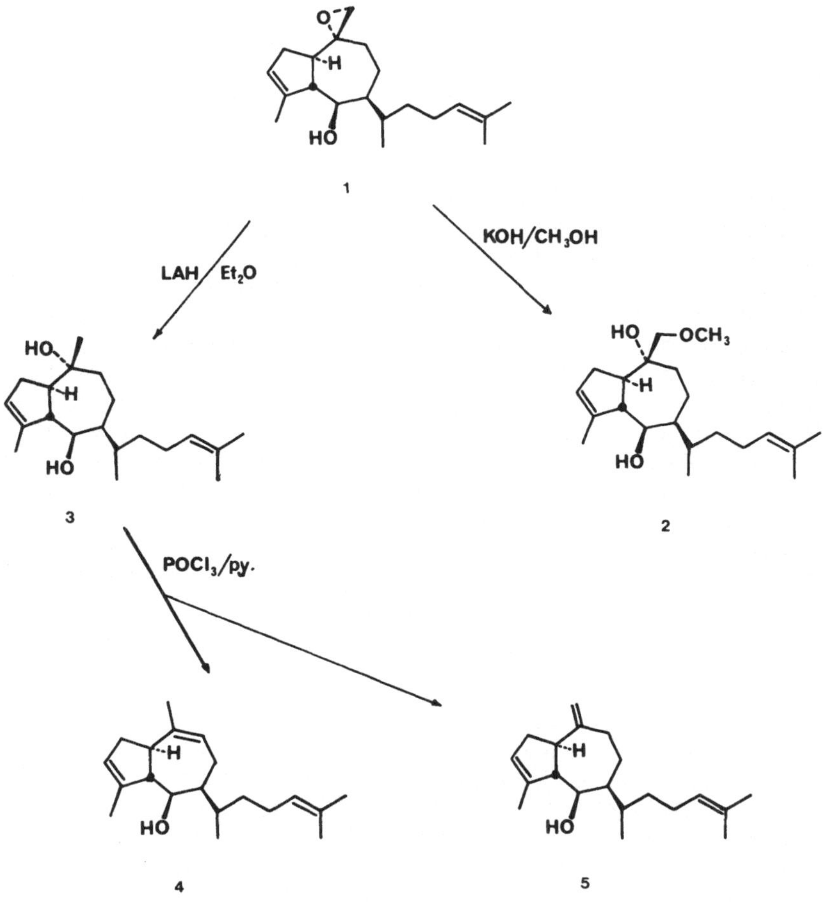

Figure 4.   Reactions of pachydictyol A epoxide (1).

Table. Physical data for acetoxycrenulatin (6).

$C_{22}H_{32}O_4$ $M^+ = 360.2308$ (calc. 360.2300)

ir: 1765, 1735, 1225, 1040, 1030 cm^{-1} (CCl$_4$)

nmr: ^{1}H (220 MHz, CDCl$_3$) ^{13}C (CDCl$_3$)

^{1}H (220 MHz, CDCl$_3$)	^{13}C (CDCl$_3$)	
0.39 (1H, m)	8.36 (t)	35.68
0.89 (1H, m)	10.28	43.87 (t)
0.99 (3H, d, J = 6 Hz)	17.10	47.35 (d)
1.02 (3H, d, J = 7.5 Hz)	17.66	71.48 (t)
1.59 (3H, s)	21.28	72.20 (d)
1.70 (3H, s)	23.42	123.51 (d)
2.03 (3H, s)	25.47	128.76 (s)
3.23 (1H, d, J = 7.5 Hz)	25.66	132.44 (s)
4.74 (1H, dd, J = 16, 2.5 Hz)	25.91	166.48 (s)
4.85 (1H, dd, J = 16, 2.5 Hz)	29.25	169.71 (s)
5.07 (1H, dd, J = 7.5, 7.5 Hz)	32.79	174.21 (s)
5.48 (1H, bt, J = 4 Hz)		

In addition to the collections near Puerto Peñasco, we have studied various Pachydictyon and Dictyota species collected in the central and southern areas of the Gulf of California. Both P. coriaceum, collected near Isla Turner, and D. crenulata, collected near Cabo San Lucas, contain an interesting new bicyclic diterpene. This new substance analyzed as $C_{22}H_{32}O_4$ and was recognized by infrared analysis as an α,β-unsaturated-γ-lactone and an acetate ester. We suggest the name acetoxycrenulatin for this molecule and have assigned the structure (6) based upon spectral analysis and chemical modification. The proton and carbon nmr characteristics (see Table) of acetoxycrenulatin were initially mistakenly believed to suggest its relatedness with pachydictyol A. This idea was quickly dispelled, since the compound is bicyclic and one of the rings is clearly a cyclopropane. Accounting for the degree of unsaturation, ms and cmr data show that a disubstituted α,β-unsaturated-γ-lactone moiety, cyclopropane, and one large ring are present.

Acetoxycrenulatin (6) underwent reactions which support the structure we propose (Figure 6). Catalytic hydrogenation under a variety of conditions resulted only in saturation of the side-chain olefin, yielding (7). Catalytically, neither the cyclopropane nor α,β-unsaturated lactone olefin appear reactive toward reduction. Lithium in ethylamine, however, afforded complete reduction of the lactone olefin and also reduced the ester to give (8). The nmr spectrum of (8) showed the lone bridgehead proton, alpha to the ester carbonyl, as a four-line pattern at δ 2.84 (J = 14, 10 Hz),

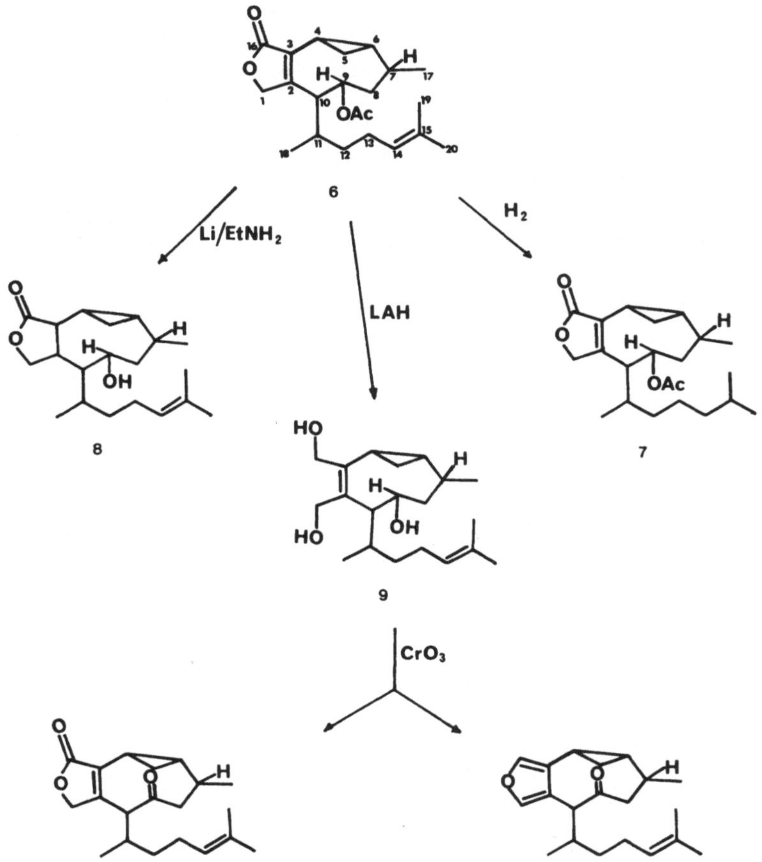

Figure 5.   Reactions of acetoxycrenulatin (6).

suggesting that C-4 and C-2 are methine carbons.   Lithium aluminum
hydride in diethyl ether, as expected, opened the lactone and
reduced the acetate, yielding the corresponding triol (9) in good
yield.   Compound (9) readily gave a tris acetate with acetic anhy-
dride/pyridine at room temperature.

Structural information to fix the location and size of the side
chain was obtained by Sarrett oxidation of the triol (9). The major
products were the keto-lactone (10) and the keto-furan (11), formed
in equal amounts at 0°C. Analysis of the nmr spectra showed that a
methylene and a methine carbon were flanking the newly-formed ketone
moiety in both (10) and (11). The mass spectrum of (11) confirmed
the presence of a "pseudobenzyl"-substituted iso-octenyl side chain
by intense loss of C_8H_{15} from the parent ion. In separate experiments, the sequence (6)→(9)→(11) was completed, using the dihydro
derivative (7). In an analogous way, dihydro (11) was produced,
and the loss of C_8H_{17} was observed in its mass spectrum.

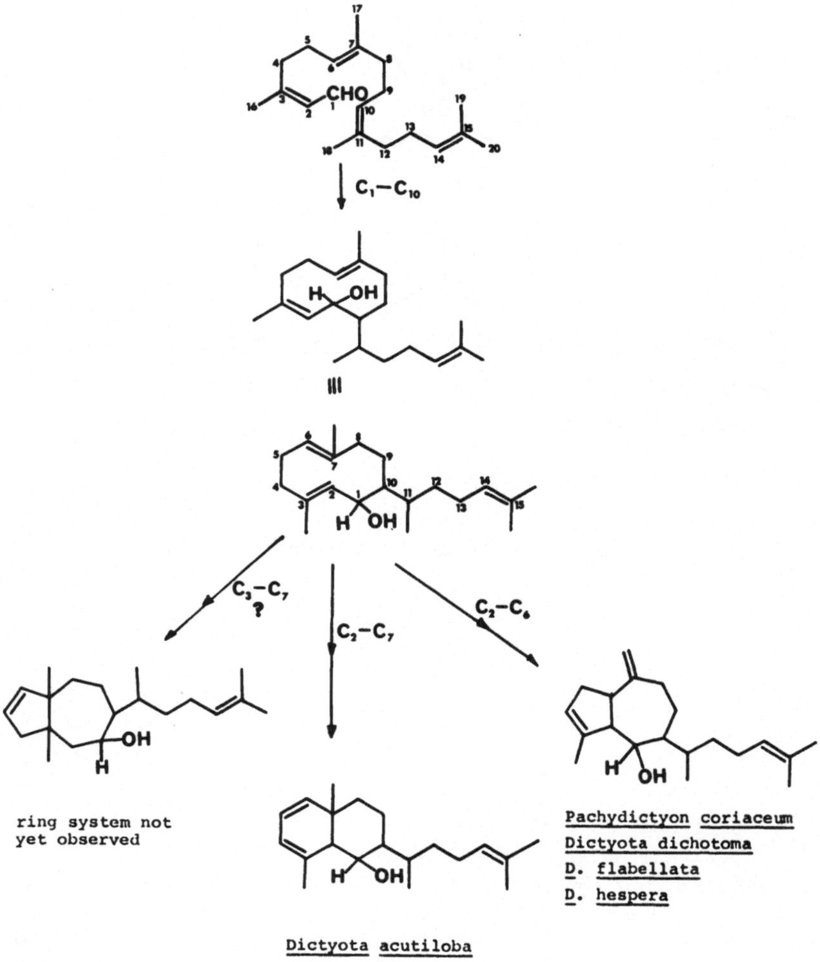

Figure 6

A series of nmr experiments involving decoupling and lanthanide-induced shift analysis with selected compounds in this series allowed the gross structure of acetoxycrenulatin to be formulated as (6). Very little information could be gathered to delineate the stereo-chemistry of this molecule as it related to C-4, C-6, C-19, and C-10. Molecular models indicate some highly favorable configurations which will require confirmation by more definitive methods.

The origin of acetoxycrenulatin and pachydictyol A-type compounds in members of the Dictyotaceae is a subject of debate.

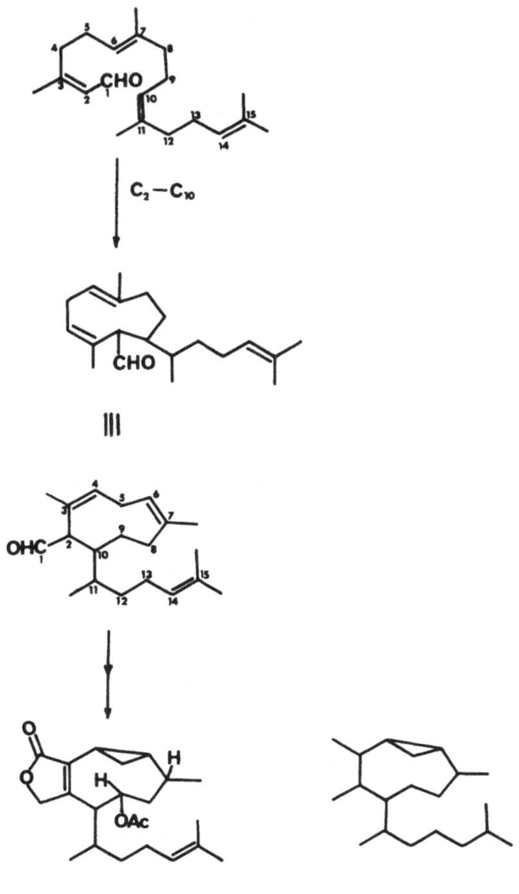

acetoxycrenulatin   crenulatane

Figure 7

Based upon the metabolites isolated, we can point out some curious
trends. As a common precursor, geranyl geranial (Figure 6) accounts
nicely for the production of "germacrene-like" diterpenoids from
C-1 to C-10 bond formation. This monocyclic diterpenoid ring
closes by standard methods at C-2 to C-6, to yield the perhydroazulene skeleton found in pachydictyol A and in dictyol A and B.
Recently[14] Erickson has isolated a eudesmane-like diterpenoid from
D. acutiloba. C-2 to C-7 bonding provides entry into this new
system. The macrocyclic intermediate also allows us to infer that
a new system resulting from C-3 to C-7 closure will be possible.
Using the same reasoning, we believe that geranyl geranial can
ring-close from C-2 to C-10. This closure generates an aldehyde
with the proper framework to be an immediate precursor to cyclopropane and lactone formation in acetoxycrenulatin[15].

REFERENCES

1. Y. Hashimoto, N. Fusetani, and K. Nozawa, Proceedings VII
    International Seaweed Symposium, 569 (1971).

2. P. R. Burkholder, L. M. Burkholder, and L. R. Almodovar,
    Bot. Mar. 11, 149 (1968).

3. I. S. Hornsey and D. Hide, Br. Phycol. J. 9, 353 (1974).

4. T. Berti, G. Fassina, and S. Pignatti, Giorn. Botanico Ital.
    70, 609 (1963).

5. T. J. Starr, M. Piferrer, and M. Kajima, Tex. Rep. Biol. Med.
    24, 208 (1966).

6. E. Kurosawa, M. Izawa, K. Yamamoto, T. Masamune, and T. Irie,
    Bull. Chem. Soc. Japan 39, 2509 (1966) and 37, 1053 (1964).

7. R. E. Moore, J. A. Pettus, Jr., and J. Mistysyn, J. Org. Chem.
    39, 2201 (1974) and references cited therein.

8. W. Fenical, J. J. Sims, D. Squatrito, R. M. Wing and P.
    Radlick, J. Org. Chem. 38, 2383 (1973).

9. A. G. Gonzalez, J. Darias, J. D. Martin, and C. Pascual,
    Tetrahedron 29, 1605 (1973).

10. D. R. Hirschfeld, W. Fenical, G. H. Y. Lin, R. M. Wing, P.
     Radlick, and J. J. Sims, J. Amer. Chem. Soc. 95, 4049 (1973).

11. E. Fattorusso, S. Magno, L. Mayol, C. Santacroce, D. Sica, V. Amico, G. Oriente, M. Piattelli, and C. Tringali, Chem. Commun. <u>14</u>, 575 (1976).

12. L. Minale and R. Riccio, Tetrahedron Lett., 2711 (1976).

13. T. Levring, M. A. Hoppe, and O. J. Schmidt, <u>Marine Algae: A Survey of Research and Utilisation</u>, Cram, DeGruyter and Co., Hamburg, 1969.

14. K. Erickson, Clark University, personal communication.

15. This research was supported by the National Institutes of Health (postdoctoral fellowship to F.J.M.) and the Department of Commerce, Sea Grant Program.

HIGHLY HYDROXYLATED PHENOLS OF THE PHAEOPHYCEAE

K.-W. Glombitza

Institut für Pharmazeutische Biologie

der Universität Bonn, BR-Deutschland

INTRODUCTION

Vesicles of high optical refractivity are to be found in many cells of Phaeophyceae. They are present in nearly all Laminariales, Fucales and Ectocarpales. Occasionally, they have been seen also in the classes of seaweeds other than the Phaeophyceae. They were called "physodes", "Hansteen's fucosan granules" or "fucosan vesicles". Their content was named fucosan, or was summarized under "phaeophycean tannins"[33].

The vesicles are unstable to hypotonic solutions; they burst. As proven by Bouck of the Yale University in 1965[1] by electron microscopical observations, this means that the vesicles are membrane enclosed spaces. It seems that the cell inclusions, blackly coloured by osmium tetroxide, migrate in the living cell to the wall and are excreting their contents through the wall to the outside or into intercellular spaces.

This is the electron optical evidence of the fact that algae excrete, at least partly, their tannins into the marine water[2,18,20]. Already in 1964, the occurrence of UV-light absorbing substances in the marine water had been observed by Craigie and McLachlan[5]. There, the phenols are developing yellow-coloured complexes called "Gelbstoff" in the presence of proteins, carbohydrates and oxygen[28-30]. Expectedly, these excretion products are acting as strong antibacterials[3,4] and algicides[5,31].

There have been, up to now, only poor investigations on their chemical structures. From more or less specific colour reactions, Crato[6] concluded that the contents of the physodes were phloroglucinol (la) or its derivatives. Indeed Takahashi[32], Shirahama[27], Ogino[21] and Craigie[5] found phloroglucinol as a product of alkaline degradation of Ecklonia cava, Cystophyllum hakodatense, Sargassum ringgoldianum and Fucus vesiculosus extracts. Haug and Larsen[16] reported that the substances of Ascophyllum nodosum were strong reductants and that their equivalent weight was 40.

In ethanolic or aqueous solutions UV spectra similar to many leucocyanidins were found. The substances, therefore, were supposed to be tannins of the catechin type[5,21]. This would be surprising, as never before had flavonoids or other secondary phenylpropanes been found in brown algae.

RESULTS

Detection of Phloroglucinol

When separating pre-purified extracts of various Phaeophyceae by thin layer chromatography and detecting the free phenols by spraying with diazonium salt solution, we got a large series of red or orange-red spots. The spot patterns are very different not only in various genera, but also within the same genus[33].

Only one of these substances could be identified rather easily. It was scratched out, purified by sublimation and recrystallisation and identified with certainty as phloroglucinol (la)[14].

During a systematic screening of 26 different species, phloroglucinol was identified unambiguously in 17 species of the genera, Cladostephus, Dictyota, Fucus, Himanthalia, Cystoseira, Bifurcaria, Halidrys, Laminaria, Chorda and Saccorhiza. In a certain stage of development in Cystoseira it amounts to almost 50% of the total phenolic fraction, that is, 1 to 2% of thallus dry weight.

The presence of phloroglucinol was not strictly proved, even after enrichment, in Dictyopteris and Laminaria. No phloroglucinol was detected in Pilayella littoralis, Colpomenia peregrina, Ascophyllum nodosum, Pelvetia canaliculata, Laminaria digitata and Laminaria saccharina[14]. However, there are various indications which suggest that at least the Fucales and the Laminariales are capable of synthesizing phloroglucinol.

The other substances present in the seaweed extracts are extremely unstable usually. They can be handled and separated only after methylation or acetylation.

Identification of Fucols

When a mixture of the acetylated phenols from <u>Fucus vesiculosus</u>, prepurified if necessary, was examined by thin layer chromatography on silica gel, approximately fourteen more or less distinguishable bands became visible[23].

The first substance below phloroglucinol triacetate (1b) was isolated in a yield of 15 mg out of 500 g freeze dried <u>Fucus</u>. The molecular weight was established by mass spectrometry to be 502, corresponding to the molecular formula $C_{24}H_{22}O_{10}$.

1a   R = H

1b   R = Ac

During fragmentation induced by electron collision it loses ketene stepwise up to six times, to an ion $C_{12}H_{10}O_6$ at m/e 250.

This is evidence that all of the oxygen is located in phenol acetate groups. After a partial ketene fision water is eliminated giving at last very strong peaks. This indicates the presence of o,o'-dihydroxylated biphenyls. Such compounds, as Riedl[24] has reported, readily undergo a water elimination to form dibenzofurans.

The NMR spectrum taken in hexadeuteroacetone shows a singlet for four aromatic protons at 7.04 ppm and at upper field at 1.97 and 2.28 ppm signals for four and two acetyl groups each. According to the results of mass spectrometry and NMR, the substance must be a biphenyl derivative with a symmetrical structure. There are two possible alternatives: biphenyl derivative of pyrogallol, or of phloroglucinol. Owing to the easy elimination of water in the mass spectrometer, the surprising similarity of the IR spectrum of the unknown substance to that of phloroglucinol triacetate and the significant differences from pyrogallol triacetate it must be 2,4,5,2',4',5'-hexaacetoxybiphenyl (2).

Another substance could be identified as a homologous ter- phenyl derivative which is also built up of phloroglucinol moieties (3)[10].

2

3

In a third compound we determined a molecular weight of 1002. From the spectroscopic data we must consider it to be a homologous quaterphenyl derivative. For this substance we have to propose two isomeric formulae. Due to the intensity pattern of the NMR signals we suppose it to be a mixture of two possible isomers, tetrafucol A (4A) and B (4B). We call this new type of compounds fucols, and want to define under this name substances containing phloroglucinol moieties connected by aryl-aryl bonds. Amongst the brown algae we have found them in the Dictyotales as well as in various families and genera of Fucales. We could not find them in Laminariales. Their occurrence in Fucus vesiculosus, F. serratus, F. spiralis, Bifurcaria rotunda and Dictyota dichotoma has been proved.

Identification of Fucophloroethols

The fucols are found readily detected by a red colour upon spraying with vanillin - H_2SO_4, which does not change when the TLC plates are kept for longer time, whereas other spots gradually turn brown.

Another difference becomes visible when UV spectra are taken in acetonitrile. The fucol acetates present a strong absorption at 238 nm. This is shifted to the longer wavelength in a second group of compounds, named by us fucophloroethols, where an additional slight shoulder between 270 and 275 nm, was observed. According to our experiences in this class of substances, such an absorption indicates a benzene nucleus which is substituted by a fourth oxygen function, mostly a phenyl ether grouping.

Molecular weights, determined by mass spectrometry, of four isolated compounds were found to be 710, 918, 1126 and 1334, indicating sum formulae of $C_{34}H_{30}O_{17}$, $C_{44}H_{38}O_{22}$, $C_{54}H_{46}O_{27}$ and $C_{64}H_{54}O_{32}$ respectively. 8, 10, 12 or 14 times ketene is eliminated by these compounds to yield ions at 374, 498, 622, 746 with sum formulae of $C_{18}H_{14}O_9$, $C_{24}H_{18}O_{12}$, $C_{30}H_{22}O_{15}$ and $C_{36}H_{26}O_{18}$ respectively.

4A

4B

R = H₃CCO

The difference of $C_6H_4O_3$ of the sum formulae of the terminal
members of the splitting series is evidence of a homologous series
whose members differ from one another by one trihydroxy benzene.
A part of the benzene nuclei ought to be connected by ether bridges,
which are broken easily in the mass spectrometer, to yield the ion
of monomeric trihydroxy benzene at m/e 126. The differences between
the total number of oxygen atoms and the number of phenolic hydroxyls,
calculated from the number of ketene eliminations is conclusive proof
of the number of ether bridges present in each molecule.   There is
one in compound (5), two in compound (6) and three in compound (7),
and four in the fourth compound.

5

6

7

A common feature in the mass spectra of both fucols and fuco-
phlorethols is the elimination of water, leading to benzofurans.
In the fucols this was seen once to several times; in fucophloroethols
never more often than once. Here, evidently, the structural element
of an o,o'-dihydroxylated biphenyl is to be expected only once in
each compound. This biphenyl moiety is linked by an ether bridge
to the next trihydroxy benzene and, by elimination of H_2O, it is
converted to the ion at m/e 232.

In full agreement herewith, there is in all spectra a fragmentation at m/e 374 which is due to a five-fold hydroxylated biphenyl
linked by an ether bridge with a trihydroxy benzene. In parallel
with ether splittings induced by electronic impact, a series of
disintegrations with regular ketene eliminations were observed, well
known from other polyhydroxy phenyl ethers, in the higher molecular
weight compounds, starting from the molecular ion.

As is to be expected, there are many similarities of the ^{1}H-NMR spectra taken in CDCl$_3$. A common feature with all compounds is a signal of two aromatic protons at 6.98 ppm, which is characteristic of protons in terminal phenyls of the biphenyl type. Compounds (5-7) gave a signal of two protons at 6.93 - 6.97 ppm, the fourth compound a signal of four protons, corresponding to one or two phenyl rings respectively, connected by ether oxygen. Signals in the region of 6.7 ppm are correlated with the protons marked b, which are present twice in compound (5), four times in compound (6), six times both in compound (7) and in the fourth one.

Methyl group signals of terminal acetoxy groups not substituted in o,o'-position were seen to be at 2.26 ppm. When the acetoxy groups are in o,o'-position to a biphenyl binding, the methyl signal is at 1.95 ppm approximately. The signals, characteristic of the six protons of acetyls in o,o'-position of the ring of the biphenyl moiety linked by an ether bridge is at about 2.05 ppm.

Surprisingly, here a signal of four methyls was found with the fourth compound.

When an ether bridge in o,o'-position is present, resonance signal is given at 2.12 ppm, as illustrated by two, four and six acetyls in compounds (5), (6) and (7) respectively. Instead of the signal of eight acetyl groups to be expected, in the fourth compound an irregularity was found in this respect.

It can therefore be concluded that compound (5) is a fucophoroethol octaacetate, that compound (6) is fucodiphlorethol decaacetate, and that compound (7) fucotriphlorethol dodecaacetate.

It seems that in the fourth compound, presumed to be fucopentaphlorethol tetradecaacetate, the rings are no longer in a linear arrangement, but rather linked by a binding point not yet elucidated, or that it is a mixture of several isomers.

The correctness of the structure proposal of (2), published by us in 1975[10], and of (5), presented by us at Bangor in 1974[8], has meanwhile been confirmed by Ragan and Craigie[22] by their paper published early in 1976. They had isolated the same substances from the Canadian Fucus vesiculosus.

In Cystoseira baccata isomers of fucophlorethol (8) and diphlorethol (9) are found, built up in angular but not in straight form. The same structure principle was evident in Laminaria ochroleuca. In fucodiphlorethol from Laminaria ochroleuca the phenyl ether groups are, of course, not arranged ortho but meta (compound 10).

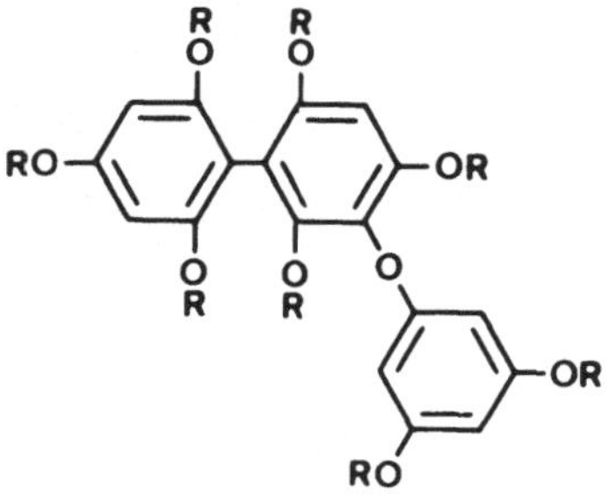

8

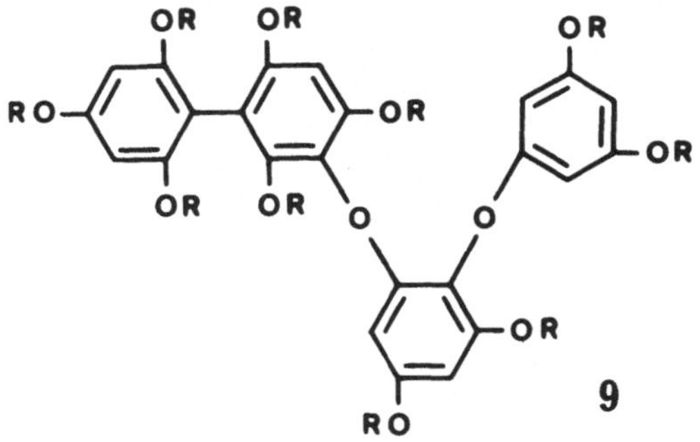

9

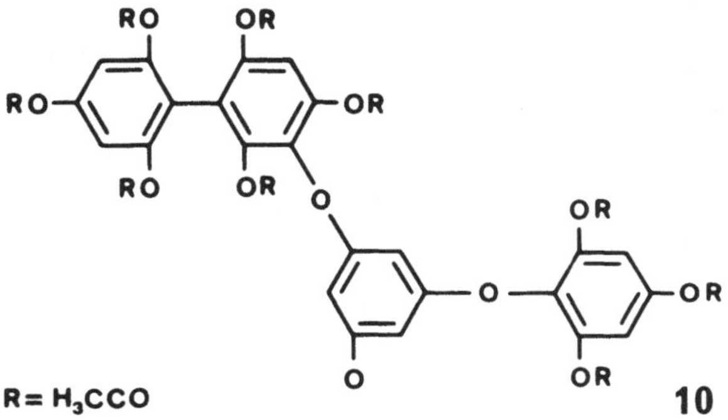

R= H₃CCO

10

Identification of Phlorethols

In Cystoseira tamariscifolia we detected a substance with two
benzene rings connected by one ether bridge and substituted with
five acetoxy groups. From the spectroscopic data it was considered
to be a 2,4,5,3',5'-pentaacetoxydiphenyl ether. Meanwhile, the
structure has been confirmed by total synthesis. We propose the
name "diphlorethol pentaacetate" (11)[13].

It occurs also in Dictyota dichotoma[7] and Laminaria ochroleuca[17].
Here also higher homologs like triphlorethol-C-acetate (12)[9] appear.
It is accompanied by series of oligophenyl ethers with 4 benzene
rings. The separation and the structure elucidation of the various
isomeric tetraphlorethols is not ready as yet. According to the
data received up to now, the structures (13), (14) and (15) are in
discussion.

In one of the proposals the ether bridges of the phloroglucinol
moieties are obviously set up m,p-, in another one m,m- and in
the third m,o-[17].

In Laminaria ochroleuca halogenated members of this class of
compound had been detected for the first time. In a mass spectrum
of a diphlorethol acetate a small amount of monochlorodiphlorethol

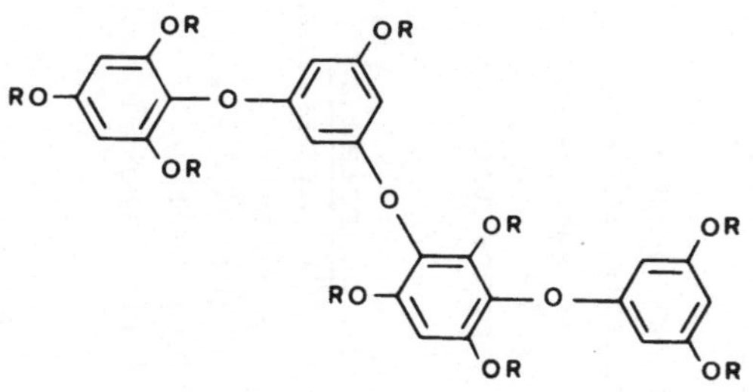

R = Ac

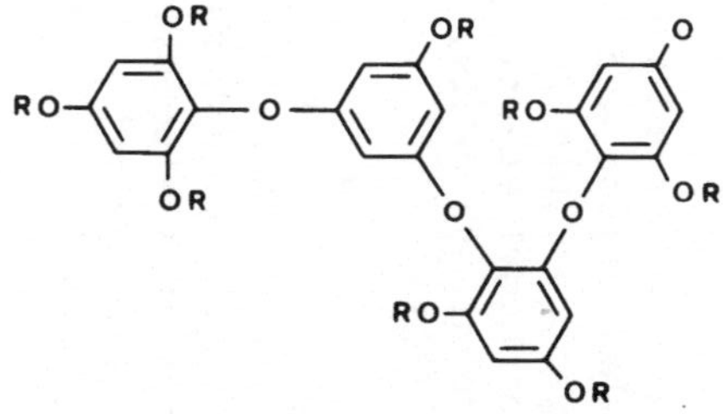

R=H₃CCO

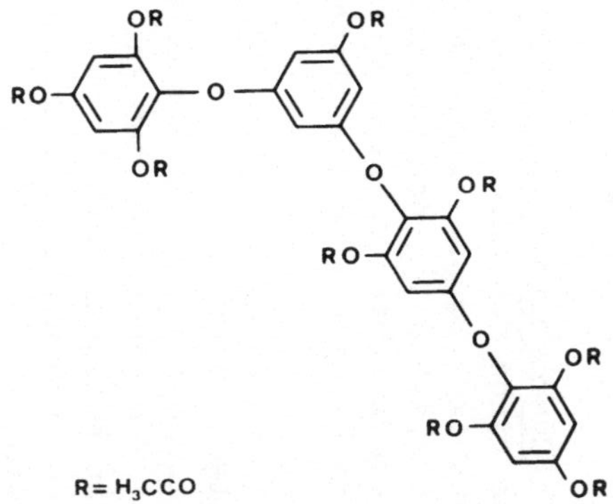

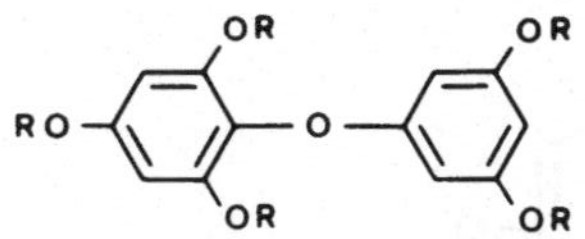

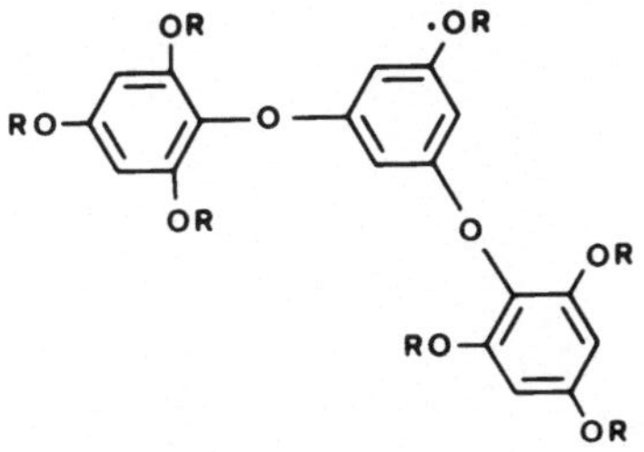

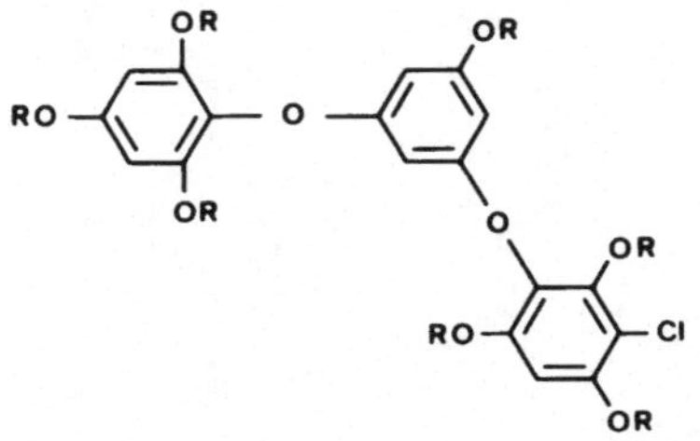

Structure proposals 13-15 not yet proven; working hypothesis

acetate could be seen, and from triphlorethol acetate a mono-
chlorotriphlorethol acetate (16) could be separated and cleaned
up. The fraction of the higher polymers also contains halogenated
compounds, which means that every third benzene nucleus is bearing
one chlorine atom.

Identification of the Fuhalols

The fourth class of substances was called fuhalols. They had
been found for the first time in Halidrys siliquosa[15] and
Bifurcaria rotunda[11]. From the phlorethols they differ by the
presence of one or more additional hydroxyl groups. The structures
of bifuhalol acetate (17) and trifuhalol (18) were established by
spectroscopic examination and total synthesis[11,25]. As to tetra-
fuhalol acetate, we have to discuss two alternative formulae (19A,
19B). The site of branching might be at the terminus or at the
first ring. The high molecular weight fuhalols of Halidrys
siliquosa are characterized by the non-linearity of the molecule
and a five-fold substitution at the branching ring[12].

From both algae we have been able to isolate a pentafuhalol
acetate (20) with a side chain elongated by one further ring. Up
to now the fuhalols have been found only in the order of the
Fucales, specifically in the Cystoseiraceae. There it seems, however, that other possible isomeric tetrafuhalols can be found, for
in spite of nearly identical R_f and the same molecular weight, the
PMR spectra of some of the isolated compounds are slightly different.

By ^{13}C-NMR spectroscopy the structure of a heptafuhalol acetate
(21) from Halidrys siliquosa could be determined. The assignment
of the signals had been achieved mainly by spectral correlation
with bifuhalol hexaacetate and trifuhalol octaacetate[26].

Insufficient chemical shift dispersion does not allow one to
assign the methyl resonances at δ 20-21 nor those due to the carbonyl carbons between δ 167 and 169. More favourable are the
tertiary aromatic carbon resonances in the region between δ 101 and
116, in particular those of the quaternary aromatics which cover a
very large range (δ 130 to 158).

Because of the large size of the molecule the rotational correlation time is long enough to shorten the transverse relaxation
times of the proton-bearing carbons to the extent where appreciable
broadening results. By contrast the quaternary carbons exhibit
considerably larger T_2's and consequently give rise to narrow signals. Almost all of the quaternary aromatic carbons which are
magnetically non-equivalent therefore appear as separate resonances.

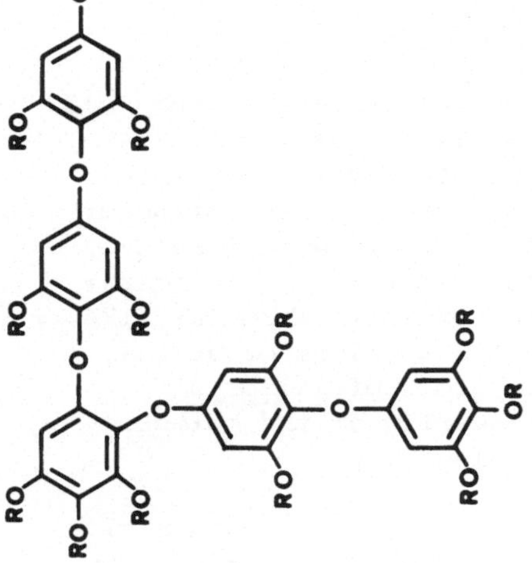

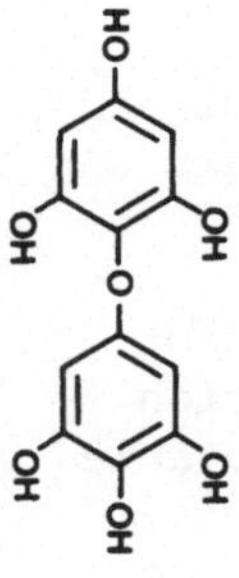

17

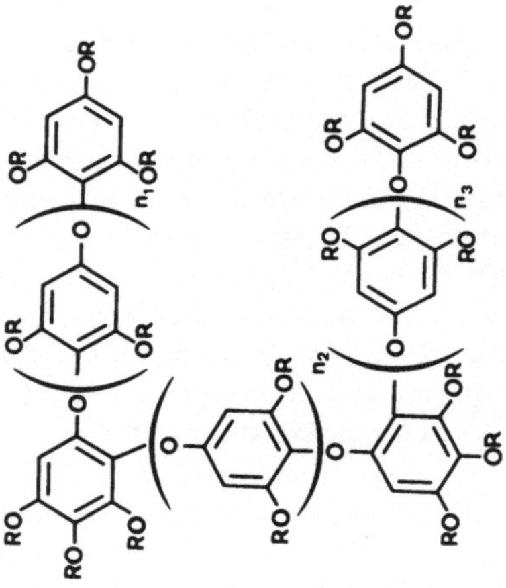

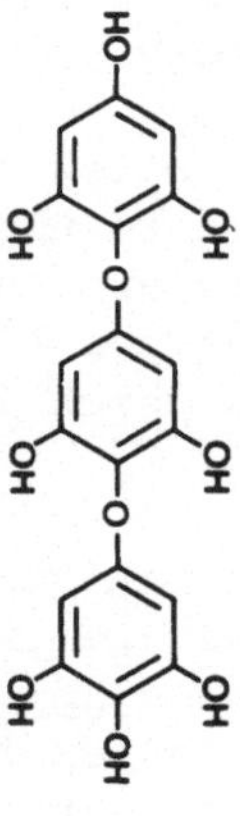

18

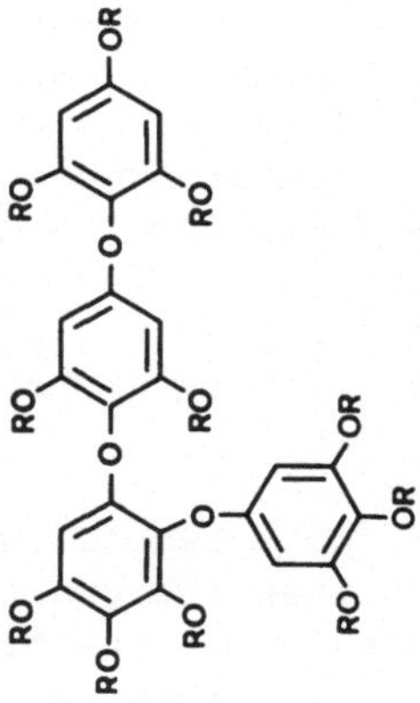

19A

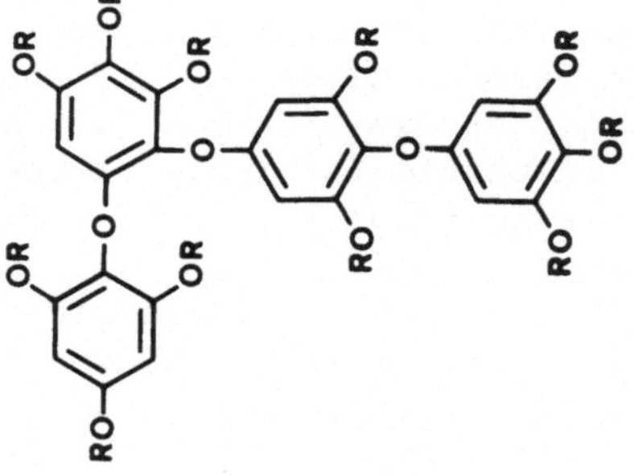

19B

20

21

A partial assignment of the aromatic carbon signals is achieved with the aid of the correlation diagram, Figure 1. The resonances in bifuhalol acetate and trifuhalol acetate were designated by means of chemical shift arrangements based on the fact that oxygens in ortho- and para-positions strongly shield the carbon in question, while those in meta-positions give rise to slight deshielding. Using the known chemical shift increments, it could be predicted that C-5^7 and 5^1 in bifuhalol acetate should be the least shielded and are thus assigned to the resonances at 154.8 and 146.9 ppm, respectively. A one-to-one assignment is possible by comparison with the spectrum of trifuhalol acetate, where an additional line appears at 154.0 ppm. In heptafuhalol acetate we find signals for four carbons at lower field and three at higher field. Analogous arguments permit us to assign the higher field quaternary carbons C-2^1 and C-2^7 as well as those belonging to the proton bearing carbons 4^1, 6^1, 4^7, 6^7 and 4^2, 6^2. The pairs of isochronous carbons 1^7, 3^7 and 1^1, 3^1 and 1^2, 3^2 resonating near 144 ppm cannot unambiguously be assigned.

It is not surprising that the predicted shifts deviate considerably from the experimental values. This is to be expected in view of the fact that the substituent parameters were derived from the corresponding monosubstituted benzenes. Generally, too large a value is calculated, but the trends are in qualitative agreement.

To shorten the explanations, we can conclude that the spectrum of heptafuhalol acetate contains all resonances observed for the other two, and must, therefore, contain the same rings, partly in multiple versions.

Though this structure proposal is in agreement with all spectroscopic data, a different sequence of the middle rings cannot completely be excluded. In any case, heptafuhalol acetate has to be in line with the building principle shown in (21), whereby for n_1, n_2, n_3 values of 0, 1 and 2 are conceivable, but the sum of n_1, n_2, $n_3 = 3$ has to be satisfied.

The fundamental pattern valid for heptafuhalol acetate holds true also for a mixture of polymer compounds isolated from Halidrys and Bifurcaria[26].

SUMMARY

Most of the substances mentioned, at least those of higher molecular weights, have the properties of tannins. We have named this class of phenols "phlorotannins", referring to tannins constructed of phloroglucinol moieties linked by aryl-aryl or aryl-ether bonds or of a mixed type. They may be substituted by one or more additional hydroxyl groups or chlorine atoms.

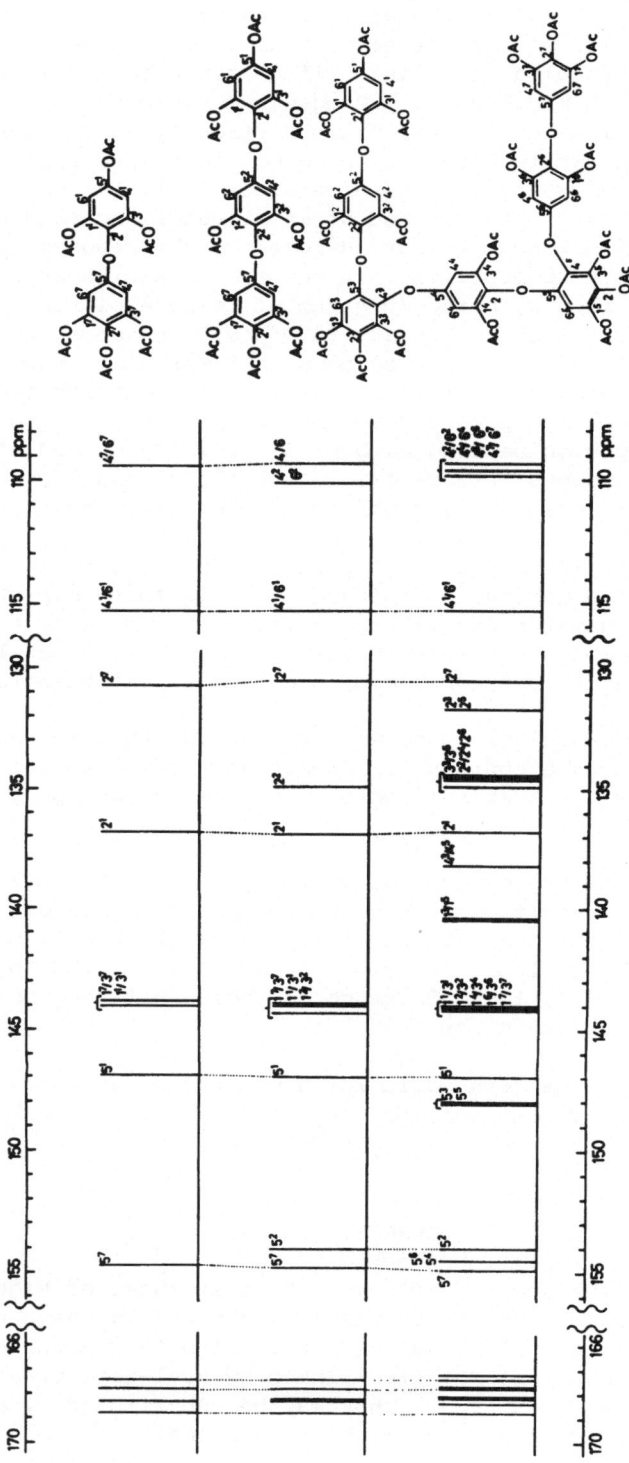

Figure 1. CMR spectra of compound (21) and model compounds

ACKNOWLEDGEMENTS

I want to thank my co-workers Dr. Sattler (Halidrys), Dr. Rosener (Bifurcaria, Cystoseira tamariscifolia), Dr. Rauwald (Fucus), Mrs. Wiedenfeld (Cystoseira baccata), Miss Koch (Laminaria ochroleuca), Mrs. Geisler (Dictyota dichotoma) and Mr. Vilter (screening), as well as the Deutsche Forschungsgemeinschaft for the grant.

REFERENCES

1. G. B. Bouck, J. Cell. Biol. 26(2), 523 (1965).

2. M. Chadefaud, 1ere Thèse pour obtenir le Grade de Docteur en Sciences naturelles, Paris, 1935.

3. J. T. Conover, Proc. Vth Int. Seaweed Symp., 99 (1965).

4. J. T. Conover and J. M. Sieburth, Bot. Mar. 6, 147 (1964).

5. J. S. Craigie and J. McLachlan, Can. J. Bot. 42, 23 (1964).

6. E. Crato, Ber. Dtsch. Bot. Ges. 10, 295 (1892).

7. C. Geisler, Dissertation Bonn, in preparation.

8. K.-W. Glombitza, Proc. VIIth Int. Seaweed Symp., 1974, in press.

9. K.-W. Glombitza, M. Koch, and G. Eckhardt, Phytochem. 15, 1082 (1976).

10. K.-W. Glombitza, H. W. Rauwald, and G. Eckhardt, Phytochem. 14, 1403 (1975).

11. K.-W. Glombitza and H. U. Rosener, Phytochem. 13, 1245 (1974).

12. K.-W. Glombitza, H. U. Rosener, and M. Koch, Phytochem. 15, 1279 (1976).

13. K.-W. Glombitza, H. U. Rosener, and D. Möller, Phytochem. 14, 1115 (1975).

14. K.-W. Glombitza, H. U. Rosener, H. Vilter, and H. W. Rauwald, Planta Med. 24, 301 (1973).

15. K.-W. Glombitza and E. Sattler, Tetrahedron Lett., 4277 (1973).

16. A. Haug and B. Larsen, Acta Chem. Scand. 12, 650 (1958).

17. M. Koch, Dissertation Bonn, in preparation.

18. M. H. Le Touze, Rev. Gen. Bot. 24, 33 (1912).

19. T. Levring, K. Vet. O Vitterh. Sarnh. Handl, FG Ser. B 3, 11 (1945).

20. M. McCully, J. Cell. Sci. 3, 1 (1968).

21. C. Ogino and Y. Taki, Tokyo Univ. Fish. 43, 1 (1957).

22. M. Ragan and J. S. Craigie, Can. J. Biochem., (1976).

23. H. W. Rauwald, Dissertation Bonn, 1976.

24. W. Riedl, Justus Liebigs Ann. Chem. 597, 148 (1955).

25. E. Sattler and K.-W. Glombitza, Arch. Pharm. 308, 813 (1975).

26. E. Sattler, K.-W. Glombitza, F. W. Wehrli, and G. Eckhardt, submitted for publication, 1976.

27. K. Shirahama, J. Fac. Agric. Hokkaido Imp. Univ. 49, 57 (1942).

28. J. M. Sieburth, J. Exp. Mar. Biol. Ecol. 3, 290 (1969).

29. J. M. Sieburth and A. Jensen, J. Exp. Mar. Biol. Ecol. 2, 174 (1968).

30. J. M. Sieburth and A. Jensen, J. Exp. Mar. Biol. Ecol. 3, 275 (1969).

31. S. Suneson, K. Fysiogr. Saellsk. Lund Foerh. 12, 183 (1942).

32. T. Takahashi, Tokyo Kogyo Shikensho Hokoku 26, 1 (1931).

33. H. Vilter, Staatsarbeit Bonn, 1973.

34. G. Wiedenfeld, Dissertation Bonn, in preparation.

NATURAL PRODUCTS FROM THE RED SEAWEED DELISEA FIMBRIATA

J. J. Sims, J. A. Pettus, Jr., and R. M. Wing

Departments of Chemistry and Plant Pathology

University of California, Riverside, California 92502 USA

The exceptional feature of marine natural products chemistry which sets it apart from the more traditional natural products chemistry is the wide-spread incorporation of the halogens into organic compounds. It seems obvious, after the fact, that this rich halogen metabolism is a result of the high concentration of halide ions in sea water. The seaweeds appear to be the primary site of halogen incorporation. Among the seaweeds, the Rhodophyta have yielded the largest number of halogen metabolites. The Bonnemaisoniaceae family is, as yet, the only one which incorporates all three of the common halogens, bromine, chlorine, and iodine, into organic metabolites. The compounds found are usually simple molecules which are often volatile[1,2,3]. Another feature of the compounds of Bonnemaisoniaceae is their toxicity. Many of them are alkylating agents. This chemical armament is probably the reason for our observation of Asparagopsis taxiformis growing in the open on Pacific coral reefs, where most algae are heavily grazed by the diverse animal life.

Our investigation of a number of algae collected near Palmer Station, Antarctic Peninsula, turned up one antibiotic-producing alga, Delisea fimbriata (Bonnemaisoniaceae). The samples were air-dried and, among the drying algae, Delisea fimbriata was the only one giving an odor different enough to be noticed. Presumably the more volatile compounds responsible for this smell were lost in drying. We have not found any particularly odoriferous volatiles in this study. We are trying to obtain further frozen samples to investigate this point. One further fact aroused our interest in D. fimbriata. In the area, it is one of a few species which are mostly free of the normal organisms (micro and macro) which colonize other algae.

Extraction of <u>D</u>. <u>fimbriata</u> with methylene chloride and then methanol gave a good yield of organic solubles. After silica gel chromatography, two fractions were of interest. The more polar of these was found to contain a series of lactones which had anti-microbial activity. High pressure liquid chromatography of the lactone fractions gave seven closely related lactones differing only slightly in their spectral properties. The major compound was shown by high resolution mass spectrometry to have the empirical formula $C_{11}H_{12}Br_2O_4$. The other lactones were a dibromoisomer, a pair of bromoiodo isomers, one bromochloro compound, and a tribromo compound. All of the lactones had the same carbon skeleton, differing only in halogen content and, at this point, an unknown isomerism.

The major isomer had ultraviolet absorption in methanol at 291 nm (ϵ = 8900). The infrared spectrum gave evidence for a conjugated lactone structure, with strong bands at 5.58, 5.77, 6.08 and 6.20 μ. The proton nmr spectrum (100 MHz, CCl_4, δ) showed a substituted n-butyl side chain (CH_3, 0.97t; CH_2, 1.39m; CH_2, 1.86m; CHOAc, 5.44dd, J = 7.4, 6.3) an acetate methyl (2.01,s), and an olefinic hydrogen (=CHBr, 6.24,s). The ^{13}C spectrum (CFT-20, $CDCl_3$, ppm) gave further evidence for acetate (170,s) and lactone (162,s) carbonyls, and tetrasubstituted (149,s; 131,s) and trisubstituted (130,s; 92,d) enolic double bonds; other absorptions were at 68d, 33t, 21q, 19t, and 14q.

Possible structures considered were of two types, α-pyrones and γ-lactones. The ultraviolet absorption was close to a number of α-pyrones, but the infrared absorption, particularly at 5.58 μ, pointed to a γ-lactone structure.

When the major lactone was dissolved in methanol and treated with a few drops of KOH solution, a purple color immediately appeared. Rapid acidification of the solution with mineral acids caused the solution to go colorless, and work-up gave a high yield of two diastereomers corresponding to addition of methanol to the lactone. Spectral changes indicated loss of the enolic double bond with generation of a -CH_2Br group. This made the γ-lactone structure the more likely structure

The two methanol adducts were separated by HPLC, and one of them was crystalline (mp 80-80.5^0). Its structure and absolute configuration were established by X-ray diffraction analysis as (1). Working backwards, the original lactone had to be (2).

The isomers of (2) are the E and Z isomers at the exocyclic double bond. Assignment of stereochemistry about the double bond in (2) was possible when we observed that the minor isomer was converted completely to the major one on heating. The change was not affected by solvent or light. From molecular models, it seems

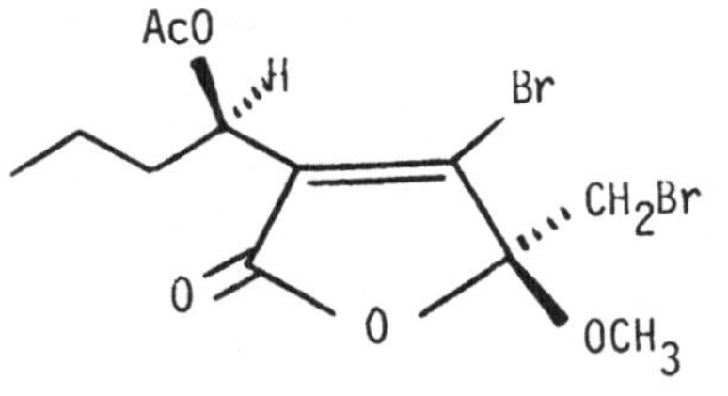

1

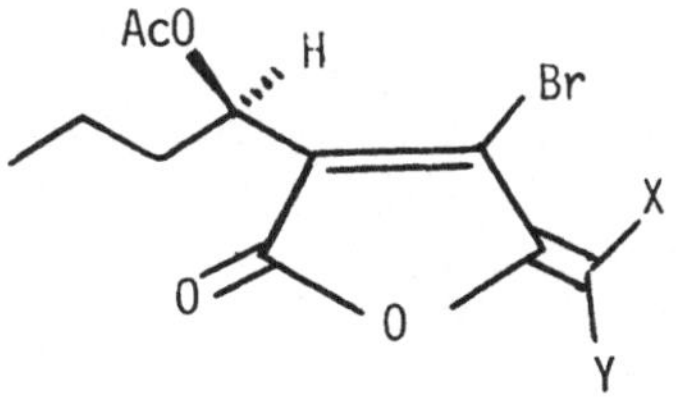

2

X	Y	% Dry Wt.
a	H	Br	0.20
b	Br	H	0.075
c	I	H	0.02
d	H	I	0.05
e	H	Cl	0.007
f	Br	Br	0.002

that the two bromine atoms in the E-isomer (2b) would crowd each other and that no crowding is evident in the Z-isomer (2a). Thus, the major isomer from the plant is (2a), the Z-isomer. The structures of the other lactones followed from their spectral properties as (2c-f).

3 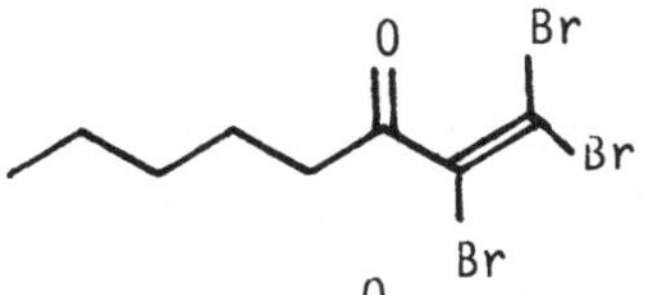   0.14 %

4 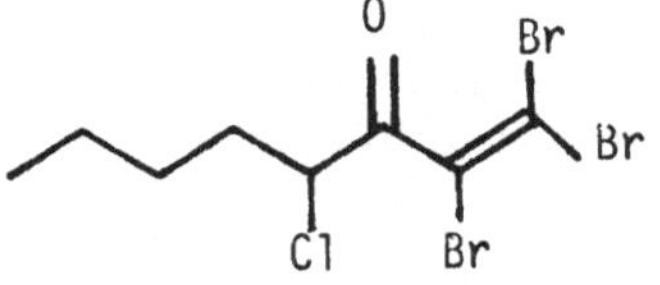   0.014 %

5 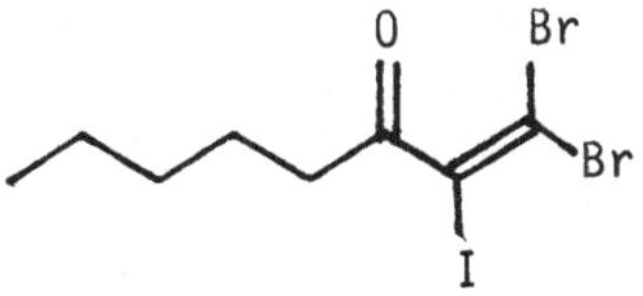   0.003 %

6 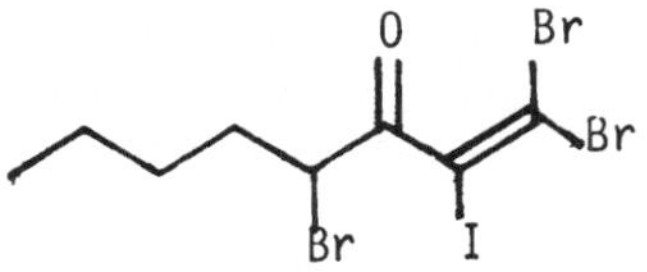   0.003%

7 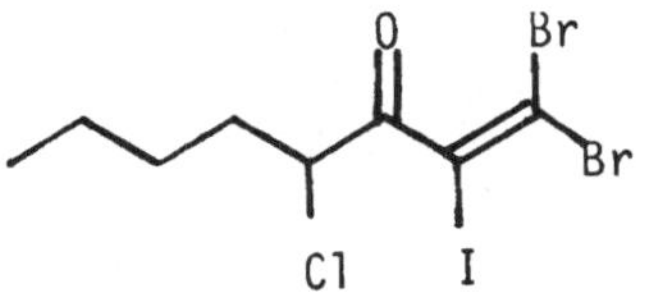   Trace

The less polar fractions from the original extract were separated by HPLC into five separate compounds. All were highly
halogenated ketones whose structures were deduced mainly from
their mass spectral fragmentations. The major ketone had strong
carbonyl absorption in its infrared spectrum at 5.82 μ and UV
absorption (MEOH) λ_{max} 220 (7200), 282 (1080). Its molecular ion
required an empirical formula of $C_8H_{11}Br_3O$. The proton nmr spectrum (60 MHz, CCl_4, δ) had peaks at 0.88 (3H,t), 1.4 (6H,m), 2.74
(2H,t). The ^{13}C spectrum showed a carbonyl (197.8), a tetrasubstituted double bond (122.8, 91.8), and five peaks of an apparent n-pentyl chain (41.3,t; 31.7,t; 23.8,t; 23.0,t; 14.5,q). The
mass spectral fragmentation revealed a McLafferty rearrangement
leading to ion a (m/e 360). The ketone undergoes α-cleavage,
giving ions b (m/e 289) and c (m/e 99, base peak). This spectral
information requires that the major ketone be formulated as (3)
(1,1,2-tribromo-oct-1-ene-one). The structures of (4-7) followed
from their spectral properties. For example, (4) and (6) gave the
expected McLafferty and α-cleavage ions for α-haloketones. The
location on the double bond of the iodine in compounds (5) and (7)
is not certain.

Both the infrared absorption and the ultraviolet absorption
of the ketones (3-7) are uncharacteristic of α,β-unsaturated
ketones. The observed absorption is what one would expect for the
carbonyl of a saturated acyclic or six-ring ketone, while the
ultraviolet absorption is that expected for the tribromo double
bond isolated from the carbonyl group. These anomalous observations
can be explained if the carbonyl group and the double bond are not
co-planar and there not conjugated. Apparently the size of the
Br atom α to the carbonyl is great enough to disrupt the normally
more stable co-planarity of the two groups, thus giving rise to the
spectral properties of the isolated chromophores.

Figure. Mass spectral fragmentations of (3).

Synthetic Scheme

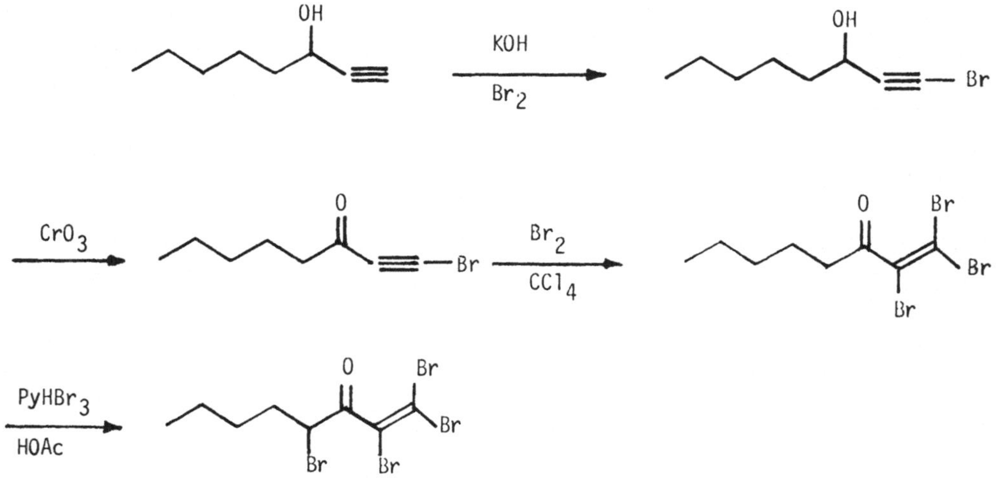

Scheme. Synthesis of compounds (3) and (6).

Confirmation of the structures of compounds (3) and (6) was obtained by a synthesis from commercially-available 1-octyne-3-ol according to the scheme. The compounds obtained had properties identical to those of the natural compounds.

ACKNOWLEDGEMENTS

This research was supported by National Science Foundation Grant No. CHE74-13938. We thank Dr. T. R. De Laca for collecting the algal samples.

REFERENCES

1. W. Fenical, Tetrahedron Lett., 4463 (1974).

2. B. J. Burreson, R. E. Moore, and P. P. Roller, Agri. Food Chem. 24, 856 (1976).

3. J. F. Siuda et al., J. Am. Chem. Soc. 97, 937 (1975).

THE BREADTH OF MONOTERPENE SYNTHESIS BY MARINE RED ALGAE:

POTENTIAL DIFFICULTIES IN THEIR APPLICATION AS TAXONOMIC MARKERS

P. Crews

Thimann Laboratories, University of California

Santa Cruz, California 94064 USA

Our interest in the organic chemistry of toxic marine organisms has led us to study a number of species of marine red algae. Several years ago our attention was drawn to the family Plocamiaceae, and this has resulted in our discovery of an interesting series of polyhalogenated monoterpenes[1-3].

Five species within Plocamiaceae are native to the North American Pacific coast[4a,b] and have been directly available to us. Plocamium cartilagineum (Dixon) is the most abundant, and it is also world-wide in occurrence. Its habitat in the Monterey Bay region is unusual in that it ranges from the middle intertidal zone to a depth of approximately 125 feet. Plocamium violaceum (Farlow), whose type locality is Santa Cruz, is moderately abundant. Two other species, P. oreganum (Doty) and P. tenue (Kylin), are rarely found on the central California coast but are moderately abundant in the Pacific Northwest. Plocamiocolax pulvinata (Setchell) is an epiphyte specific to P. cartilagineum and is moderately abundant.

During the last three years we have examined each of these species, except for P. tenue, from a variety of geographical areas. We have found them all to be rich in halogenated monoterpenes, and a great deal of our energy has been devoted to total structural characterizations of these monoterpenes[1-3,5-6]. Concurrently with our isolation and structure work, we have, in several cases, been able to initiate studies directed at understanding the mechanisms which control both the production and fate of algal monoterpenes[7,8]. Our close proximity to Monterey Bay and its diversity of both seaweed flora and rocky intertidal areas has greatly facilitated this aspect of our research.

While not complete, our chemical study of monoterpene synthesis by red seaweeds has reached a stage where many trends of structural types, biogenetics, and relationships between component compositions and species can be clearly identified. Three major aspects of this chemistry will be discussed below: the monoterpene architectures observed to date, intraspecies variations in monoterpene occurrence, and algal monoterpene production as a function of phyletic and morphological considerations.

Type	Examples	Reference
1,5,7-octatriene	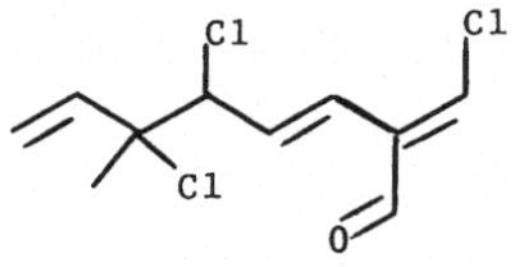 cartilagineal (1)	1
	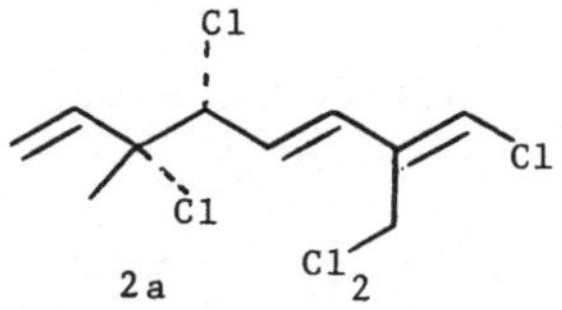 2a	9
	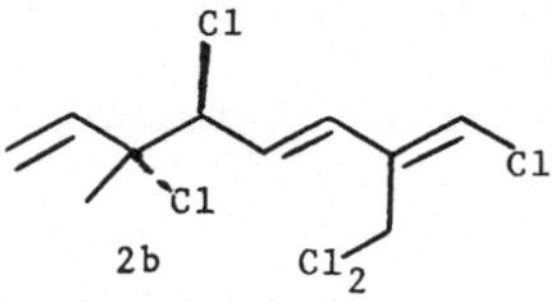 2b	9
1,5-octadiene	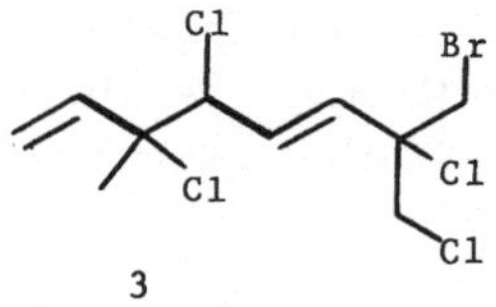 3	10a
	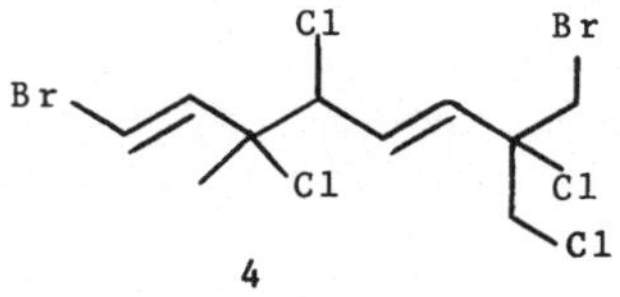 4	
2,7-octadiene	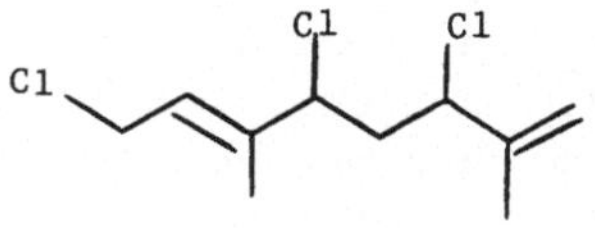 5	10b

Figure 1. Summary of halogenated monoterpene constitutions from Plocamiaceae.

MONOTERPENE STRUCTURAL TYPES

All of the polyhalogenated monoterpenes from Plocamiaceae observed to date can be divided into four major structural families. The first group contains the common acyclic head-to-tail arrangement, and further structural distinctions can be made within this group according to the extent of unsaturation. The structures of cartilagineal (1) and pentachloride (2) in Figure 1 illustrate the most unsaturated type. Numerous other acyclic 1,5,7-octatriene halocarbons have been reported from P. cartilagineum[9], while cartilagineal (1) represents one of the few oxygenated monoterpenes observed and can be obtained only from P. cartilagineum found at certain geographical locations. Two examples of the acyclic 1,5-octadiene skeleton are shown in Figure 1[10a]. We have encountered several examples of the 2,7-octadiene frame, including structure (5) and halogen isomers of (5) having mixed halogens Cl and Br[10b].

1-vinyl-1,5-dimethyl-
cyclohexane

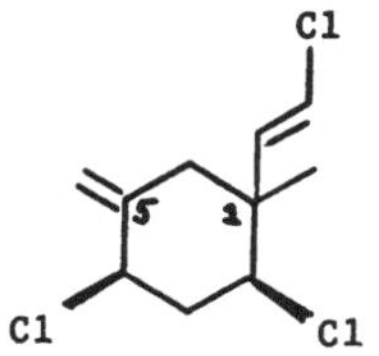

5

plocamene-D (6)

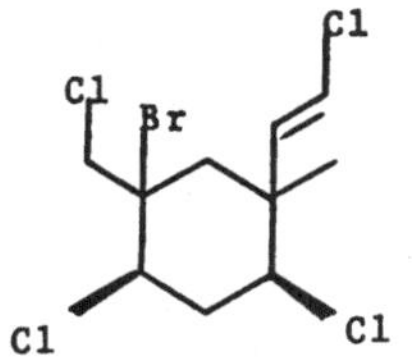

11

violacene (7)

2-vinyl-1,5-dimethyl-
cyclohexane

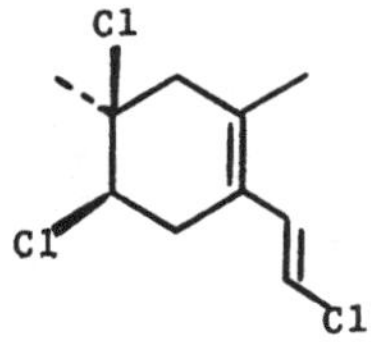

2

plocamene-B (8)

dihydropyran

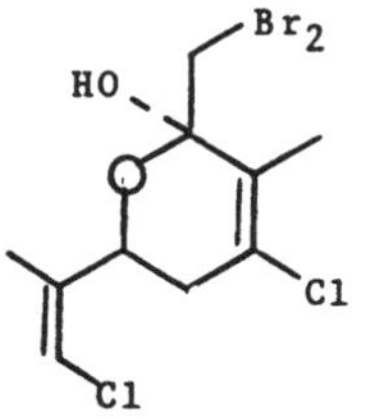

12

9

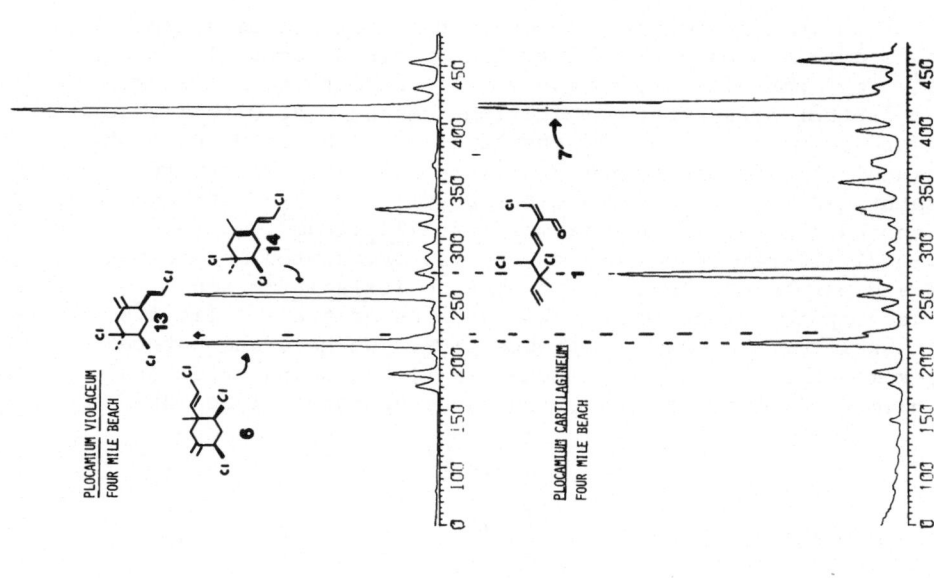

Figure 3. GLC/MS trace of P. violaceum and P. cartilagineum from the northern Monterey Bay area.

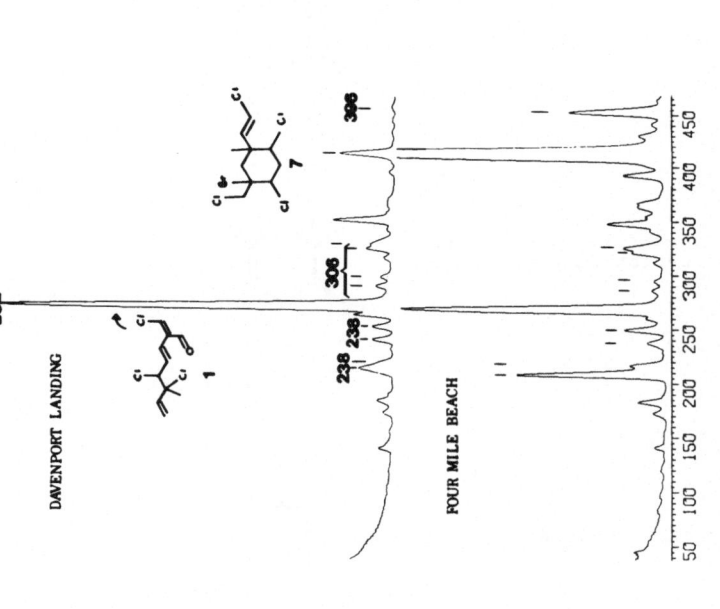

Figure 2. GLC/MS trace of P. cartilagineum from the northern Monterey Bay area.

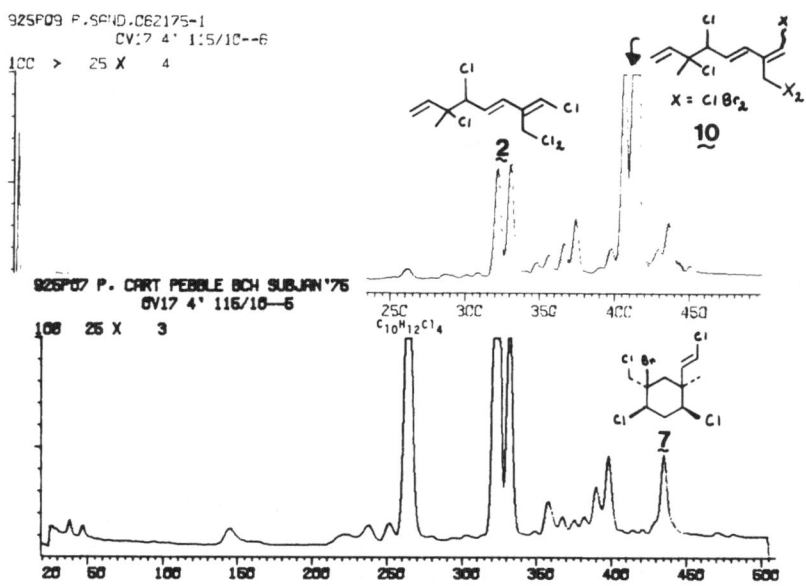

Figure 4.   GLC/MS trace of P. sandvicense from Hawaii and P. cartilagineum from the southern Monterey Bay area.

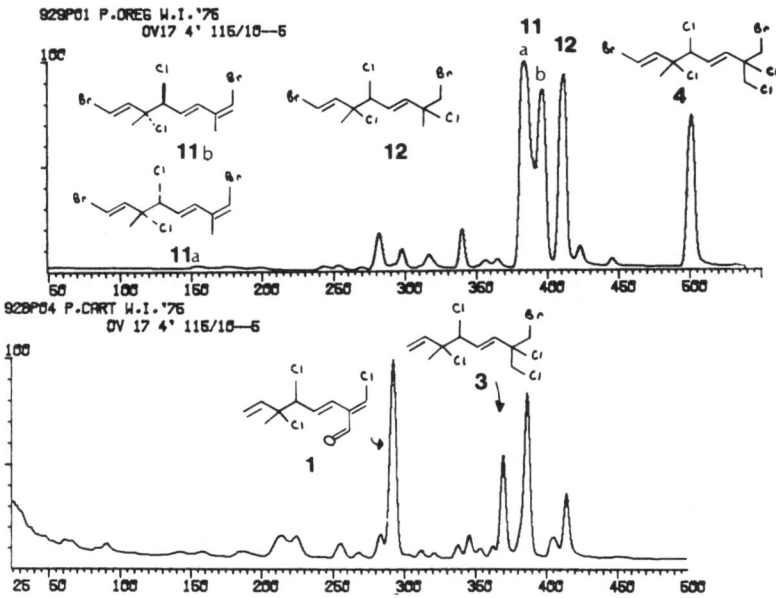

Figure 5.   GLC/MS trace of P. oreganum and P. cartilagineum from Partridge Point on Whidbey Island, Washington.

Turning to the monocyclic types, plocamene-D (6)[5] and violacene (7)[11] provide two examples of the 1,1,5-trialkyl cyclohexane ring. These two compounds are closely related, but they are not artifacts of isolation. Instead, (7) is undoubtedly derived from (6) by enzyme-directed addition of BrCl. Both (6) and (7) can be related back to the acyclic head-to-tail skeletons above by a cleavage of the C_1 to C_6 bond. The second type of cyclohexane ring system is represented by plocamene-B (8)[2]. This rearranged isoprenoid is probably formed via a _trans_-chlorovinyl migration from a suitable precursor in the 1-vinyl-1,5-dimethyl cyclohexane series. A final monocyclic structural type recently described from an Australian _Plocamium_ is represented by dihydropyran (9), from _Plocamium costatum_[12].

INTERSPECIES VARIATIONS IN HALOGENATED MONOTERPENES

At a very early stage in our isolation work on _P. cartilagineum_, we observed that several monoterpene frameworks were elaborated by this species and that their relative concentrations changed when the collection location was varied by only a few miles. As an example, Figure 2 shows that net concentrations of components (1) and (7) vary considerably in _P. cartilagineum_ specimens collected within a four-mile radius. In connection with our broad study of _Plocamium_, we have obtained _P. cartilagineum_ from various sites within California and Washington and from west of Santiago, Chile. Figures 3-6 compare specific monoterpenes from these various collections, including: (Figure 4) 1,5,7-octatrienes (2a-b), which are common to _Plocamium sandvicense_ (J Agardh) from Maui, Hawaii and _P. cartilagineum_ from Pebble Beach (Monterey County, subtidal); (Figure 5) 1-bromo-1,5-dienes (4) and (12), which are common to _P. oreganum_ from Whidbey Island, Washington[10a,13] and the 1-debromo-1,5-diene (3), which is common to _P. cartilagineum_ at the same location; (Figure 3) 1-vinyl-1,5-dimethyl cyclohexanes (6) and (7) and 2-vinyl-1,5-dimethyl cyclohexanes (13) and (14), which are common to _P. violaceum_ and _P. cartilagineum_, both from Four Mile Beach, Santa Cruz County; and (Figure 6) several 1-vinyl-1,5-dimethyl cyclohexanes, especially (7), which are common to _P. violaceum_ from Davenport Landing, Santa Cruz County, and _P. cartilagineum_ from Chile.

These comparisons show that the breadth of monoterpene constitutions elaborated by _P. cartilagineum_ almost equals the total range observable from all other _Plocamium_ species that we have examined to date. In addition, among the _Plocamium_ that we have studied, _P. cartilagineum_ exhibits the largest variation in composition of monoterpenes as a function of geographical location.

Interestingly, _P. cartilagineum_ has not yet been observed to elaborate representatives of the acyclic 2,7-diene type. Rather, these metabolites have been available from _P. violaceum_ at collection sites in the southern Monterey Bay region (Figure 7)[10b].

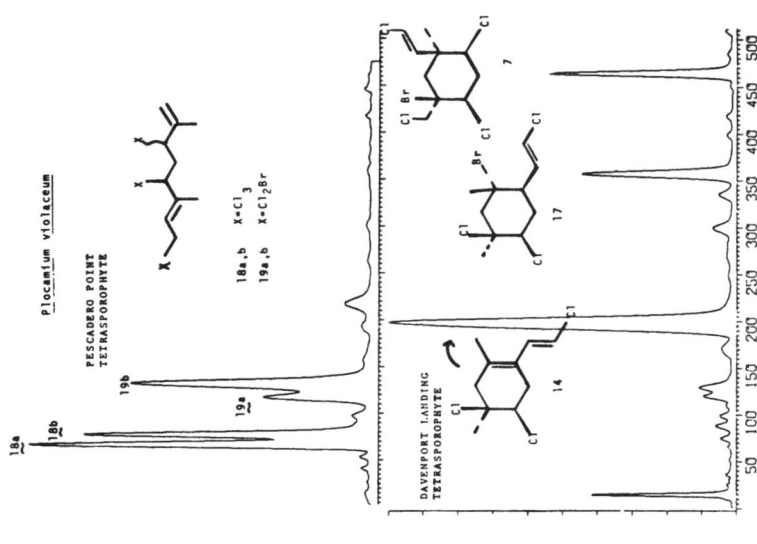

Figure 7. GLC/MS trace of P. violaceum from the northern and southern Monterey Bay areas.

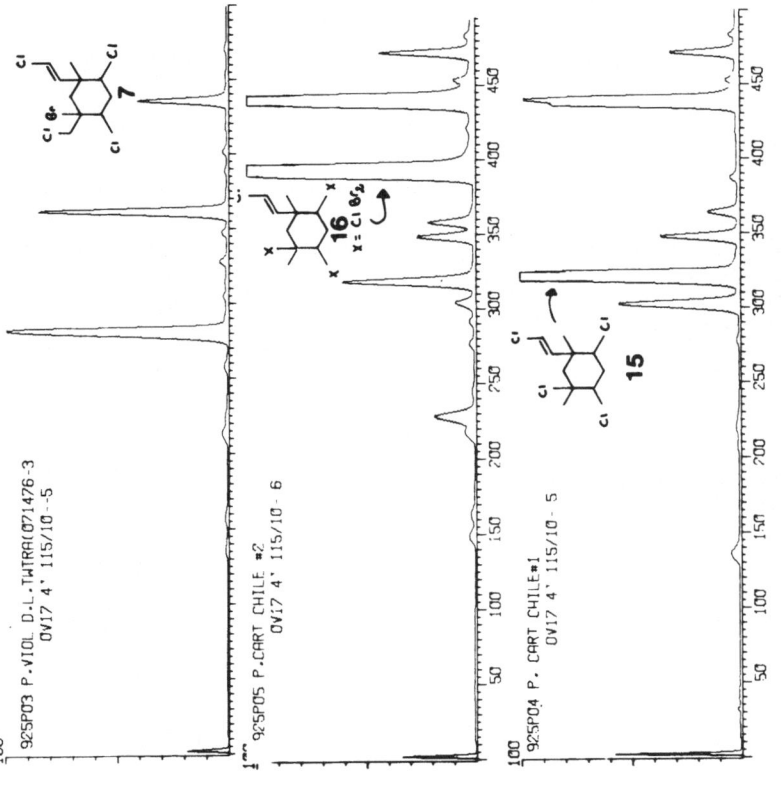

Figure 6. GLC/MS trace of P. violaceum from Davenport Landing, Santa Cruz County, and P. cartilagineum from Montemar, Chile.

Table 1. Halogenated terpene composition of P. violaceum from
northern (n) and southern (s) Monterey Bay regions.

#	Season	Form[*]	(7)	(14)	(17)	(6)	(13)
Davenport Landing[n]							
1	1974, Nov	U	30.0	29.7	26.0	3.8	1.5
2	1976, July	U	26.0	42.1	21.5	5.6	4.8
3	1976, July	T	17.0	59.2	16.4	3.2	4.2
4	1976, July	F	26.3	42.1	17.6	8.7	5.3
5	1976, July	S	17.3	54.3	21.0	3.1	4.2
Pescadero Point[s]			(18a)	(18b)	(19a)	(19b)	
6	1975, Nov	U	30.6	16.9	9.5	43.0	
7	1976, June	T	30.6	25.4	11.9	32.1	

*U=unsorted; T=tetrasporophyte; F=female; S=sterile

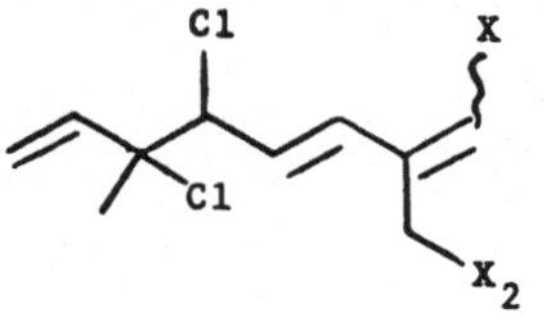

10   X=ClBr$_2$ 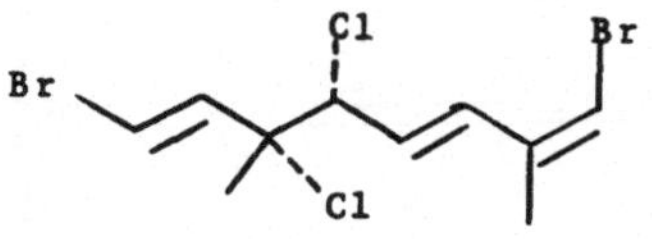   11a

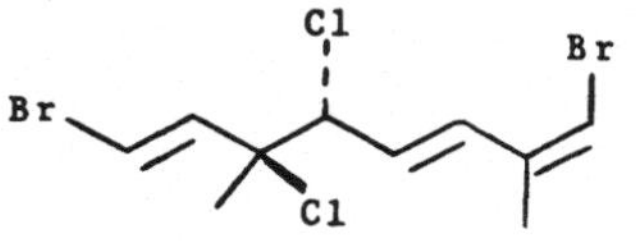

11b 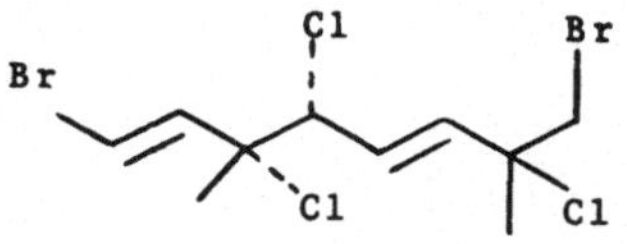   12

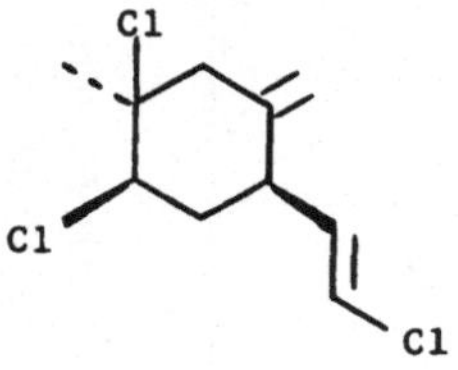

13 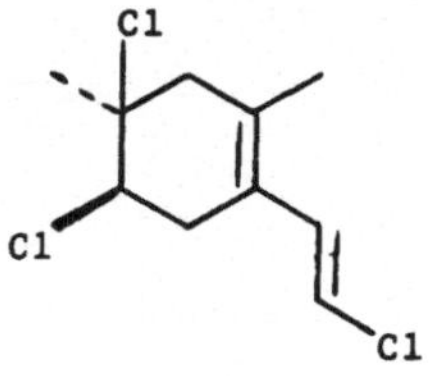   14 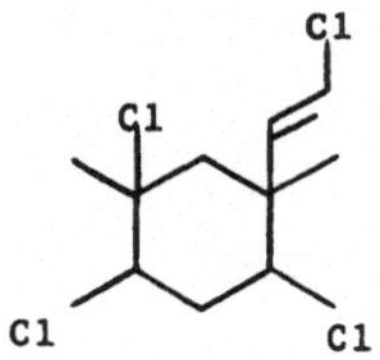   15

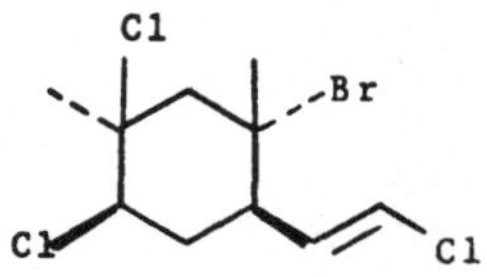

17 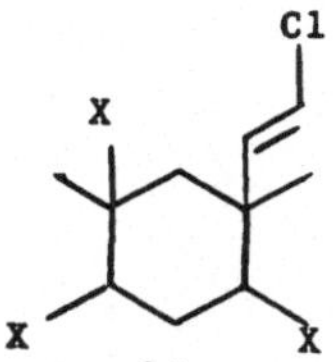   16

The marked difference between the terpene composition
observed in P. violaceum from northern and southern Monterey Bay
locations was intriguing and stimulated the additional experiments
described below. The two simplest but somewhat different rational-
izations that could account for this particular phenomenon are
either the operation of different biosynthetic pathways to the
acyclic and alicyclic families as a function of algal life-history
type or the existence of multiple chemical forms[14] of P. violaceum.
In order to distinguish between these two possibilities, we under-
took a coordinated chemical study of the various life-history stages
of P. violaceum from different locations in the Monterey Bay region.
Table 1 provides some examples of the data collected in this study,
and the full experimental details and discussion will be published
elsewhere[8]. A comparison of Table 1 entries 1-5 for specimens from
Davenport Landing shows that the same sets of major alicyclic components, (7), (14), and (17), are present, although slight
variations were observable in their relative percentages. A
similar consistency in acyclic components can also be observed via
entries 6 and 7 in Table 1. It is apparent, based upon these data,
that the occurrence of alicyclic halogenated terpenes from P.
violaceum in the northern Monterey Bay region is invariant, regard-
less of the life-history form or the season, and an exactly parallel
situation exists in the synthesis of acyclic halogenated terpenes
by P. violaceum in the southern Monterey Bay region.

PHYLETIC AND MORPHOLOGICAL CONSIDERATIONS

An additional input on the breadth of monoterpene synthesis
by red algae comes from our investigations on the red algae
Microcladia (Ceramiaceae)[7]. In part, our attention was drawn to
this genera because of the striking similarity in appearance
between P. cartilagineum and Microcladia coulteri (Harvey), and
between P. violaceum and M. borealis (Ruprecht). In addition, in
the Monterey Bay area each of these respective pairs of algae
co-exists in exactly the same intertidal and/or subtidal region.

Our suspicions of similar terpene chemistry in Plocamium and
Microcladia were confirmed beyond our expectations[7]. GLC examination of the individual extracts from subtidal collections of M.
coulteri and P. cartilagineum from Pebble Beach, Monterey County,
revealed similar profiles of halogenated monoterpenes (Figure 8).
In fact, this phenomenon was not limited to the above case;
further exploration by GLC revealed similar haloterpenes, including
(6), (7), (14), and (17), from M. borealis and P. violaceum from
several northern Monterey Bay sites[15]. Still another demonstration
of this trend is shown for monoterpenes (4), (11), and (12) in
Figure 9, from M. borealis and P. oreganum, which are contiguous
at Partridge Point on Whidbey Island, Washington. Finally, while

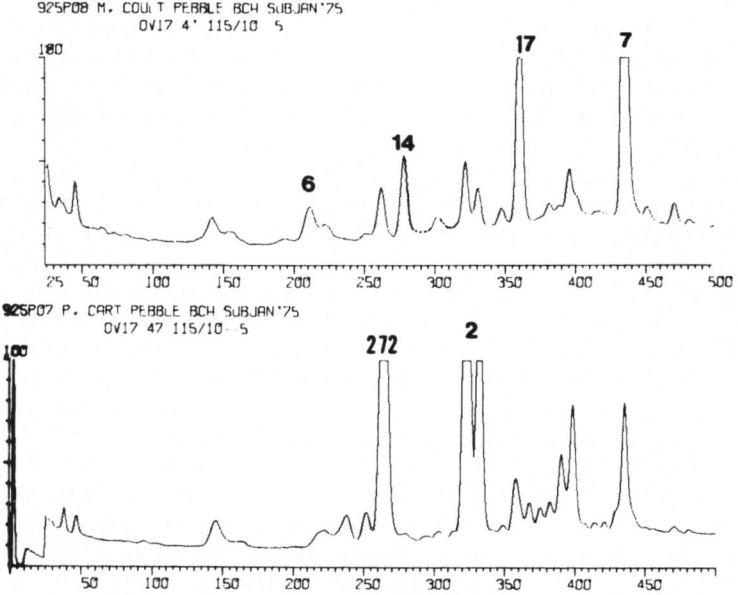

Figure 8.   GLC/MS trace of M. coulteri and P. cartilagineum from
Pebble Beach, Monterey County.

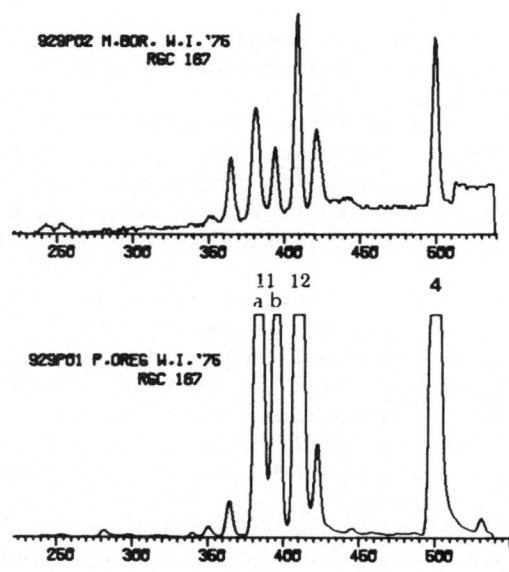

Figure 9.   Specific ion GLC/MS trace at m/e=167 (C_4H_5ClBr) for
M. Borealis and P. Oreganum from Whidbey Island, Washington.

the absolute yields of halogenated monoterpenes from Microcladia
are uniformly much lower than those from Plocamium, the monoterpene
concentrations are sufficiently high to allow their isolation and
structural verification by pmr analysis[7].

DISCUSSION AND CONCLUSIONS

Several of the results mentioned above document the fact that
intraspecies variations of monoterpene components are common for
both P. cartilagineum and P. violaceum. An interesting parallel is
that the major halogenated monoterpenes from Chondrococcus hornemanni
(Rhizophyllidaceae) have also been observed to vary markedly between
different collection sites[16]. On the other hand, little or no
component variation has been either observed or suggested for the
array of halogenated secondary metabolites which have been reported
from the more than one dozen other genera of Rhodophyta which carry
out organic halogenation[17].

There has been prior debate over the taxonomy of P. cartilagineum, owing to the wide variation in the morphological features used
to distinguish the Pacific coast plants from the European specimens[4a].
However, in a recent revision, Dixon has dropped the varietal dis-
tinction previously applied to this species[18]. The observations of
the large chemical variations within P. cartilagineum and P. violaceum reported in this paper suggest that taxonomic uncertainty may
still exist for both species and that a coordinated chemical-
morphological study could be of use. In order to illustrate this
potential uncertainty, we have adopted the designation of chemotype
to express the situation in which apparently morphologically
indistinguishable specimens of P. violaceum can be regularly dif-
ferentiated by chemical analysis. Thus, based upon our observations
(Table 1 and Figure 7), we have designated the northern Monterey
Bay form of P. violaceum as chemotype α and the southern form as
chemotype β.

In spite of the above, our present view of the general utility
of halogenated monoterpenes as a chemosystematic aid at the species
level is quite guarded. In particular, major difficulties are
presented by the large diversity of halogenated monoterpenes from
P. cartilagineum as a function of geography and by the large over-
lap of structural types among the various Plocamium species we have
examined. An additional complication is provided by the similarity
in constituents from Microcladia and Plocamium specimens. The
simplest rationalization of this latter phenomenon is that Microcladia species are able to concentrate the secondary metabolites
which are produced by adjacent Plocamium plants. The strong
parallels in halogenated terpene profiles discussed above and
shown in Figures 8 and 9 provide reasonable support for this
hypothesis. It should be possible to minimize the general impact

of the latter phenomenon on chemosystematics by noting the absolute percentage yields of the secondary metabolite under consideration or by a comparative study of the typé illustrated in Figures 8 and 9.

ACKNOWLEDGEMENTS

I wish to thank the UCSC Committee on Research for support of this research and the NSF Chemical Instrumentation Program for their financial assistance in the purchase of the GC/MS system used in this work. The efforts of my co-workers E. Kho-Wiseman, P. Ng, and L. Campbell are greatly appreciated.

REFERENCES AND NOTES

1.   P. Crews and E. Kho, J. Org. Chem. <u>39</u>, 3303 (1974).

2.   P. Crews and E. Kho, J. Org. Chem. <u>40</u>, 2568 (1975).

3.   P. Crews, E. Kho, and P. Ng, Abstract 178, 30th Northwest Regional A.C.S. Meeting, June, 1975.

4.   (a) G. M. Smith, <u>Marine Algae of the Monterey Peninsula</u>, 2nd edition, Stanford University Press, 1969; (b) R. F. Scagel, <u>A Guide to the Common Seaweeds of British Columbia</u>, British Columbia Provincial Museum Handbook, Victoria, B. C., 1967; (c) E. Y. Dawson, Pacific Sci. <u>15</u>, 370 (1961).

5.   P. Crews and E. Kho-Wiseman, Abstract 122, 31st Northwest A.C.S. Meeting, June, 1976.

6.   P. Crews and E. Kho-Wiseman, J. Org. Chem., submitted.

7.   P. Crews, P. Ng, E. Kho-Wiseman, and C. Pace, Phytochemistry, in press.

8.   P. Crews, L. Campbell, and E. Heron, J. Phycol., submitted.

9.   J. S. Mynderse and D. J. Faulkner, Tetrahedron <u>31</u>, 1963 (1975).

10. (a) P. Crews, Pacific Conference on Chemistry and Spectroscopy, November, 1976; (b) E. Kho-Wiseman and P. Crews, ibid.

11. J. S. Mynderse and D. J. Faulkner, J. Amer. Chem. <u>96</u>, 6771 (1974).

12. J. J. Sims, personal communication.

13. Monoterpene (12) has recently been reported as a constituent of _Aplysia californica_: C. Ireland, M. O. Stallard, D. J. Faulkner, J. Finer, and J. Clardy, J. Org. Chem. __41__, 2461 (1976).

14. A number of terrestrial plants have been observed to have different chemical forms. For a general discussion, see (a) A. Abraham, D. Lavie, and I. Kirson, Israel J. Chem. __14__, 60 (1975); (b) T. J. Mabry, Pure and Appl. Chem. __34__, 377 (1973).

15. Compare Figures 1 and 3 in reference 7.

16. (a) B. J. Burreson, F. X. Woolard, and R. E. Moore, Chem. Lett., 1111 (1975) and Tetrahedron Lett., 2155 (1975); (b) N. Ichikawa, Y. Naya, and S. Enomoto, Chem. Lett., 1333 (1974).

17. (a) W. Fenical and J. N. Norris, J. Phycol. __11__, 104 (1975); (b) W. Fenical, J. Phycol. __11__, 245 (1975) and references therein.

18. P. S. Dixon, Blumea __15__, 55 (1967).

STUDIES ON THE ABSOLUTE CONFIGURATION OF PERIDININ AND DINOXANTHIN[*]

Jon Eigill Johansen and Synnøve Liaaen-Jensen

Organic Chemistry Laboratories, Norwegian Institute of

Technology, University of Trondheim, Trondheim, Norway

Gunner Borch

Chemistry Department A, The Technical University of

Denmark, Lyngby, DK-2800, Denmark

The structure of peridinin (1), the characteristic major carotenoid of dinoflagellates, has recently been elucidated[1-3]. It is
a unique C_{37}-skeletal nor-carotenoid, formally lacking three carbon
atoms in the centre of the molecule (Scheme 1) and carrying butenolide, allene, epoxy, acetoxy and hydroxy functions.

Plausible biosynthetic routes from a traditional C_{40}-carotenoid
precursor have been envisaged[1,4], involving oxidation of the C-9
methyl to carboxyl, followed by cyclization _via_ either an acetylenic
bond or a double bond followed by re-oxidation. An enzymatically
controlled, photochemically allowed electrocyclic reaction could
account for elimination of the C_3-unit as an acetylenic derivative,
as shown in Scheme 1.

The search for likely precursors among other carotenoids of
five selected dinoflagellates[5] revealed the presence of traditional
C_{40}-carotenoids such as β,β-carotene[6] (2), astaxanthin (3, minor,
occasional), the acetylenic diatoxanthin (4), diadinoxanthin (5),
the allenic dinoxanthin (6), and the C_{37}-skeletal carotenoids
peridininol (7), pyrrhoxanthin (8), and pyrrhoxanthinol (9). Of

---

[*]No. XXI in the series Algal Carotenoids; No. XX Phytochemistry,
in press

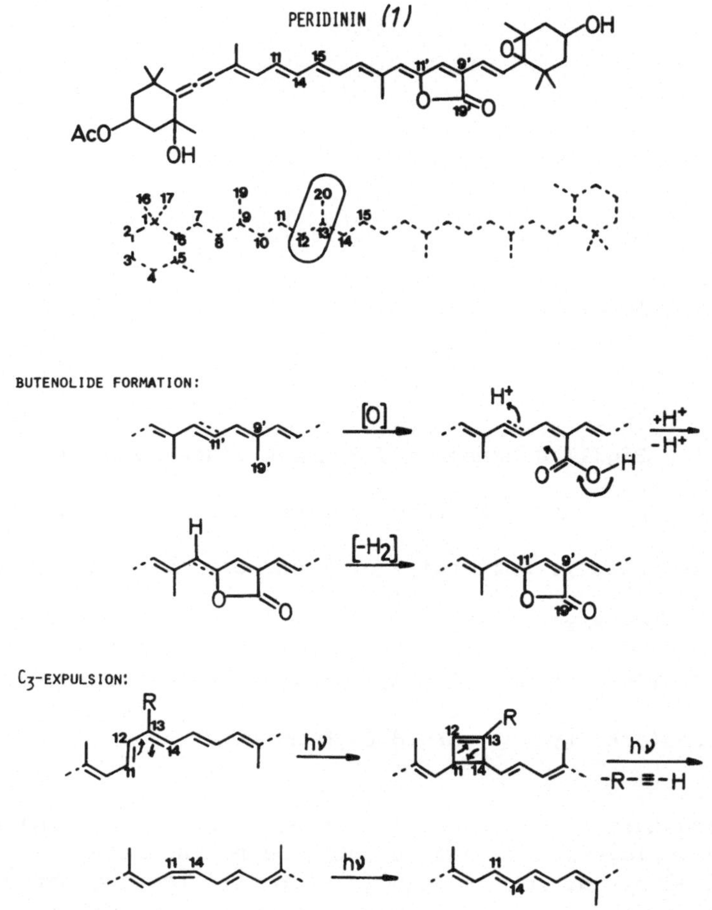
PERIDININ (1)
AcO
OH
OH
BUTENOLIDE FORMATION:
[O]
+H+
-H+
[-H2]
C3-EXPULSION:
R
hν
hν
-R-≡-H
hν

Scheme 1. Postulated biosynthesis of peridinin[1,4]

these carotenoids, dinoxanthin (6) has two end groups in common with peridinin (1) and is a likely precursor candidate.

In our previous work[1-5], the stereochemistry of peridinin (1) was not considered. Determination of the absolute configuration of peridinin with 6 chiral centres was of interest in itself and in a biosynthetic context.

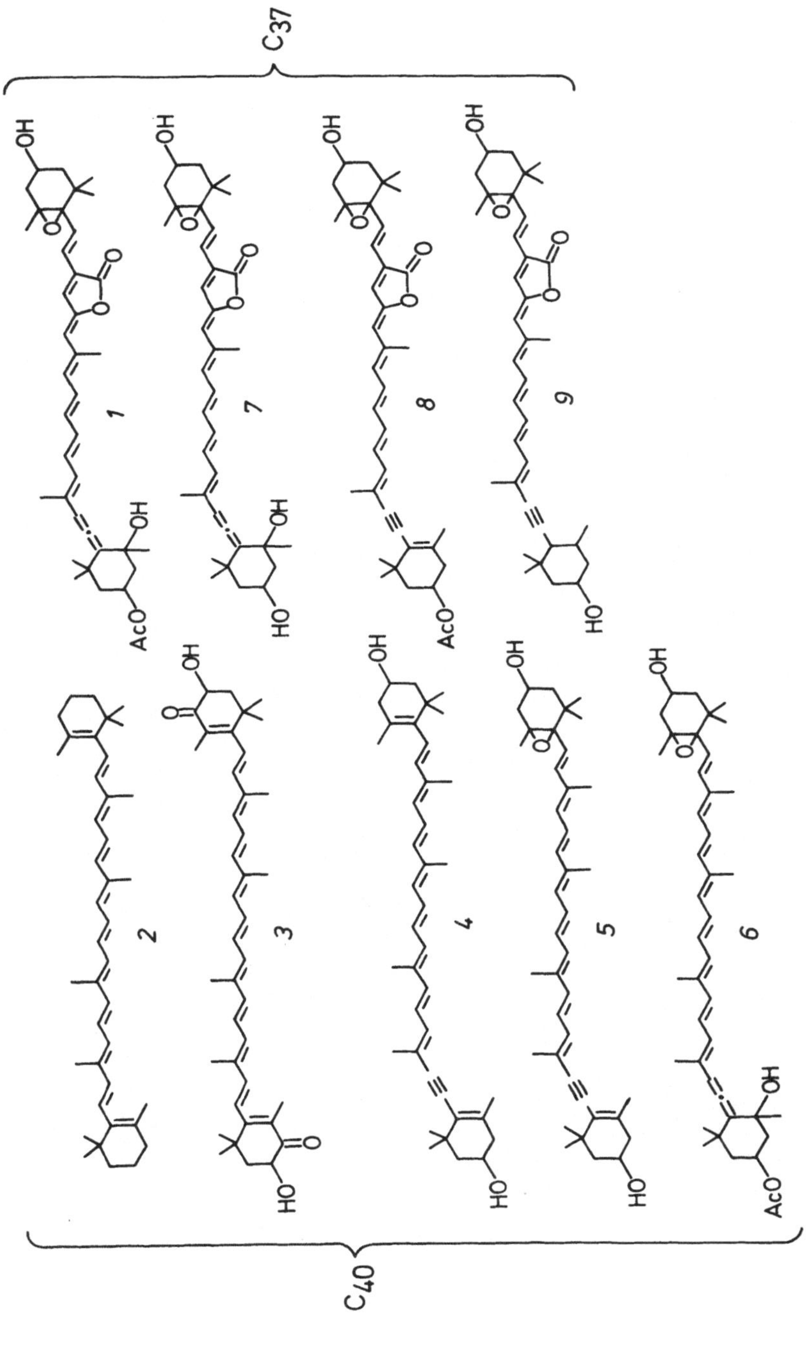

Scheme 2. Carotenoids of dinoflagellates[5].

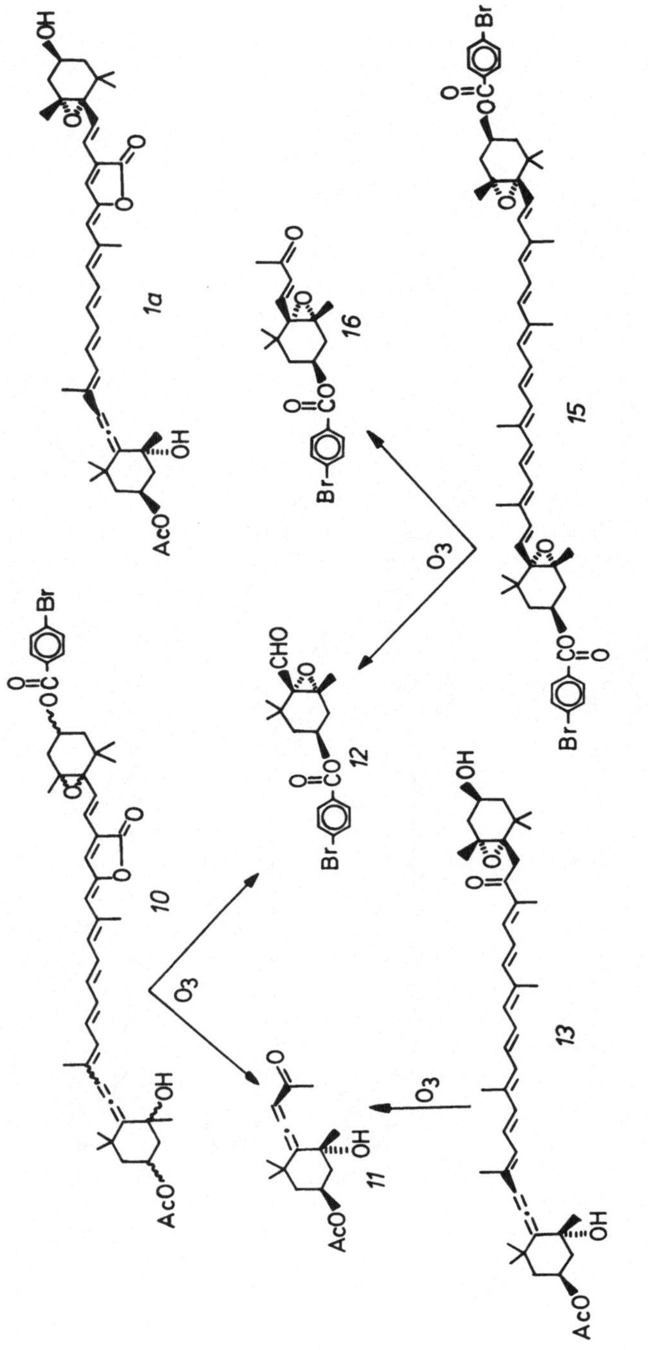

Scheme 3. Ozonolysis of peridinin p-bromobenzoate (10), fucoxanthin (13), and violaxanthin di-p-bromobenzoate (15).

The following approaches were considered:

A.   X-ray analysis of peridinin p-bromobenzoate (10).

B.   ^{1}H NMR and ^{13}C NMR correlations with stereochemically defined
      models for determination of relative configuration.

C.   CD correlations based on intact carotenoids.

D.   Oxidative degradation of peridinin p-bromobenzoate (1) for [1]H
      NMR and CD correlation with authentic derivatives of caroten-
      oids with similar end groups of known chirality.

E.   Establishment of the absolute configuration of dinoxanthin (6)
      for evaluation of a possible biogenetic relationship to
      peridinin (1).

So far, perfect crystals of peridinin p-bromobenzoate (10) for
X-ray have not been obtained. It should be commented that, to date,
few successful X-ray analyses of carotenoids have been achieved.
^{1}H NMR correlations will be discussed. ^{13}C NMR data are available
but must await future interpretation.

CD correlations based on intact carotenoids were ruled out
because of the lack of relevant, related carotenoids and the weak
Cotton effect of intact peridinin.

Consequently, oxidative degradation was required, and an
ozonolysis of peridinin p-bromobenzoate (10)[2] was performed (Scheme
3). The desired ketone (11) and aldehyde (12), together containing
all the chiral centres, could also theoretically be obtained by
oxidative degradation of fucoxanthin (13)[9] and violaxanthin (14a)
as the di-p-bromobenzoate (15), respectively. Both (13) and (14)
have known chirality[7,8].

Table 1 gives ^{1}H NMR data for peridinin (1) and relevant carotenoid models. Fucoxanthin (13) and neoxanthin (17), of known
stereochemistry[8,10], have the same allenic end group as peridinin
(1). Correspondence between signals of the lateral methyl groups,
acetoxy and allenic protons associated with this end group[2,9,10],
supports assignment of the same relative stereochemistry of the
allenic end group in (1),(13) and (17). This is also true for
vaucheriaxanthin (18)[11], which has unestablished chirality. Regarding the relative configuration of the epoxidic end group, comparison
of the chemical shifts of the ring methyl groups of peridinin (1)
with other carotenoids containing this end group is less conclusive.
Fucoxanthin (13), neoxanthin (17), and natural violaxanthin (14a)
all have 3,5(3',5')-trans configuration. For comparison, semi-
synthetic violaxanthin (14b), with 3,5(3',5')-cis configuration
was prepared by epoxidation of (3R,3'R)-zeaxanthin diacetate with

Table 1. ^{1}H NMR signals (δ) of peridinin (1) and relevant models in CDCl$_3$

Compound	Allenic end group							Epoxidic end group					
	Me-1	Me-5	Me-9	Me-13	H-3	Ac	H-8	Me-1	Me-5	Me-9	Me-13	H-3	H-7
Peridinin (1)[2]	1.07, 1.35	1.37	1.81	--		2.01	6.05	0.97, 1.20	1.20	--	2.20		
Fucoxanthin (13)[2,9]	1.07, 1.33	1.37	1.79	1.96		2.00	6.05	0.97, 1.20	1.20	1.92	1.96		
Neoxanthin (17)[10]	1.06, 1.33	1.33	1.79	1.94		--	?	0.96, 1.14	1.17	1.91	1.94		
Vaucheriaxanthin (18)[11]	1.04, 1.35	1.35	1.78	1.96		--	6.00	0.98, 1.16	1.21	--	1.96		
Nat. violaxanthin (14a)[8,12]								0.95, 1.12	1.16	1.89	1.92		
Semisynth. violax. (14b)[8]								1.00, 1.14	1.17	1.90	1.93		
Present Work													
Ketone 11 from 1a[13]	1.15, 1.45	1.43	2.18	--	5.37	2.03	5.85	1.01, 1.15	1.19	1.92	1.96		
from 13[13]	1.15, 1.42	1.42	2.17	--	5.39	2.03	5.84						
synth. racem.[13]	1.16, 1.43	1.43	2.18	--	5.39	2.03	5.84						
Aldehyde 12 from 1a								1.16, 1.35	1.46	--	--	5.16	9.77
from 14a								1.16, 1.35	1.46	--	--	5.16	9.77
Ketone 16 from 14a								1.04, 1.23	1.27	2.29	--	5.16	7.06*

*H-8 δ 6.31 $\underline{J}_{7,8}$ = 16 Hz

m-chloroperbenzoic acid, followed by alkaline hydrolysis[13]. In
the 3,5-cis case (14b), the chemical shift of one of the gem
dimethyl groups differs markedly from that of the 3,5-trans compounds (13, 17, 14a) (cf. previous data for 14b[8]). However, the
stereochemistry of peridinin (1) cannot be safely assigned from
these data.

Ozonolysis of peridinin p-bromobenzoate (10) (Scheme 3) provided several products, the eight least polar of which were
investigated. Seven of these products contained the benzoate
function. The acetylated allenic ketone 11 was identical in all
respects to the same allenic ketone obtained by ozonolysis of fucoxanthin (13)[2, 9, 13b]. This includes ^{1}H NMR data given in Table 1
for (11) from peridinin (1) and from fucoxanthin (13)[14], including
racemic 11[14], as well as CD data (Figure 1), mass spectra, and
electronic spectra[9, 13b]. The absolute configuration of the three
chiral centres in the allenic end group of peridinin (1a, Scheme 3)
is thereby proved.

The second key ozonolysis product of peridinin p-bromobenzoate
(10) was the previously undescribed epoxidic aldehyde (12, Scheme 3),
characterized here by ^{1}H NMR, CD, and electronic and mass spectra.
Ozonolysis of violaxanthin di-p-bromobenzoate (15), of known chirality, gave the same product (12), together with its vinylogue (16)
and three other products.

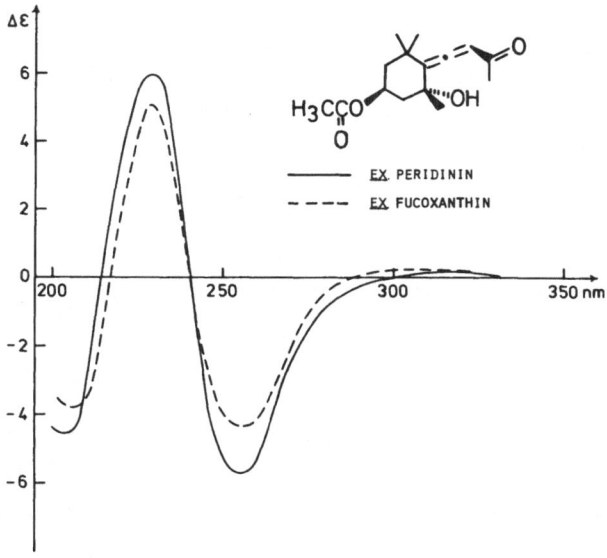

Figure 1. CD spectra in EPA solution (diethyl ether, isopentane,
ethanol 5:5:2) of the allenic ketone (11) ⎯⎯ from peridinin (1a)
and --- from fucoxanthin (13).

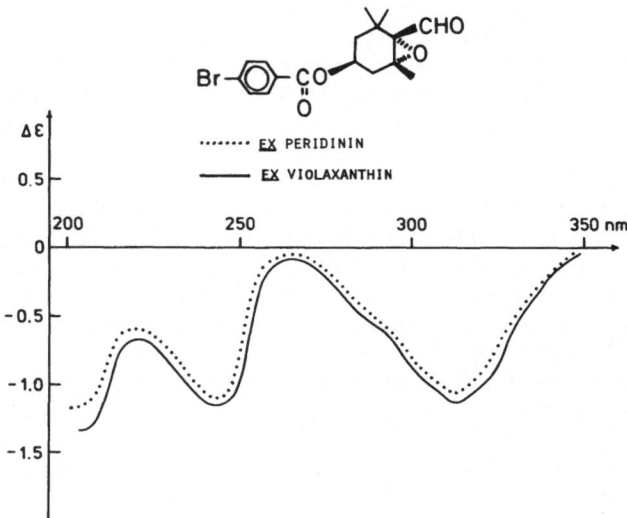

Figure 2.   CD spectra in EPA solution of the epoxy-aldehyde (12)
··· from peridinin (1a) and —— from violaxanthin di-p̲-bromo-
benzoate (15).

Again, ^{1}H NMR data compiled in Table 1 for product (12) from
peridinin p̲-bromobenzoate (10) and from violaxanthin di-p̲-bromo-
benzoate (15) support the same relative stereochemistry for the
epoxy end group in peridinin (1) and violaxanthin (14a).   The
3,5-c̲i̲s̲ model is not available.   However, considering the differences in chemical shifts of natural (14a) and semi-synthetic
violaxanthin (14b) (Table 1), the 3,5-c̲i̲s̲ relationship for product
(12) from peridinin is disregarded.   CD data (Figure 2) prove the
same absolute configuration of the epoxidic aldehyde (12) from
peridinin (1') and from violaxanthin (14a).

In conclusion, the absolute configuration of peridinin (1a)
is as given in Scheme 3, that is, (3S̲,5R̲,6R̲,3'S̲,5'R̲,6'S̲)-5',6'-
epoxy-3,5,3'-trihydroxy-6,7-didehydro-5,6,5',6'-tetrahydro-12,13,
20-trinor-β,β-caroten-19',11'-olide 3-acetate.   The allenic end
groups have the same chirality as all other carotenoids containing
this end group, including fucoxanthin (13) and neoxanthin (17),
and the epoxidic end group has the same chirality as in all other
3-hydroxy-5,6-epoxy-carotenoids.   These statements are, of course,
restricted to carotenoids having established stereochemistry.

The absolute configuration of dinoxanthin (6), the gross
structure of which was recently shown to be neoxanthin 3-acetate,
has now been studied.

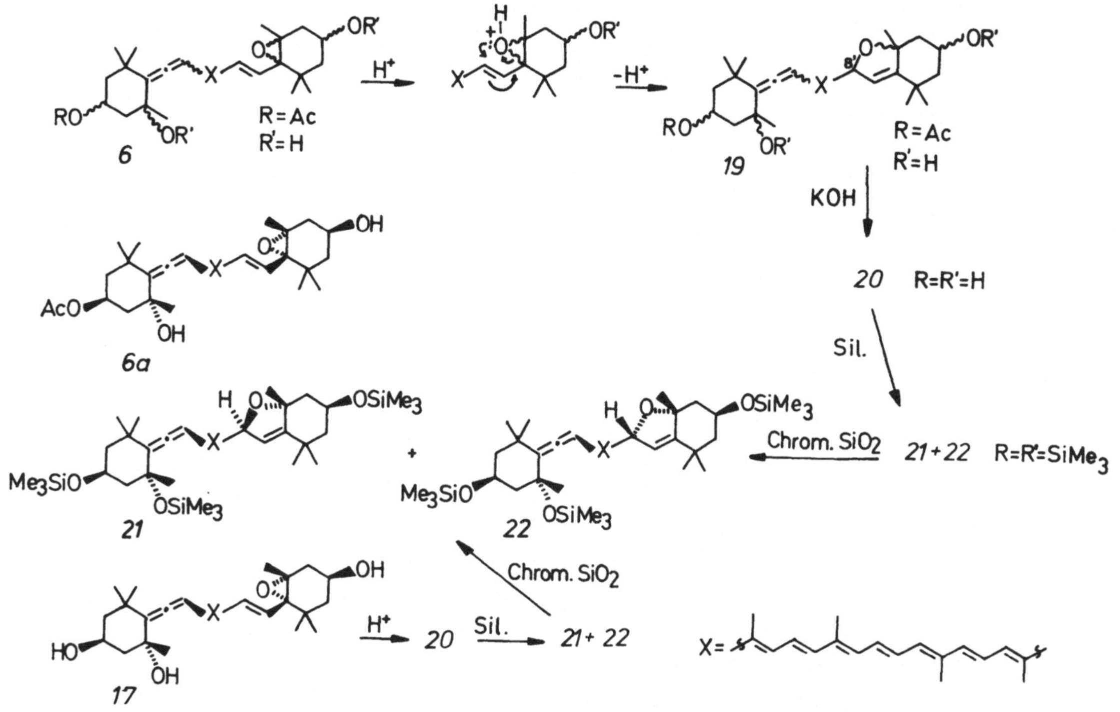

Scheme 4. Reactions relating dinoxanthin (6) and neoxanthin (17).

Amphidinium carterae, left from our previous study[5], contained
the C-8 epimeric dinochromes ((19),Scheme 4) presumably formed by
acid-catalyzed epoxide-furanoxide rearrangement of natural dinoxanthin (6).   The dinochromes (19) were hydrolyzed with alkali to
the triol (20), which, on silylation, provided the tri(trimethyl)-
silyl ether (21) (Scheme 4).   These reactions are known to retain
the stereochemistry at C-3,5,8,3',5'[15].   Chromatography on silica
gave the individual C-8 epimers (22) and (23).

Neoxanthin (17) of known chirality[10], isolated from maple
leaves, was converted to the silylated furanoxides (22) and (23)
via the neochromes (24) by the same sequence, and the two silylated
C-8' epimeric neochromes were separated.

The R_f-values of the silylated furanoxides (22) and (23) from
dinoxanthin (6) and from neoxanthin (17) were identical.   Which is
which, in terms of structure, has not been established[15], but this
is irrelevant for the present argument.   Furthermore, CD data
(Figure 3) for the all-trans isomers of (22) and (23) from the two
sources agree so well that the identical absolute configuration of
neoxanthin (17) and dinoxanthin ((6a), Scheme 4), (3S,5R,6R,3'S,
5'R, 6'S)-5',6'-epoxy-6,7-didehydro-5,6,5',6'-tetrahydro-β,β-
carotene-3,5,3'-triol 3-acetate is tentatively assigned.   ^{1}H NMR

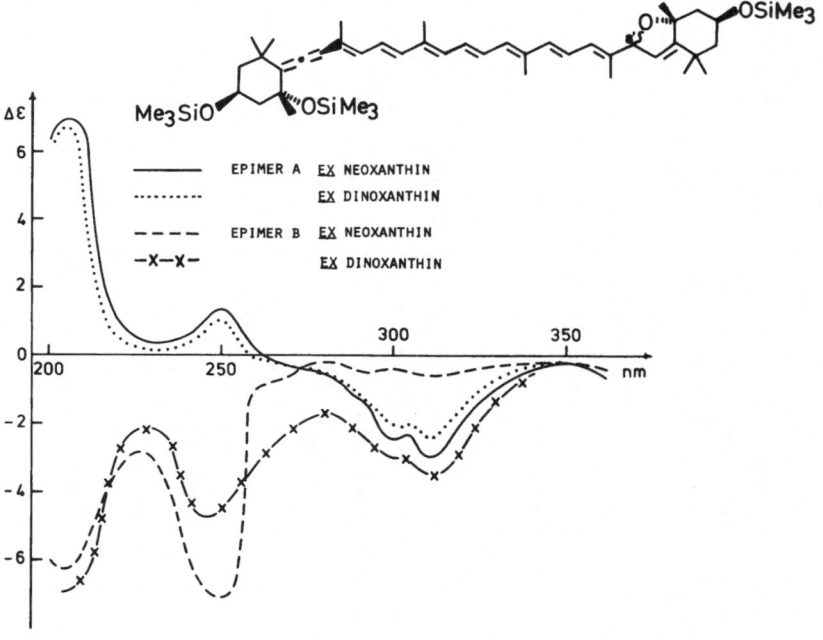

Figure 3.   CD spectra in EPA solution of the C-8' epimeric neochrome
tri(trimethyl)silyl ethers (22) and (23); epimer A (least polar) ——
from neoxanthin (17), --- from dinoxanthin (6) and epimer B (most
polar) --- from dinoxanthin (17), -x-x- from dinoxanthin (6).

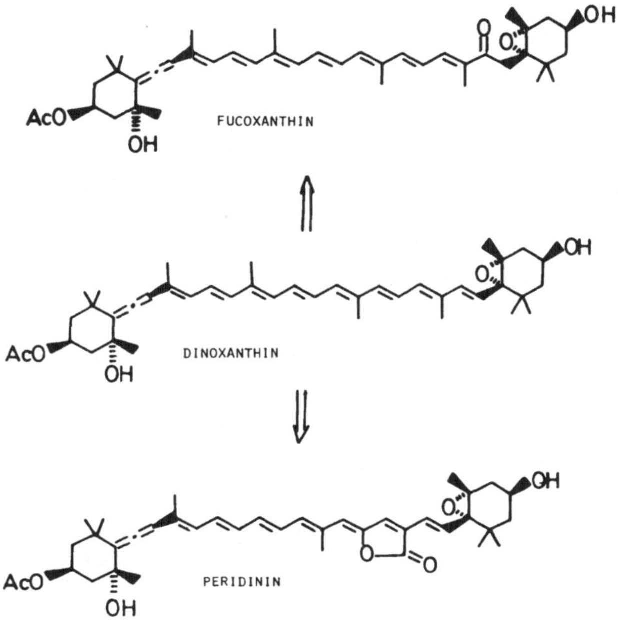

Scheme 5. Hypothetic biogenetic relation between dinoxanthin (6a), peridinin (1a) and fucoxanthin (13).

data for neoxanthin (17) and its furanoid derivatives are available[7,10,15,18], but such data for dinoxanthin (6), a minor constituent in dinoflagellates, are not. Relative configurations therefore cannot be checked by ^{1}H NMR. Not all chiral centres in the silylated neochromes (22) and (23) necessarily influence the Cotton effect, but diastereomers are not expected to be chromatographically identical.

The established chirality of peridinin (1a) and the probable stereochemistry of dinoxanthin (6a) support a close biogenetic relationship (Scheme 5).

In some dinoflagellates, peridinin (1a) is replaced by fucoxanthin (13)[16]. Some of the fucoxanthin-producing dinoflagellates have two nuclei, ascribed to an endosymbiont of the Crysophyta. However, other fucoxanthin-producing dinoflagellates have no second nucleus[17]. Dodge[17] has advanced the hypothesis that dinoflagellates originally were heterotrophic organisms and that the fucoxanthin-containing species represent an early stage of the photosynthetic apparatus, prior to the development of peridinin.

Consideration of the chemical structures of fucoxanthin (13) and peridinin (1a), stereochemistry included, supports a close biosynthetic relationship between these carotenoids. However, a branched scheme with a common precursor such as dinoxanthin (6a) seems more likely than a linear route (Scheme 5). However, so far, the co-occurrence of dinoxanthin (6a) and fucoxanthin (13) has not been unequivocally established[16,18].

Experimental details of this work will be available elsewhere[19].

ACKNOWLEDGEMENTS

The peridinin used was provided by Dr. H. H. Strain, formerly at the Chemistry Division, Argonne National Laboratory, Argonne, USA. Violaxanthin was a generous gift from Dr. J. Szabolcs, Chemical Institute of the Medical University, Pecs, Hungary.

This work was supported by grants from The Norwegian Research Council of Science and the Humanities and from Hoffman-La Roche, Basel, to S.L.J.

REFERENCES

1. H. H. Strain, W. A. Svec, K. Aitzetmüller, M. C. Grandolfo, J. J. Katz, H. Kjøsen, S. Norgård, S. Liaaen-Jensen, F. T. Haxo, P. Wegfahrt, and H. Rapoport, J. Am. Chem. Soc. 93, 1823 (1971).

2. H. H. Strain, W. A. Svec, P. Wegfahrt, H. Rapoport, F. T. Haxo, S. Norgård, H. Kjøsen, and S. Liaaen-Jensen, Acta Chem. Scand. B 30, 109 (1976).

3. H. Kjøsen, S. Norgård, S. Liaaen-Jensen, W. A. Svec, H. H. Strain, P. Wegfahrt, H. Rapoport, and F. T. Haxo, Acta Chem. Scand. B 30, 157 (1976).

4. H. Kjøsen, Thesis, Synthetic and Spectroscopic Studies on Carotenoids, Norw. Inst. Technology, University of Trondheim, 1972.

5. J. E. Johansen, W. A. Svec, S. Liaaen-Jensen, and F. T. Haxo, Phytochem. 13, 2661 (1974).

6. IUPAC and IUB, Nomenclature of Carotenoids (Rules approved 1974), Butterworths, London, 1976.

7. K. Bernhardt, G. P. Moss, G. Tóth, and B. C. L. Weedon, Tetrahedron Lett., 115 (1976).

8. L. Bartlett, W. Klyne, W. P. Mose, P. M. Scopes, G. Galasko, A. K. Mallams, B. C. L. Weedon, J. Szabolcs, and G. Tóth, J. Chem. Soc. C, 2527 (1969).

9. R. Bonnett, A. K. Mallams, A. A. Spark, J. L. Tee, B. C. L. Weedon, and A. McCormick, J. Chem. Soc. C, 469 (1969).

10. L. Cholnoky, K. Györgyfy, A. Rónai, J. Szabolcs, G. Tóth, G. Galasko, A. K. Mallams, E. S. Waight, and B. C. L. Weedon, J. Chem. Soc. C, 1256 (1969).

11. H. Nitsche, Z. Naturforsch. 28c, 641 (1973).

12. B. C. L. Weedon, Fortschr. Chem. Org. Naturstoffe 27, 81 (1969).

13. H. Cadosch and C. H. Eugster, Helv. Chim. Acta 57, 1466 (1974).

13b. S. Hertzberg, T. Mortensen, G. Borch, H. W. Siegelman, and S. Liaaen-Jensen, Phytochem, to be published.

14. J. R. Hlubucek, J. Hora, S. W. Russell, T. P. Toube, and B. C. L. Weedon, J. Chem. Soc. C, 848 (1974).

15. D. Goodfellow, G. P. Moss, J. Szabolcs, G. Tóth, and B. C. L. Weedon, Tetrahedron Lett., 3925 (1973).

16. S. W. Jeffrey, M. Sielicki, and F. T. Haxo, J. Phycol. 11, 374 (1975).

17. J. D. Dodge, Phycologia 14, 253 (1975).

18. J. P. Riley and T. R. S. Wilson, J. Mar. Biol. Ass. U.K. 45, 583 (1965).

19. J. E. Johansen, Thesis, Chemical Studies on Selected Algal and Bacterial Carotenoids, Norw. Inst. Technology, University of Trondheim, 1977.

ALGAL CAROTENOIDS AND CHEMOSYSTEMATICS

Synnøve Liaaen-Jensen

Organic Chemistry Laboratories, Norwegian Institute of

Technology, University of Trondheim, Trondheim-NTH, Norway

May I first remind you that carotenoids are usually yellow-red, isoprenoid, polyene pigments, widely distributed in nature, where they serve important functions such as protection against photo-dynamic damage and auxiliary light absorbtion for photosynthesis and phototaxis[1-3].

In the marine environment we meet the highest structural diversity of carotenoid pigments. Carotenoids are synthesized de novo by all photosynthetic organisms and by certain non-photosynthetic bacteria and fungi[4]. Many higher organisms have the ability to structurally modify the carotenoids taken in through the diet.

Figure 1 gives a few selected examples of carotenoids encountered in the marine environment: β-carotene (5), or rather β,β-carotene by the new IUPAC nomenclature[1], is a common bicyclic C_{40}-carotene present in all algae. Its skeleton is formally made up of eight head-to-tail linked isoprenoid units, tail-to-tail linked in the centre of the molecule. Bacterioruberin (59) is a C_{50}-tetrol, characteristic of halophilic bacteria[5]. Here, two extra isopentenyl units are attached to 2,2' positions. Actinioerythrin (60), found in sea anemones, is an esterified 2,2'-bisnor C_{38}-carotenoid α-ketol where two carbon atoms from the rings are formally removed. Peridinin (23) is a unique C_{37} skeleton carotenoid restricted to dinoflagellates[7,8]. Here, three carbon atoms from the polyene chain are formally expelled, and several functional groups including butenolide, epoxy, acetoxy, hydroxy and allenic functions are present.

From functional aspects, wide distribution and structural diversity, the potential use of carotenoids as chemosystematic

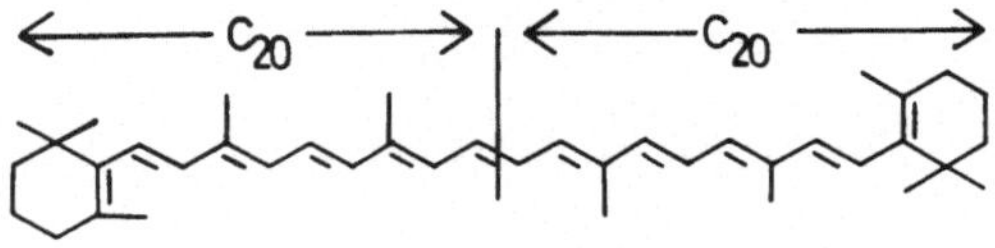

β,β-CAROTENE (5)

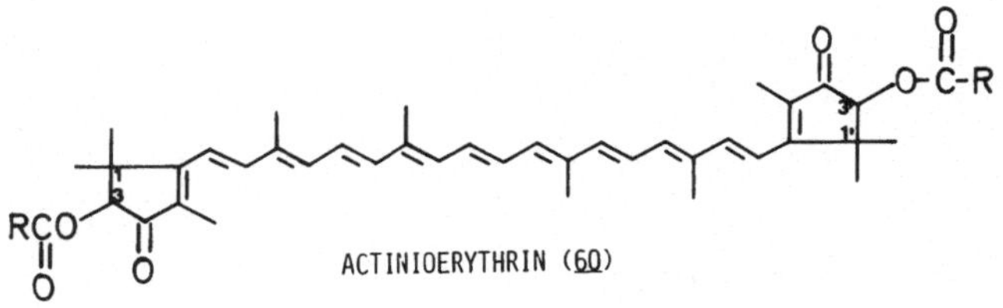

ACTINIOERYTHRIN (60)

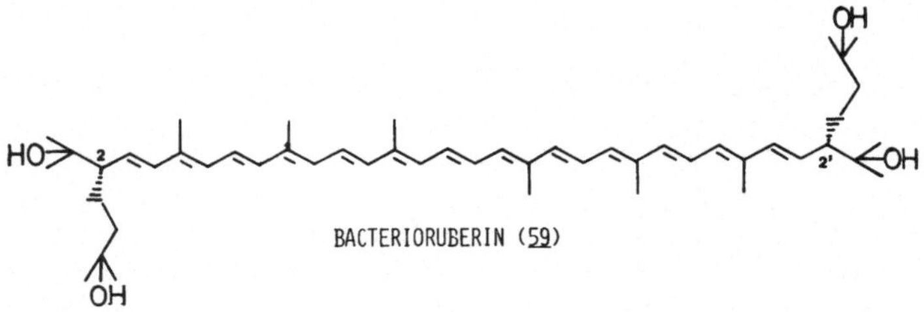

BACTERIORUBERIN (59)

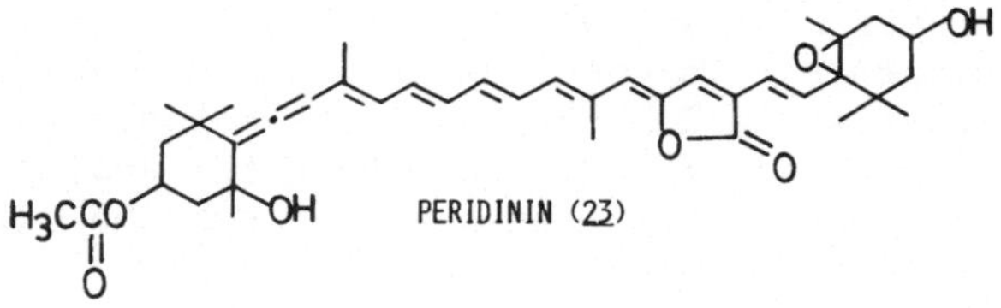

PERIDININ (23)

Figure 1.   Examples of marine carotenoids

markers seems obvious. Together with morphological, physiological, and biochemical criteria, chemosystematics may serve classifica- tion[9]. Consideration of biosynthetic pathways, in addition to ultimate products, is a still better criterion.

Available evidence suggests that in carotenogenesis the path- way from mevalonic acid <u>via</u> isopentenyl pyrophosphate to C_{10}, C_{15}, C_{20}, and C_{40} units is a general one. This seems also to be the case for the sequential dehydrogenation of the colourless triene

·Figure 2. Use of chirality in biosynthetic considerations

C_{40} precursor phytoene (1) to coloured aliphatic carotenes. Less
evidence is available concerning the terminal steps in carotenoid
synthesis involving cyclization, introduction of oxygen functions,
etc.[4]

If biosynthetic data are lacking, consideration of absolute
configuration may provide indirect biosynthetic information. For
example, the proposed conversion of lutein to astaxanthin in gold-
fish[10] was cast into doubt by the chirality of lutein (57) and
astaxanthin (46), which were subsequently established[11-13] (Figure 2).
An epimerization at C-3', not included in the route suggested,
would be required.

The following presentation will be restricted to algal carotenoids, with some emphasis on our own work. A consideration of
the carotenoids encountered in fourteen algal classes, following
the classification of Christensen[14] (Table 1), will be made. The
most primitive algae are shown on the top and the more highly
developed ones at the bottom.

Many workers, including those in our group in Trondheim, have
been engaged in studies of algal carotenoids. Around sixty different carotenoids, mostly having well established structures, are
encountered within these algal classes. From three to twenty-one
different carotenoids are isolated from each class, with an average
of ca. eleven carotenoids per class. Space will not permit a
detailed treatment of each class. Hopefully a more generalized
consideration will be equally beneficial for the non-carotenoid
specialist. From their characteristic structural features, algal
carotenoids may be grouped into carotenes (hydrocarbons) and
acetylenic, allenic, glycosidic, epoxidic, ketonic and hydroxylic
carotenoids. Two or more of these functions may be present in a
single carotenoid.

Table 1. Classification of algae, mainly according to Christensen[14]

	Division	Class	Subclass	No. Orders
Procaryotes	Cyanophyta	Cyanophyceae		5
Eucaryotes	Rhodophyta	Rhodophyceae	Bangiophyceae	5
			Florideophycideae	6
	Chromophyta	Cryptophyceae		2
		Dinophyceae		6
		Rhapidophyceae		1
		Chrysophyceae		11
		Haptophyceae		2
		Bacillariophyceae		2
		Xanthophyceae		5
		Eustigmatophyceae		1
		Phaeophyceae		10
	Chlorophyta	Euglenophyceae		2
		Loxophyceae		1
		Prasinophyceae		1
		Chlorophyceae		11

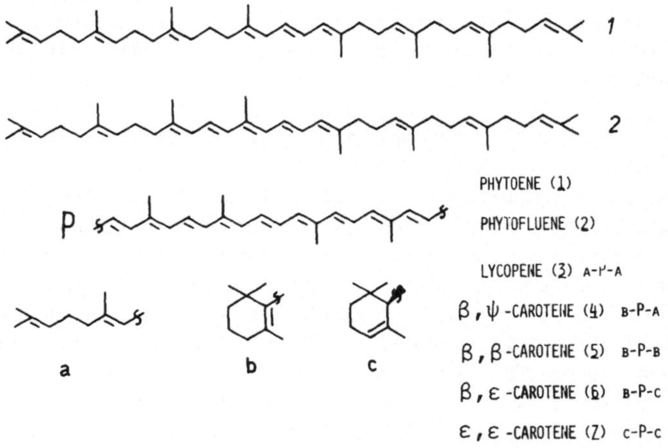

PHYTOENE (1)
PHYTOFLUENE (2)
LYCOPENE (3) A-P-A
β,ψ-CAROTENE (4) B-P-A
β,β-CAROTENE (5) B-P-B
β,ε-CAROTENE (6) B-P-C
ε,ε-CAROTENE (7) C-P-C

Figure 3. Algal carotenes (1 - 7)

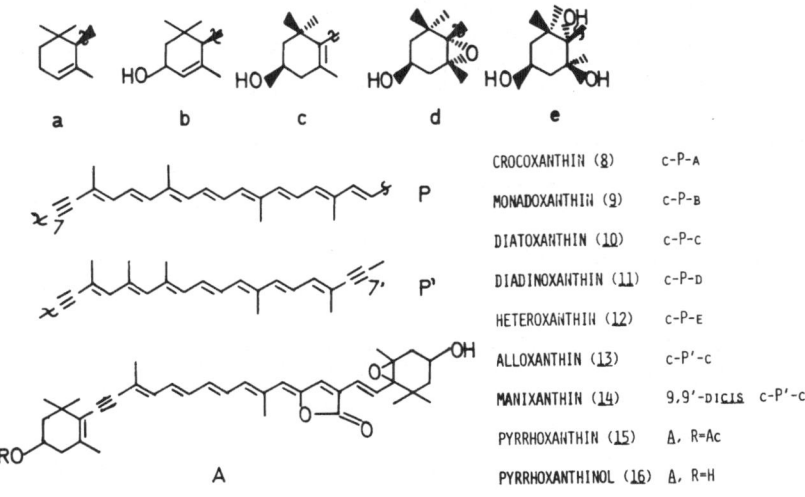

CROCOXANTHIN (8)	c-P-a
MONADOXANTHIN (9)	c-P-b
DIATOXANTHIN (10)	c-P-c
DIADINOXANTHIN (11)	c-P-d
HETEROXANTHIN (12)	c-P-e
ALLOXANTHIN (13)	c-P'-c
MANIXANTHIN (14)	9,9'-DICIS c-P'-c
PYRRHOXANTHIN (15)	A, R=Ac
PYRRHOXANTHINOL (16)	A, R=H

Figure 4. Acetylenic algal carotenoids (8 - 16)

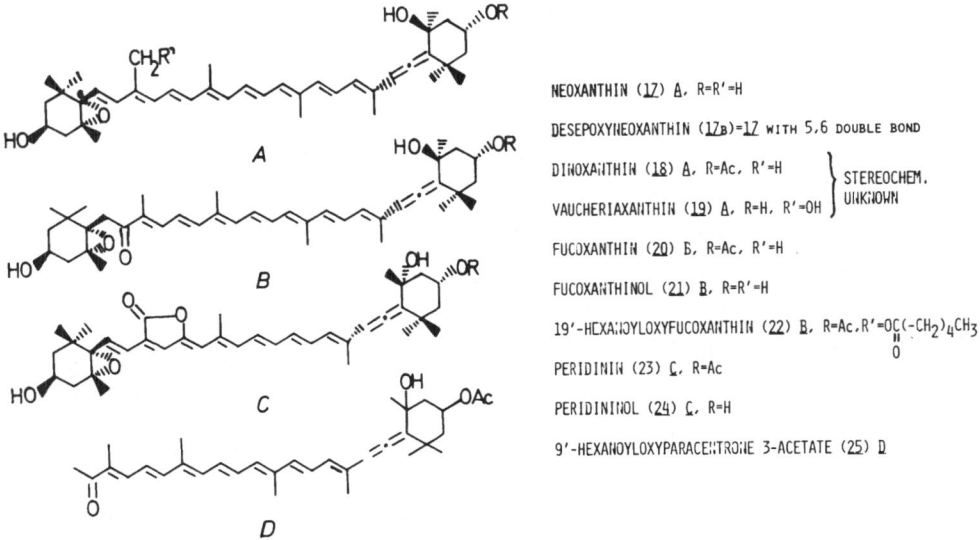

NEOXANTHIN (17) A, R=R'=H

DESEPOXYNEOXANTHIN (17B)=17 WITH 5,6 DOUBLE BOND

DINOXANTHIN (18) A, R=Ac, R'=H } STEREOCHEM.

VAUCHERIAXANTHIN (19) A, R=H, R'=OH } UNKNOWN

FUCOXANTHIN (20) B, R=Ac, R'=H

FUCOXANTHINOL (21) B, R=R'=H

19'-HEXANOYLOXYFUCOXANTHIN (22) B, R=Ac,R'=OC(-CH$_2$)$_4$CH$_3$

PERIDININ (23) C, R=Ac

PERIDININOL (24) C, R=H

9'-HEXANOYLOXYPARACENTRONE 3-ACETATE (25) D

Figure 5. Allenic algal carotenoids (17 - 25)

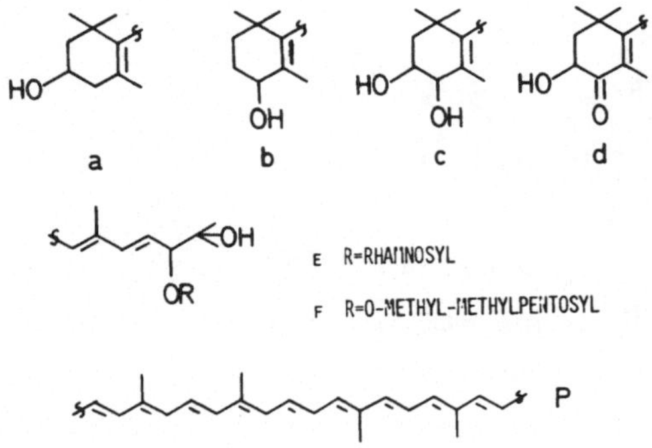

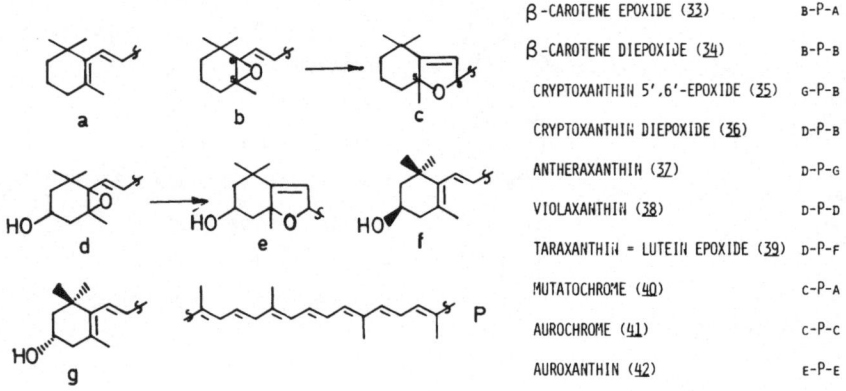

MYXOXANTHOPHYLL (26) A-P-E

4-KETOMYXOXANTHOPHYLL (27) C-P-E

APHANIZOPHYLL (28) C-P-E

OSCILLAXANTHIN (29) D-P-E

MYXOL-2'-O-METHYL-METHYLPENTOSIDE (30) A-P-F

4-KETO-MYXOL-2'-O-METHYL-METHYLPENTOSIDE (31) D-P-F

OSCILLOL-2,2'-DI(O-METHYL)-METHYLPENTOSIDE (32) D-P-F

E R=RHAMNOSYL

F R=O-METHYL-METHYLPENTOSYL

Figure 6. Glycosidic algal carotenoids (26 - 32)

C_{40} carotenes (Figure 3) include the aliphatic precursors phytoene (1) and phytofluene (2) and coloured monocyclic and bicyclic carotenes with ψ-, β-, and ϵ- end groups.

Acetylenic bonds are invariably located in the 7(7') position. Several monoacetylenic and one diacetylenic derivative are known with ϵ-, 3-OH-ϵ, 3-OH-β, 3-OH-epoxy and triol end groups, as shown in Figure 4.

β-CAROTENE EPOXIDE (33)   B-P-A

β-CAROTENE DIEPOXIDE (34)   B-P-B

CRYPTOXANTHIN 5',6'-EPOXIDE (35) G-P-B

CRYPTOXANTHIN DIEPOXIDE (36)   D-P-B

ANTHERAXANTHIN (37)   D-P-G

VIOLAXANTHIN (38)   D-P-D

TARAXANTHIN = LUTEIN EPOXIDE (39) D-P-F

MUTATOCHROME (40)   C-P-A

AUROCHROME (41)   C-P-C

AUROXANTHIN (42)   E-P-E

Figure 7. Non-allenic, non-acetylenic, epoxidic algal carotenoids (33 - 42)

All carotenoid allenes (Figure 5) have the same hydroxylated, acetylated end group, stereochemistry included. Variations in the rest of the molecule are encountered; note for instance the neo-xanthin-related vaucheriaxanthin (19) with 19-hydroxy group, fucoxanthin (20) with 8-keto group, and peridinin (23).

Glycosidic algal carotenoids (Figure 6) are either aliphatic or monocyclic, with the aliphatic end group (e,f) in common. They are all secondary allylic rhamnosides or O-methyl-methyl pentosides.

Epoxides containing none of the previously mentioned functions are encountered (Figure 7). 5,6-Epoxides (b,d) readily rearrange to the 5,8-furanoid isomers (d, f), which may be artifacts.

Carotenoid ketones not already mentioned under allenes (Figure 8) include 4-keto compounds like echinenone (43) and astaxanthin (46) and 8-keto carotenoids like siphonaxanthin (47).

Finally, some algal carotenoids (Figure 9) bear hydroxy functions only in the 3(3') or 2(2') positions.

Being an organic chemist, I would at this point like to give an example from our own work on structure elucidation and chirality studies on algal carotenoids (Figure 10). For nostoxanthin from *Anacystis nidulans*, other workers[15] had suggested the bisallenic structure (61), a structure incompatible with the electronic spectrum, which clearly demonstrated a β,β-carotene type chromophore[16]. The mass spectrum showed a molecular ion at *m/e* 600, consistent with

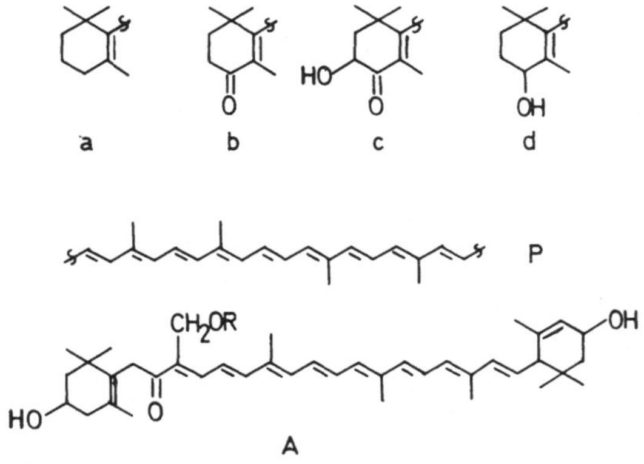

ECHINENONE (43)	B-P-A	
CANTHAXANTHIN (44)	B-P-B	
3-HYDROXY-ECHINENONE (45)	C-P-A	
4'-HYDROXY-ECHINENONE (45B)	B-P-D	
ASTAXANTHIN (46)	C-P-C	
SIPHONAXANTHIN (47)	A	R=H
SIPHONEIN (48)	A	R=ACYL

Figure 8. Non-allenic, ketonic algal carotenoids (43 - 48)

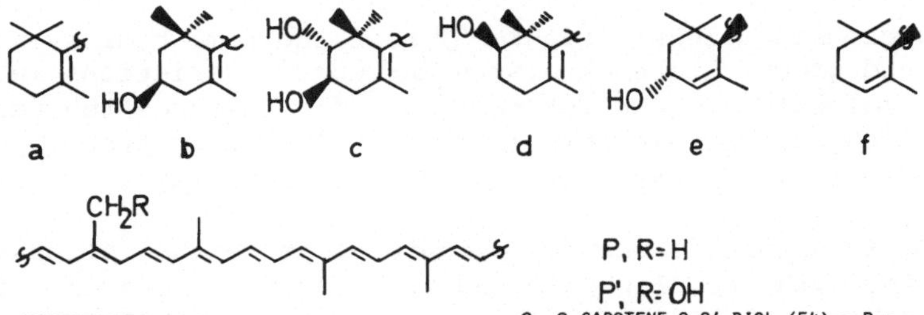

a   b   c   d   e   f

CRYPTOXANTHIN (49)   B-P-A   P, R=H
                                         P', R=OH
ZEAXANTHIN (50)   B-P-B   β,β-CAROTENE-2,2'-DIOL (54) D-P-D

CALOXANTHIN (51)   C-P-B   β,ε-CAROTEN-2-OL (55)   D-P-F

NOSTOXANTHIN (52)   C-P-C   α-CRYPTOXANTHIN (56)   E-P-A

β,β-CAROTENE-2-OL (53)   D-P-A   LUTEIN (57)   B-P-E

                                         LOROXANTHIN (58)   B-P'-E

Figure 9. Algal carotenoids with hydroxy functions only (49 – 58)

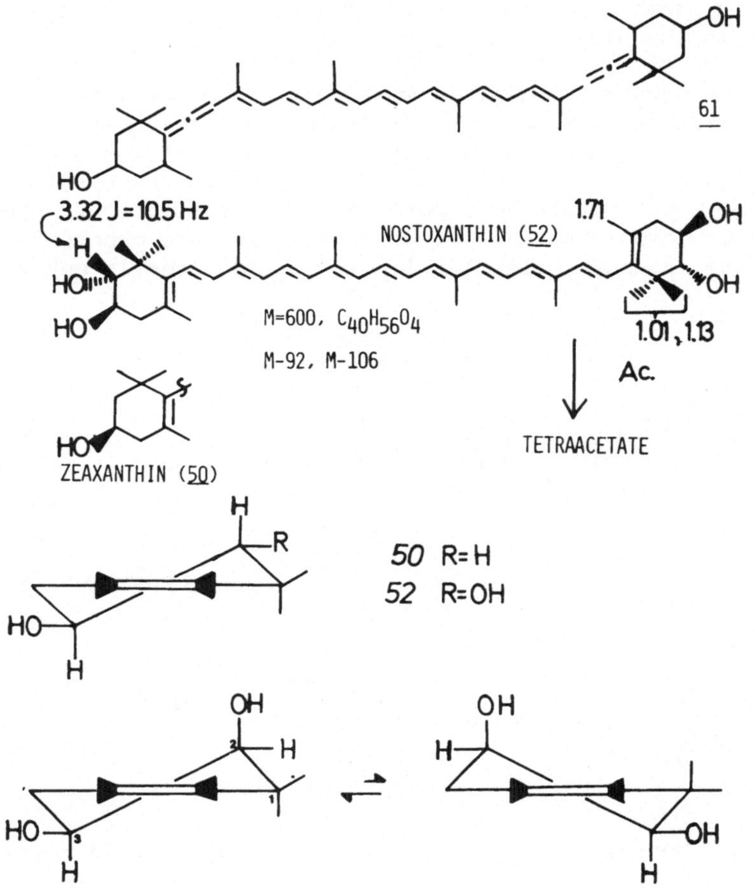

Figure 10. Example of structure elucidation and chirality studies

$C_{40}H_{56}O_4$. The formation of a tetraacetate demonstrated the presence of four primary or secondary hydroxy groups, and the ^{1}H NMR data defined the β,β-carotene-2,3,2',3'-tetrol structure (52). A two-proton doublet at δ 3.32 (J = 10.5 Hz) revealed a _trans_ diaxial coupling for the C-2(2') protons. 3,4-Diol substitution was ruled out from IR and ^{1}H NMR evidence[17].

Assignment of absolute configuration followed from the ^{1}H NMR and the CD spectrum, identical with that of β,β-carotene-3,3'-diol (50 = zeaxanthin). From the conformational rule[23], the preferred helicity of the cyclohexene half-chair dictates the Cotton effect. Both 2R,3R (52) and 2S,3R (52b) substitution would give a preference for the same half-chair as for zeaxanthin (3R, 50) to avoid 1,3-diaxial interaction. However, the 2,3-_trans_ relationship is defined by the ^{1}H NMR, and hence nostoxanthin is 2R,3R,2'R,3'R-β,β-carotene-2,3,2',3'-tetrol (52)[17].

Let us now return to our general consideration of algal carotenoids. Table 2 is a summary of the distribution pattern of carotenoids in algae. Trivial names have been used except for the carotenes.

Table 2. Distribution pattern of carotenoids in algae[27]. For structures, see Figure 4 (1-7), Figure 5 (8-16), Figure 6 (17-25), Figure 7 (26-32), Figure 8 (33-42), Figure 9 (43-48) and Figure 10 (49-58).

Algal Class		Carotenoids
Cyanophyceae	Glycosides:	myxoxanthophyll (26)
		4-keto-myxoxanthophyll (27)
		aphanizophyll (28)
		oscillaxanthin (29)
		myxol-2'-O-methyl-methylpentoside (30)
		4-keto-myxol-2'-O-methyl-methylpentoside (31)
		oscillol-2,2'-di(O-methyl)-methylpentoside (32)
	Other:	lycopene (3)
		β,β-carotene (5)
		β,ψ-carotene (4)
		cryptoxanthin (49)
		zeaxanthin (50)
		echinenone (43)
		canthaxanthin (44)
		3'-hydroxy-echinenone (45)
		caloxanthin (51)
		nostoxanthin (52)
		mutatochrome (40)

Table 2. (continued)

Algal Class		Carotenoids
Rhodophyceae	Common:	β,β-carotene (5)
		β,ε-carotene (6)
		zeaxanthin (50)
		lutein (57)
	Questionable or occasional:	
		α-cryptoxanthin (56)
		cryptoxanthin (49)
		antheraxanthin (37)
		violaxanthin (38)
		auroxanthin (42)
		aurochrome (41)
		neoxanthin (17)
		fucoxanthin (20)
		taraxanthin (39)
Cryptophyceae		β,β-carotene (5)
		β,ε-carotene (7)
		ε,ε-carotene (7)
		crocoxanthin (8)
		monadoxanthin (9)
		alloxanthin (13)
		manixanthin (14)
Dinophyceae	Common:	β,β-carotene (5)
		diatoxanthin (10)
		diadinoxanthin (11)
		dinoxanthin (18)
		peridinin (23)
	Occasional:	phytoene (1)
		phytofluene (2)
		β,ε-carotene (6)
		astaxanthin (46)
		pyrrhoxanthin (15)
		pyrrhoxanthinol (16)
		peridininol (24)
		fucoxanthin (20)
Raphidophyceae		β,β-carotene (5)
		antheraxanthin (37)
		lutein epoxide (39)
Chrysophyceae	Common:	β,β-carotene (5)
		fucoxanthin (20)
	Other:	zeaxanthin (50)
		antheraxanthin (37)
		violaxanthin (38)
		cryptoxanthin (49)
		cryptoxanthin epoxide (35)
		neoxanthin (17)

Table 2. (continued)

Algal Class		Carotenoids
Haptophyceae	Common:	β,β-carotene (5)
		diatoxanthin (10)
		diadinoxanthin (11)
		fucoxanthin (20)
	Other:	β,ε-carotene (6)
		β-carotene epoxide (33)
		cryptoxanthin diepoxide (36)
		cryptoxanthin (49)
		echinenone (43)
		canthaxanthin (44)
		fucoxanthinol (21)
		desepoxyneoxanthin (17b)
		19'-hexanoyloxyfucoxanthin (22)
		19'-hexanoyloxyparacentrone 3-acetate (25)
Bacillariophyceae		β,β-carotene (5)
		ε,ε-carotene (7)
		diatoxanthin (10)
		diadinoxanthin (11)
		fucoxanthin (20)
		neoxanthin (17)
Xanthophyceae	Common:	β,β-carotene (5)
		diatoxanthin (10)
		diadinoxanthin (11)
		heteroxanthin (12)
		vaucheriaxanthin (19)
		neoxanthin (17)
	Other:	β-carotene diepoxide (34)
		cryptoxanthin 5',6'-epoxide (35)
		cryptoxanthin diepoxide (36)
Eustigmatophyceae	Common:	β,β-carotene (5)
		diatoxanthin (10)
		diadinoxanthin (11)
		heteroxanthin (12)
		vaucheriaxanthin (19)
	Other:	neoxanthin (17)
		β-carotene diepoxide (34)
		cryptoxanthin 5',6'-epoxide (35)
		cryptoxanthin diepoxide (36)
Phaeophyceae	Common:	β,β-carotene (5)
		violaxanthin (30)
		fucoxanthin (20)
	Other:	ε,ε-carotene (7)
		zeaxanthin (50)
		antheraxanthin (37)
		diatoxanthin (10)

Table 2. (continued)

Algal Class		Carotenoids
		diadinoxanthin (11)
		fucoxanthinol (21)
		neoxanthin (17)
Euglenophyceae	Common:	β,β-carotene (5)
		diatoxanthin (10)
		diadinoxanthin (11)
		heteroxanthin (12)
		neoxanthin (17)
	Other:	cryptoxanthin (49)
		cryptoxanthin 5',6'-epoxide (35)
	Eye-spot:	echinenone (43)
		3-hydroxy-echinenone (45)
		canthaxanthin (44)
		astaxanthin (46)
Prasinophyceae/Loxophyceae		lycopene (3)
		β,ψ-carotene (4)
		β,β-carotene (5)
		β,ε-carotene (6)
		zeaxanthin (50)
		lutein (57)
		violaxanthin (38)
		neoxanthin (17)
		siphonaxanthin (47)
		siphonein (48)
Chlorophyceae	Common:	β,β-carotene (5)
		β,ε-carotene (6)
		lutein (57)
		zeaxanthin (50)
		antheraxanthin (37)
		violaxanthin (38)
		neoxanthin (17)
	Occasional:	β,ψ-carotene (4)
		ε,ε-carotene (7)
		cryptoxanthin (49)
		cryptoxanthin 5',6'-epoxide (35)
		lutein epoxide (39)
		loroxanthin (58)
		β,β-caroten-2-ol (53)
		β,β-carotene-2-2'-diol (54)
		β,ε-caroten-2-ol (55)
		siphonaxanthin (47)
		siphonein (48)

Ketonic carotenoids produced during nitrogen starvation:
echinenone (43)
4'-hydroxy-echinenone (45b)
canthaxanthin (44)
3-hydroxy-canthaxanthin (45)
astaxanthin (46)

Table 3. Characteristics of the carotenoid, chlorophyll and biliprotein distribution pattern in various algal classes.

Class (Christensen)	Carotenoid Major Carotenes	Characteristic Xanthophylls	Chlorophyll	Biliprotein
1. Cyanophyceae	β,β	Monocyclic rhamnosides ⎫ no	a	+
2. Rhodophyceae	β,β β,ε	Simple ⎬ epoxides	a	+
3. Cryptophyceae	β,β β,ε	Acetylenic ⎭	a, c_2	(+)
4. Dinophyceae	β,β	Acetylenic; peridinin	a, c_2	
5. Raphidophyceae	β,β β,ε	Simple epoxides	a, c_2	
6. Chrysophyceae	β,β	Fucoxanthin	a, c_2, c_1	
7. Haptophyceae	β,β β,ε	Fucoxanthin; acetylenic	a, c_2, c_1 ?	
8. Bacillariophyceae	β,β β,ε	Fucoxanthin, acetylenic	a, c_2, c_1	
9. Xanthophyceae	β,β	Vaucheriaxanthin; acetylenic	a, c	
10. Eustigmatophyceae	β,β	Vaucheriaxanthin; violaxanthin	a	
11. Phaeophyceae	β,β	Fucoxanthin; violaxanthin	a, c_2, c_1	
12. Euglenophyceae	β,β	Acetylenic	a, b	
13. Prasinophyceae/ Loxophyceae	β,β β,ε	Violaxanthin	a, b	
14. Chlorophyceae	β,β β,ε	Violaxanthin; neoxanthin	a, b	

Table 4. Structural features of major carotenoids encountered in various algal classes.

Structural feature	1. Cyanophyceae	2. Rhodophyceae	3. Cryptophyceae	5. Raphidophyceae	6. Chrysophyceae	11. Phaeophyceae	7. Haptophyceae	8. Bacillariophyceae	4. Dinophyceae	10. Eustigmatophyceae	9. Xanthophyceae	12. Euglenophyceae	13. Prasinophyceae/Loxophyceae	14. Chlorophyceae
2-OH											+			
O-Acyl									+	+	+			
5,6-glycol										+	+			
19-OH							+		+		+			+
Butenolide; C_3-expelled									+					
-OAc					+	+	+	+						
8-keto				+	+	+	+	+	+					+
=C=				+	+	+	+	+	+		+			+
O				+	+	+	+	+	+	+	+	+		+
C≡C			+		+		+	+	+	+	+			
ε-rings		+	+				+	+					+	+
4-keto	+									+				+
Bicyclic xanthophyll	+	+	+	+	+	+	+	+	+	+	+	+		+
Monocyclic glycosides	+													

Algal Class

1. Cyanophyceae
2. Rhodophyceae
3. Cryptophyceae
5. Raphidophyceae
6. Chrysophyceae
11. Phaeophyceae
7. Haptophyceae
8. Bacillariophyceae
4. Dinophyceae
10. Eustigmatophyceae
9. Xanthophyceae
12. Euglenophyceae
13. Prasinophyceae/ Loxophyceae
14. Chlorophyceae

Table 3 gives a simplified evaluation of the characteristics of the carotenoid pattern in each algal class, following the order of Christensen[14]. Glycosides, generally monocyclic carotenoid rhamnosides, are restricted to blue-green algae, which can readily be identified from their characteristic carotenoid composition. Carotenoid epoxides are usually not encountered in the three most primitive algal classes. Acetylenic carotenoids are encountered in Cryptophyceae, dinoflagellates, Haptophyceae, diatoms, Xanthophyceae and Euglenophyceae. The allenic fucoxanthin (20) occurs in Chryso-phyceae, diatoms, Haptophyceae and brown algae. Peridinin (23) is unique to dinoflagellates. Raphidophyceae have simple epoxides. The diepoxide violaxanthin (38) is found in the higher algal classes. Small amounts of 4-keto-carotenoids are produced by the Chlorophyta and blue-green algae. The allenic vaucheriaxanthin (19) is restricted to Eustigmatophyceae and Xanthophyceae. There are significant dif-ferences in carotenoid composition between the algal classes, nicely paralleled by systematic change in the chlorophylls, a only at the top of Table 1, a and b at the bottom, and a and c in the centre[18,19]. This is also the case for biliproteins[20].

In terms of the biosynthetic capability of algae to produce particular structural features of their carotenoids, Table 4 may be more instructive. As already indicated, monocyclic carotenoid glycosides are peculiar to blue-green algae, although they are found in various bacteria. Bicyclic xanthophylls are found in all algal classes and are of no chemosystematic value. 4-Keto-carotenoids are encountered in blue-green algae and are minor con-stituents in Euglenophyceae (eye-spot pigments) and green algae grown under nitrogen starvation. Cyclization to ε-rings is typical of certain algal classes (Table 4). Acetylenic carotenoids are encountered in several classes. Epoxidic carotenoids are synthe-sized by all except the three most primitive algal classes. Allenic carotenoids, 8-keto-carotenoids, and carotenol acetates are found as indicated. Butenolide formation and C_3-expulsion in the caro-tenoid synthesis are unique properties of dinoflagellates. However, oxidation of in-chain methyl groups to 19-ols is also effected by some other classes (Table 4). 5,6-Glycol formation seems to be restricted to Xanthophyceae and Euglenophyceae. Higher fatty acid esters of carotenoids are encountered in three classes and, finally, the 2-hydroxy-β-type is found in Chlorophyceae.

To reach the final products in algal carotenoid synthesis from an aliphatic C_{40} precursor, a minimum number of some twenty step-reactions are required. Probable sequences, based on general knowledge of carotenogenesis[4], are indicated in Figure 11. The glycosidic end group a could be reached by hydration, dehydrogena-tion, allylic oxidation, and glycosidation, cf. Reference 21. The 3-hydroxy-ε-end group c requires cyclization to the ε-ring b and allylic hydroxylation. Following cyclization to the β-ring d,

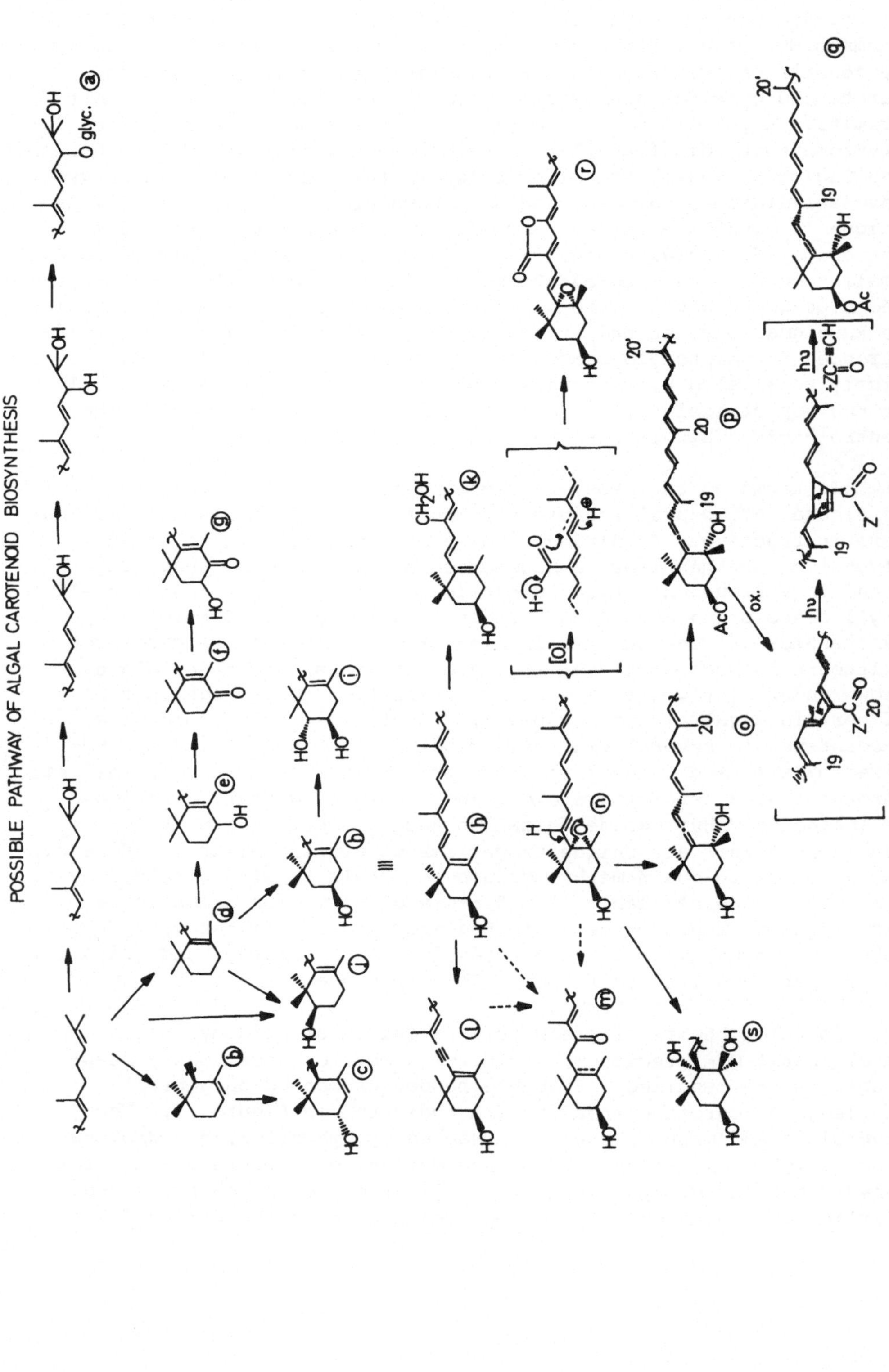

Figure 11. Postulated terminal step-reactions required in algal carotenogenesis.

allylic oxidation via e to f to the α-ketol g is plausible. Direct
3-hydroxylation of d, known to occur in <u>Chlorella</u>[22], would give h,
which by <u>trans</u> hydroxylation could provide i. The 2-hydroxy-β-end
group j could be formed from an aliphatic epoxidic precursor or
from d[23]. By allylic 19-hydroxylation, the 3-hydroxy-β-end group h
would give k and is one possible precursor of the acetylenic l; the
two latter ones are possible precursors of the 8-keto group m by
hydration and keto-enol tautomerization or hydration followed by
allylic oxidation, respectively. Epoxidation of h would give n,
which could provide the allenic end group by rearrangement to o,
followed by acetylation to p. From p, allylic oxidation, eventually
followed by C_3-expulsion, could give rise to g, whereas allylic
oxidation of h could ultimately give the butenolide r. Opening of
the epoxide could give the triol s. Stereochemical knowledge helps
to define such a scheme.

In Table 4 the structural features of the carotenoids are
arranged, not arbitrarily but in an order which reflects the
sequences of step reactions shown in Figure 12, as far as it is
possible to transfer a branched scheme to a linear system.

The order of the algal classes has been changed somewhat in
the centre of Table 4 to obtain a systematic change in the carotenoid pattern; the number to the left refer to the order in which
the classes are treated by Christensen[14].

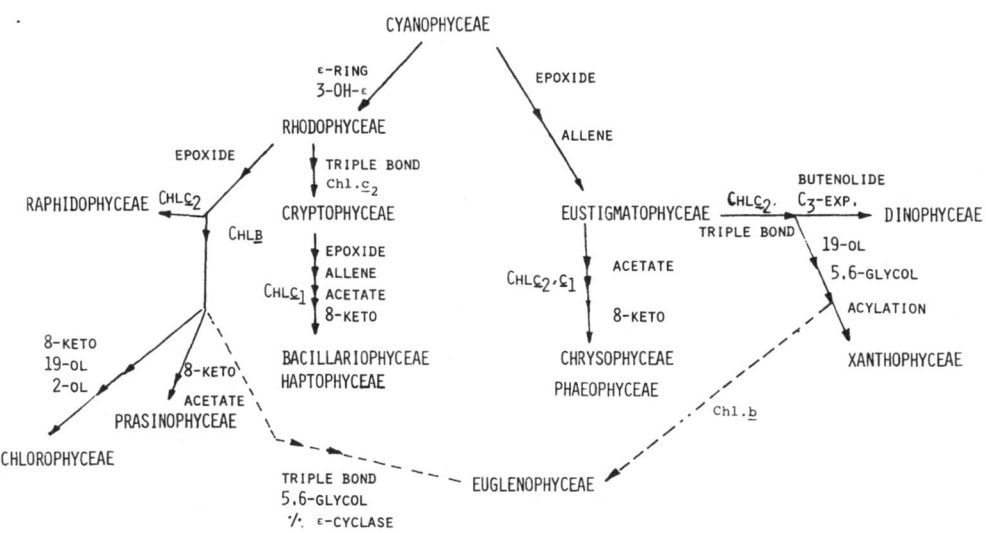

Figure 12. Hypothetic evolution of algae based on consideration of
photosynthetic pigments with emphasis on the detailed structures
of algal carotenoids

Assuming that evolution is caused by development of new enzymes, our knowledge of chloroplast pigments, with some emphasis on the detailed carotenoid structures, is used for construction of Figure 12, which suggests possible events in algal chloroplast pigment evolution. Repression of the original enzymes has been kept to a minimum, but parallel invention of enzymes is required.

Cyanophyceae with biliproteins, chlorophyll $\underline{a}$ alone, and aliphatic and monocyclic carotenoids are recognized as the most primitive algae.

Rhodophyceae, also with biliproteins and chlorophyll $\underline{a}$ only, could have evolved from Cyanophyceae by including development of an ε-cyclase system to produce ε-type carotenoids.

Raphidophyceae appear further developed in the sense that carotenoid epoxides and chlorophyll $\underline{c_2}$ are present. Development of chlorophyll $\underline{b}$ and allenic carotenoid synthesis lead to the missing link from which Prasinophyceae/Loxophyceae could have evolved after acquiring the enzymes to effect introduction of the 8-keto group and acetylation, and to the Chlorophyceae, some of which produce 19-ols and 2-ols and therefore require enzymes.

Cryptophyceae, still containing some biliproteins, chlorophyll $\underline{a}$ and $\underline{c_2}$ could have evolved from the Rhodophyceae after evolving the triple-bond-forming enzyme. Bacillariophyceae and Haptophyceae could follow by introduction of chlorophyll $\underline{c_1}$ synthesis and the synthesis of epoxidic, allenic and 8-keto carotenoids.

In this scheme, algal classes without ε-type carotenoids are presumed to have evolved by a different route from the Cyanophyceae.

The Eustigmatophyceae with chlorophyll $\underline{a}$ only may occupy an intermediate position, after the introduction of epoxidic and allenic carotenoids. To reach the Chrysophyceae and Phaeophyceae, the ability to form chlorophylls $\underline{c_2}$, $\underline{c_1}$ and 8-keto carotenoids must be acquired.

Alternatively, introduction of enzymes producing acetylenic carotenoids and chlorophyll $\underline{c_2}$ could lead to the missing link from which the Dinophyceae evolved after the invention of butenolide-forming and C_3-expulsion enzymes, and Xanthophyceae evolved after development of 19-ol and 5,6-glycol and triple-bond-forming enzymes.

The Euglenophyceae pose a particular dilemma. The left dotted line avoids invention of chlorophyll $\underline{b}$ twice, but requires total repression of the ε-cyclase.

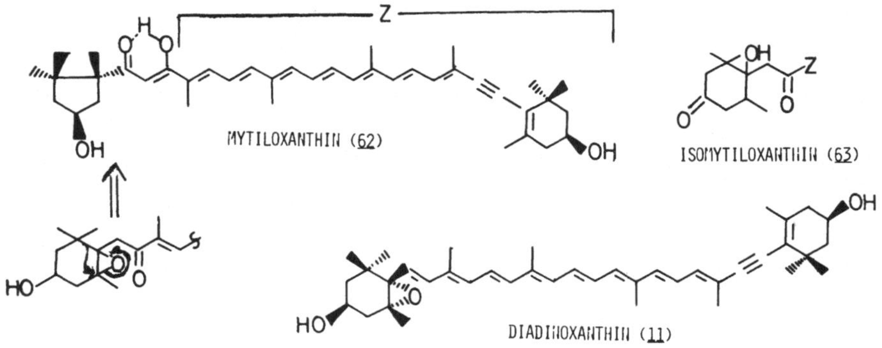

Figure 13. Possible conversion of algal carotenoids in Mytilus edilus[26]

The best lesson of Figure 12 is probably that the situation is not easy to rationalize. Several branching points are likely to have occurred in algal evolution, and the missing links complicate the picture. Similar attempts, based on less structural evidence and with different outcome, were made in 1971 by Goodwin[24].

However, as science progresses there is reasonable chance that the pigments of the photosynthetic apparatus, considered together with other important morphological and physiological (life cycle) parameters, may throw more light on these fascinating problems.

Finally, I will make a few comments on the possible use of algal carotenoids as indicators to follow the food chain in the marine environment.

It is well known that animals cannot carry out de novo carotenoid synthesis. It is also recognized that planktonic algae serve as food for various marine animals. Provided the animal resorbs the carotenoids or metabolizes them in such a way that the original structure can be recognized, our knowledge of the distribution of algal carotenoids provides a tool for following the food chain.

One example only: The correlation between the total carotenoid content in the edible sea mussel Mytilus edilus and the occurrence of dinoflagellate blooms has been demonstrated[25]. Weedon and co-workers[26] have shown that the acetylenic carotenoids mytiloxanthin (62) and isomytiloxanthin (63) are characteristic constituents of M. edilus and have suggested that the end groups may be biogenetically related to the epoxidic end group of fucoxanthin (20).

Of the C_{40} carotenoids encountered in dinoflagellates, diadinoxanthin (11) is a possible precursor candidate.

Some of the ideas discussed here will be elaborated further in a chapter being prepared on marine carotenoids[27]. Detailed references to the structure and distribution of algal carotenoids will be included there.

REFERENCES

1. IUPAC and IUB, <u>Nomenclature of Carotenoids</u> (Rules approved 1974), Butterworths, London, 1976.

2. T. W. Goodwin, <u>The Comparative Biochemistry of the Carotenoids</u>, Chapman and Hall, London, 1952.

3. N. I. Krinsky, in <u>Carotenoids</u>, ed. O. Isler, Birkhäuser, Basel, 1971, p. 669.

4. T. W. Goodwin, in <u>Carotenoids</u>, ed. O. Isler, Birkhäuser, Basel, 1971, p. 557.

5. J. E. Johansen and S. Liaaen-Jensen, Tetrahedron Lett., 955. (1976).

6. S. Hertzberg and S. Liaaen-Jensen, Acta Chem. Scand. <u>22</u>, 1714, (1968).

7. H. H. Strain, W. A. Svec, P. Wegfahrt, H. Rapoport, F. T. Haxo, S.Nordgård, H. Kjøsen, and S. Liaaen-Jensen, Acta Chem. Scand. <u>B 30</u>, 109 (1976).

8. H. Kjøsen, S. Nordgård, S. Liaaen-Jensen, W. A. Svec, H. H. Strain, P. Wegfahrt, H. Rapoport, and F. T. Haxo, Acta Chem. Scand. <u>B 30</u>, 157 (1976).

9. R. M Klein and A. Cronquist, Quart. Rev. Biol. <u>42</u>, 108 (1967).

10. W.-J. Hsu, D. B. Rodriguez, and C. O. Chichester, Int. J. Biochem. <u>3</u>, 333 (1972).

11. R. Buchecker, P. Hamm, and C. H. Eugster, Chimia <u>26</u>, 134 (1972).

12. A. G. Andrewes, G. Borch, and S. Liaaen-Jensen, Acta Chem. Scand. <u>B 28</u>, 139 (1974).

13. A. G. Andrewes, G. Borch, S. Liaaen-Jensen, and G. Snatzke, Acta Chem. Scand. <u>B 28</u>, 730 (1974).

14. T. Christensen, Systematisk Botanik, Alger, University of
     Copenhagen, Copenhagen, 1962.

15. H. Stransky and A. Hager, Arch. Mikrobiol. 72, 84 (1971).

16. S. Hertzberg, S. Liaaen-Jensen, and H. W. Siegelmann, Phyto-
     chemistry 10, 3121 (1971).

17. R. Buchecker, S. Liaaen-Jensen, G. Borch, and H. W. Siegelman,
     Phytochemistry 15, 1015 (1976).

18. A. H. Jackson, in Chemistry and Biochemistry of Plant Pigments,
     Vol. 1, ed. T. W. Goodwin, Academic Press, London, 1976.

19. L. Bogorad, in Chemistry and Biochemistry of Plant Pigments,
     Vol. 1, ed. T. W. Goodwin, Academic Press, London, 1976.

20. P. O'Carra and C. OhEocha, in Chemistry and Biochemistry of
     Plant Pigments, Vol. 1, ed. T. W. Goodwin, Academic Press,
     London, 1976.

21. S. Liaaen-Jensen, G. Cohen-Bazire, and R. Y. Stanier, Nature
     192, 1168 (1961).

22. H. Claes, Z. Naturforsch. B 14, 4 (1959).

23. H. Kjøsen, N. Arpin, and S. Liaaen-Jensen, Acta Chem. Scand.
     26, 3053 (1972).

24. T. W. Goodwin, Aspects of Terpenoid Chemistry and Biochemistry,
     Academic Press, London, 1971, p. 315.

25. A. Jensen and E. Sakshaug, J. Exp. Mar. Biol. Ecol. 5, 180,
     246 (1970).

26. A. Khare, G. P. Moss, and B. C. L. Weedon, Tetrahedron Lett.,
     3921 (1973).

27. S. Liaaen-Jensen, in Marine Natural Products. V., ed. P.
     Scheuer, to be published.

CHEMISTRY AND DISTRIBUTION OF DELETERIOUS DINOFLAGELLATE TOXINS

Y. Shimizu, M. Alam, Y. Oshima, L. J. Buckley and

W. E. Fallon, College of Pharmacy, University of

Rhode Island[*]

H. Kasai, I. Miura, V. P. Gullo, and K. Nakanishi

Department of Chemistry, Columbia University[*]

INTRODUCTION

The dinoflagellates are an important part of the marine plankton, second only the the diatoms in abundance and as primary producers of organic matter in most areas of the world's oceans. Occasionally, for reasons not fully understood, a particular dinoflagellate bloom may reach concentrations as high as 50,000 cells per ml. Blooms (red tides) of toxic dinoflagellates, particularly Gonyaulax tamarensis in the north Atlantic and G. catanella in the north Pacific, have been causing serious economic and public healtħ proglems. These problems arise when filter feeders (such as shellfish), which concentrate the poison, are consumed by organisms higher in the food chain. In man, consumption of contaminated bivalves produces paralytic shellfish poisoning (PSP), a severe and often fatal form of food poisoning. These dinoflagellate toxins are also of great pharmacological and biomedical interest because of their particular mode of action, namely, blockage of the transmission of nerve impulses by selectively blocking the passive influx of Na^+ through excitable membranes[1,2].

Prior to 1972, most of the work reported on the chemistry of PSP dealt with G. catanella and bivalves exposed to G. catanella on the Pacific coast of North America. Saxitoxin (STX,1), the active component, was first isolated from the California mussel Mytilus californianus and the Alaska butter clam Saxidomus giganteus[3] and later from cultured G. catanella cells[4]. The isolation involved

binding the toxin to a weak cation exchange resin, washing out
impurities with acetate buffer, and elution of the toxin with
dilute acetic acid. Although gram quantities of pure saxitoxin (1)
have been available since 1957, the true structure was not reported
until 1974, when crystalline derivatives of saxitoxin were
X-rayed[5,6].

Early attempts to isolate the toxin from cultured G. tamarensis
cells and shellfish exposed in the north Atlantic, using the method
developed by Schantz, et al.[3] for saxitoxin, failed because the
majority of the toxin was only weakly bound to the cation exchange
resin and was eluted from the column with the sodium acetate buffer,
along with the bulk of inert material[7,8]. However, Ghazarossian,
et al.[9] were able to identify saxitoxin in an old sample of scallops
from the Bay of Fundy. Buckley, et al.[10,11] succeeded in identifying saxitoxin as a minor component of the toxin in Mya arenaria
collected on the New England coast in 1972. They also isolated two
new toxins, later identified as gonyautoxin II and III[12], from the
sodium acetate eluate, using extraction into cold ethanol and
repeated chromatography on Bio-Gel P-2.

Meanwhile, using a different approach, we have succeeded in
isolating a total of seven toxins from G. tamarensis cells and
red-tide-infested bivalves[13,14]. This paper describes the isolation and structural elucidation of the toxins, as well as their
distribution.

ISOLATION OF NEW PSP TOXINS

Massive blooms of G. tamarensis along the New England coast
in 1972 and 1974 provided the raw material and incentive for an
extensive investigation of "east coast" PSP. The toxins were
extracted from the hepatopancreas of the exposed soft-shell clam
Mya arenaria by a procedure similar to that used for saxitoxin[15].
The extract was partitioned with chloroform to remove phospholipids,
etc., and applied to a Sephadex or Bio-Gel P-2 column. The bulk
of inert material passed through the column with water, while the
toxin was adsorbed and subsequently eluted with dilute acetic acid.
In some cases all the toxin was not bound on the first passage and
rechromatography was necessary. The adsorbed toxin was shown by
TLC to be a complex mixture consisting of saxitoxin and several new
toxins.

This mixture was separated into saxitoxin and gonyautoxin
fractions by chromatography on Bio-Rex 70. The gonyautoxin fraction
was first eluted from the column with dilute acetic acid. The saxitoxin fraction was then eluted with dilute HCl. The gonyautoxin
fraction was resolved into five pure components by careful high speed
chromatography on very fine ion exchange resin (e.g. Bio-Rex 70,

400 mesh), employing either a linear acetic acid gradient or a stepwise elution. The saxitoxin fraction was separated into pure saxitoxin and a new toxin called neosaxitoxin, using high speed chromatography on Bio-Rex 70.

Similar procedures were used for the successful isolation of the toxins from cultured G. tamarensis cells, Alaska butter clams[14] and a Japanese red tide[16].

CHEMISTRY OF PSP TOXINS

The highly hygroscopic nature of the new toxins and the minute amounts available made an exact determination of their toxicity impossible; however, they appear to possess about the same potency as saxitoxin (5500 MU/mg). The molecular weights could not be determined by analysis or by mass spectroscopy due to their scarcity, polarity or hygroscopic nature. In solution, gonyautoxin II (2) and III (3) were observed to undergo slow equilibration to form a 3:1 mixture. This rather slow equilibration was tremendously accelerated by the presence of trace amounts of base. The proton-noise-decoupled (79,729 scans) and continuous-wave-decoupled (110,499 scans) ^{13}C NMR spectra of gonyautoxin II (2) in D_2O (Figure 1) were very informative when compared to those of saxitoxin. An identical number of carbon atoms with almost identical chemical shifts was observed for both compounds. The only clear difference was the replacement of the 33.1 ppm (11-C) and 43.0 ppm (10-C) triplets in saxitoxin by a 77.6 ppm doublet and 50.9 ppm triplet in gonyautoxin II. The 44.5 ppm downfield shift of 11-C and the 7.9 ppm shift of 10-C are best explained by a hydroxyl substitution at 11-C in gonyautoxin II (2). Further evidence for this

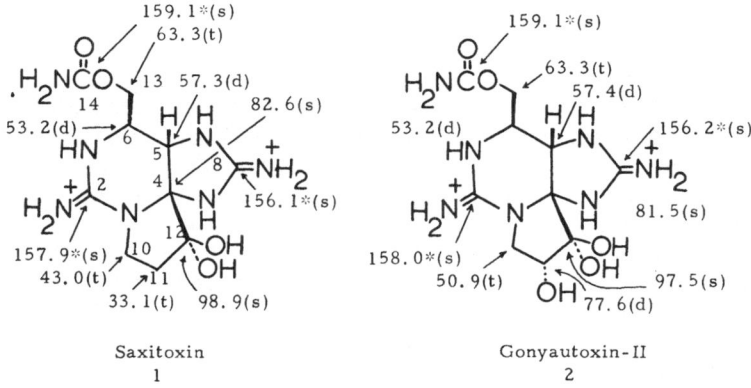

Saxitoxin
1

Gonyautoxin-II
2

Figure 1. CMR data of gonyautoxin II (2) and saxitonin (1) (*assignments may be interchanged).

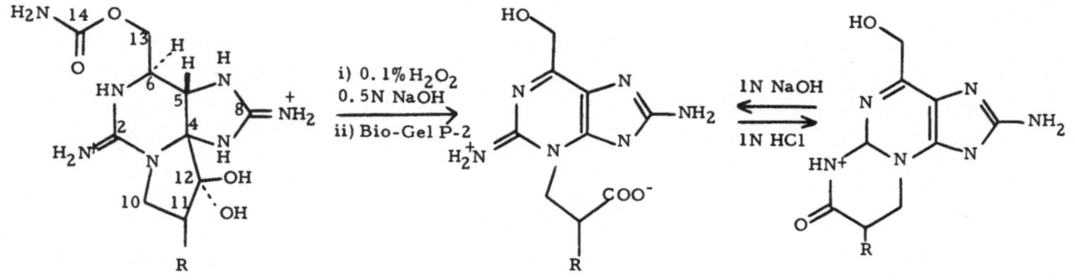

i) 0.1% H$_2$O$_2$
0.5N NaOH
ii) Bio-Gel P-2

1N NaOH
1N HCl

1 R : H Saxitoxin

2 R : ⊀ -OH Gonyautoxin-II

3 R : ⊿ -OH Gonyautoxin-III

5 R : H

4 R : ⊀ -OH

7 R : H

6 R : ⊀ -OH

Scheme 1

11-substitution was obtained by the oxidative degradation of gonyautoxin II and saxitoxin on an ultramicro scale.

Reaction of gonyautoxin II with 0.1% H$_2$O$_2$ in 0.5 N NaOH afforded, as the major product, acid (4): ^{1}H NMR (D$_2$O) 5.04 (s, 2H, 13-H),4.70 ppm (m, 3H, 10-H and 11-H) which was isolated by chromatography on Bio-Gel P-2 in about 50% overall yield. Reaction of saxitoxin under identical conditions gave acid (5): MS 234 (M-H$_2$)); ^{1}H NMR (D$_2$O) 5.06 (s, 2H, 13-H), 4.40 (t, J = 6 Hz, 2H, 10-H), 2.70 ppm (t, J = 6 Hz, 11-H). Both acids (4) and (5) formed the corresponding lactams (6) and (7) by treatment with 1 N HCl. This conversion was reversible, favoring the acid at neutral pH. The highest MS peak of acid (4) was at m/e 250, corresponding to the cyclized lactam (6). The weak Cotton effect of the CD peaks of acid (4): CD (H$_2$O) 330 (Δε-0.51), 251 (Δε-1.06), 230 nm (Δε+1.14), which was not observed in the corresponding saxitoxin derivative (5), indicated the presence of a chiral center in gonyautoxin II not at a carbon directly linked to the nucleus. Taken together, this data established the presence of 11-hydroxyl substitution in gonyautoxin II.

Table 1. ^{1}H NMR and ^{13}C NMR data for neosaxitoxin and saxitoxin

	Saxitoxin			Neosaxitoxin		
	CMR	PMR		CMR	PMR	
2	157.9*(s)	--		158.8 (s)	--	
8	156.1*(s)	--		158.1 (s)	--	
14	159.1*(s)	--		--	--	
4	82.6 (s)	--		82.2 (s)	--	
5	57.3 (d)	4.77	(J=1)	64.4 (d)	4.83	
6	53.2 (d)	3.87	(J=9,5,1)	56.9 (d)	4.15	(J=6,6)
10	43.0 (t)	3.85	(J=10)	43.3 (t)	3.80	(J=10)
		3.57	(J=10)		3.58	(J=10)
11	33.1 (t)	2.37	(m)	--	2.44	(m)
12	98.9 (s)	--		98.6 (s)	--	
13	63.3 (t)	4.27	(J=11,9)	61.1 (t)	4.43	(J=11,6)
		4.05	(J=11,5)		4.28	(J=11,6)

*Assignments may be interchanged.

The isomeric gonyautoxin III (3) appears to be an enolization product of gonyautoxin II. Although the stereochemistry at 11-C has not been definitely established, the 11α configuration appears to be less hindered and has been assigned to gonyautoxin II[17].

Neosaxitoxin possesses a chromatographic behavior very similar to that of saxitoxin except that its movement on electrophoresis is about half that of saxitoxin. The IR spectrum is also very close except for a distinct absorption at 1770 cm^{-1}, which is absent or very weak in the saxitoxin spectrum. The ^{13}C NMR spectrum of neosaxitoxin is shown in Table 1. Both chemical shifts and coupling patterns of neosaxitoxin hydrogens and carbons are very close to those of saxitoxin. The only differences are the chemical shifts of C-5 and C-6 carbons and the hydrogens attached to the carbons.

Treatment of neosaxitoxin with 7.5 N HCl gave a toxic compound which is clearly different from decarbamylsaxitoxin prepared from saxitoxin under the same conditions. Oxidation of neosaxitoxin with H_2O_2 did not afford appreciable amounts of aromatized products. However, when it was reacted in a NaOH solution in the presence of air, two fluorescent compounds with absorption maxima 334 nm and 371 nm were formed.

Table 2. Paralytic shellfish poisoning and related toxins isolated.

Sources	Purified Toxins (+ indicates approximate amounts isolated)									
	STX	GTX_1	GTX_1'	GTX_2	GTX_3	GXX	neo-STX	$APTX_1$	$APTX_2$	$APTX_3$
Gonyaulax tamarensis[1]	++	++	+	+++	+	+	+++			
Mya arenaria[2]	++	++	+	+++	+	+	+			
Mya arenaria[3]	++	+/−		+++	+		+			
Mytilus edulis[4]	+	+++	+	++	+	++				
Tapes (Amygdala) japonica[4]	+	+++	+	++	+	++				
Gonyaulax sp.[5]	+	+++	+	++	+	++				
Saxidomus giganteus[6]	+++						+			
Aphanizomenon flos-aquae[7]	++							++	++	++

[1]Cultured cells

[2]Softshell clams collected at Essex, Massachusetts in 1972 and 1974

[3]Softshell clams collected at Hampton, New Hampshire in 1972

[4]Mussels and short-necked clams collected at Oase Bay, Mie, Japan in 1975, exposed to an unidentified Gonyaulax sp. bloom

[5]A bloom collected at Oase Bay, Mie, Japan in 1975

[6]Alaska butter clams collected in Alaska

[7]Toxic blue-green alga collected at Kezar Lake, New Hampshire

DISTRIBUTION OF PARALYTIC SHELLFISH POISONS

Table 2 shows the distribution of toxins in the organisms examined to date. Seven toxins, including five gonyautoxins, saxitoxin, and neosaxitoxin, have been found in cultured Gonyaulax tamarensis cells. Gonyautoxin II and neosaxitoxin accounted for about 60% of the total toxicity in the dinoflagellate, with saxitoxin making up only about 5%. This is in contrast to the softshell clam M. arenaria, where saxitoxin accounts for 20% of the total toxicity and only trace amounts of neosaxitoxin have been detected. No GTX_5 has been detected in softshell clams. Several explanations for the change in the relative concentrations of the toxins are possible, including preferential adsorption or metabolism in the clams and differences in the stability of the toxins during storage.

Six toxins, gonyautoxins I, II, III, IV and V and saxitoxin, were isolated from shellfish and plankton samples collected during a red tide in Owase Bay, Japan[16]. Although the organism isolated at Owase Bay was identified as G. catanella on the basis of morphology[18], the toxin content more closely resembled that found in G. tamarensis.

Alaska butter clams, which are thought to acquire the toxin from G. catanella, contain about 10% neosaxitoxin in addition to saxitoxin. The presence of neosaxitoxin in Alaska butter clams is interesting because they have been the major source of saxitoxin and were previously reported to contain only saxitoxin[19]. The heterogeneity of the Gonyaulax toxins is also evident when one considers that some Gonyaulax species which are morphologically identical to G. tamarensis are nontoxic[20].

ACKNOWLEDGMENTS

The isolation and chemical work were done mostly at the University of Rhode Island and the spectroscopic work at the Department of Chemistry, Columbia University. The work at the University of Rhode Island was supported by HEW Grant FD-00619 and the Sea Grant Program, University of Rhode Island, R/D3. The work at Columbia University was supported by NIH Grant CA-11572. The authors express their deep gratitude for financial support.

REFERENCES

1. T. Narahashi, Fed. Proc., Fed. Am. Soc. Exp. Biol. 31, 1124 (1972).

2. T. Narahashi, M. S. Brodwick, and E. J. Schantz, Environ. Lett. 9, 239 (1975).

3. E. J. Schantz, J. D. Mold, D. W. Stanger, J. Shavel, F. J.
    Riel, J. P. Bowden, J. M. Lynch, R. S. Wyler, B. Riegel, and
    H. Sommer, J. Am. Chem. Soc. 79, 5230 (1957).

4. E. J. Schantz, J. M. Lynch, G. Vayvada, K. Matsumoto, and H.
    Rapoport, Biochem. 5, 1191 (1966).

5. E. J. Schantz, V. E. Ghazarossian, H. K. Schnoes, F. M. Strong,
    J. P. Springer, J. O. Pezzanite, and J. Clardy, J. Am. Chem.
    Soc. 97, 1238 (1975).

6. J. Bordner, W. E. Thiessen, H. A. Bates, and H. Rapoport,
    J. Am. Chem. Soc. 97, 6008 (1975).

7. E. J. Schantz, Ann. N. Y. Acad. Sci. 90, 843 (1960).

8. M. H. Evans, Br. J. Pharmacol. 40, 847 (1970).

9. N. E. Ghazarossian, E. J. Schantz, H. K. Schnoes, and F. M.
    Strong, Biochem. Biophys. Res. Comm. 59, 1219 (1974).

10. L. J. Buckley, M. Ikawa, and J. J. Sasner, Jr., Proc. First
    Intern. Conf. on Toxic Dinoflagellate Blooms, The Massa-
    chusetts Science and Technology Foundation, 1975.

11. L. J. Buckley, M. Ikawa, and J. J. Sasner, Jr., J. Agri. Food
    Chem. 24, 107 (1976).

12. L. J. Buckley, unpublished data.

13. Y. Shimizu, M. Alam, and W. E. Fallon, Proc. First Intern.
    Conf. on Toxic Dinoflagellate Blooms, The Massachusetts
    Science and Technology Foundation, 1975.

14. Y. Oshima, unpublished data.

15. Y. Shimizu, M. Alam, Y. Oshima, and W. E. Fallon, Biochem.
    Biophys. Res. Comm. 66, 731 (1975).

16. Y. Oshima, W. E. Fallon, Y. Shimizu, T. Noguchi, and Y.
    Hashimoto, Nippon Suisan Gakkaishi, 44, 851 (1976).

17. Y. Shimizu, L. J. Buckley, M. Alam, Y. Oshima, W. E. Fallon,
    H. Kasai, I. Miura, V. P. Gullo, and K. Nakanishi, J. Am.
    Chem. Soc. 98, 5414 (1976).

18. Y. Hashimoto, T. Noguchi, and R. Adachi, Nippon Suisan
    Gakkaishi 42, 671 (1976).

19. E. J. Schantz, V. E. Ghazarossian, H. K. Schnoes, F. M. Strong,
     J. P. Springer, J. O. Pezzanite, and J. Clardy, Proc. First
     Intern. Conf. on Toxic Dinoflagellate Blooms, The Massa-
     chusetts Science and Technology Foundation, 1975.

20. L. A. Loeblich and A. R. Loeblich, III., Proc. First Intern.
     Conf. on Toxic Dinoflagellate Blooms, The Massachusetts
     Science and Technology Foundation, 1975.

* Y. Shimizu, M. Alam, Y. Oshima, L. J. Buckley and W. E. Fallon
   Department of Pharmacognosy, College of Pharmacy
   University of Rhode Island
   Kingston, Rhode Island USA 92881

   H. Kasai, I. Miura, V. P. Gulo and K. Nakanishi
   Department of Chemistry, Columbia University
   New York, New York USA 10027

PHARMACOLOGICAL ACTIVITIES OF PURIFIED TOXINS FROM GYMNODINIUM

BREVE AND PRYMNESIUM PARVUM

G.M. Padilla, Y.S. Kim, M. Westerfield, E. Rauckman

and J.W. Moore

Duke University Medical Center, Durham, N.Carolina, USA[*]

INTRODUCTION

Biologically active natural products may be considered to be
fully characterized only when the relationship between their chemical identity and a specific activity is clearly established. This
goal has been achieved with but a few of the toxins obtained from
marine organisms, the most notable being saxitoxin[1] and tetrodotoxin[2]. Many of the bioactive compounds obtained from phytoplankton "blooms" in "red tides" have not been fully characterized
nor has their action been defined in terms of a specific activity
against a well defined physiological system. This situation arises
not from the lack of experimental acuity or resourcefulness of the
investigators, but largely from the erratic occurrence of the red-
tide outbreaks and the chemical complexity of the toxins themselves,
which occur as mixtures of several natural products[3]. The strategy
that is most often followed is to proceed with purification schemes
of one kind or another while refining the bioassay, which hopefully will reveal the exact structure-function relationship of the
toxin being isolated. We hope to show that this is probably the
most fruitful approach, since it guides the chemist pursuing his
isolation pathways while it demands of the physiologist a clearer
definition of the cellular activity with which the toxin is inter-
acting.

Our investigations have focussed on the toxins derived from
the dinoflagellate Gymnodinium breve, endemic to the west coast of
Florida and the Gulf of Mexico, and the chrysomonad Prymnesium
parvum, which inhibits estuaries and artificial brackish-water
ponds in several countries[4]. Recent studies have shown that high
pressure liquid chromatography (HPLC) will resolve crude extracts

271

of the organisms into individual toxic and non-toxic fractions.
By this procedure, the purity of the toxic components is greatly
enhanced with a minimal level of degradation which may accompany
open bed thin layer chromatography (TLC).

HIGH PRESSURE LIQUID CHROMATOGRAPHY

In a previous study[5] it was shown that the toxins from G. breve
could be separated by a combination of column and preparative TLC
into three toxic fractions (IA, IB, IC) with differing ichthyotoxic
potencies but devoid of pigments. This result was at variance with
those of Trieff et al.[6],who reported the separation of G. breve extracts into at least five toxins using a somewhat more elaborate
isolation procedure. Abbott et al.[7] isolated the neurotoxic component from a haemolytic fraction by passage through an LH_{20} column.
A variety of other analytical procedures have been used by other
investigators to isolate the toxin from these organisms, emphasi-
zing the lack of a universally accepted purification procedure which
renders a comparison of the various studies very difficult[8-11]. In
an effort to resolve this situation we decided to use HPLC to achieve
a direct and rapid isolation of the toxins, on the basis of the
highest resolving power of this technique and the varying polari-
ties of the toxic components of G. breve. Figure 1 shows the
chromatogram of an extract of G. breve eluted through a micro-
particle silica column[12]. The upper trace (Figure 1A) shows that
the separation was incomplete when an isocratic elution mode was
employed. The lower trace (Figure 1B) shows that the crude extract
was resolved into three groups of fractions when a gradient elution
(methanol in chloroform) was used. The first group contained the
major pigments, the second group the ichthyotoxic fractions, and
the third group, being the most polar, contained the haemolytic
component. It probably corresponds to the haemolytic fraction
as noted by Abbott et al.[7]. We did not analyze it further at this
time. Note that group 2 is made of of four distinct sub-fractions.
Further analysis revealed that sub-fraction 2a was the major ichthyotoxic component (see below), was unstable, and upon standing
(at 4°C, dry, in the dark) was converted into sub-fractions 2b, 2c
and 2d. Sub-fraction 2d was found to be non-toxic[12].

A direct comparison of the resolving power between TLC and
HPLC is shown in Figure 2[12]. A partially purified extract of
G. breve was subjected to preparative thin layer chromatography
(2 mm layers, silica gel 60, F-254, E. Merck) with a mixture of
chloroform and methanol (98:2) as the solvent[5]. The PTLC separation is shown on the right-hand side of Figure 2. The bands were
visualized under ultra-violet light. The material from each band
was recovered in chloroform and subjected to high pressure liquid
chromatography as in Figure 1. It is clear that with the exception
of the uppermost band,which consisted largely of pigments, all the

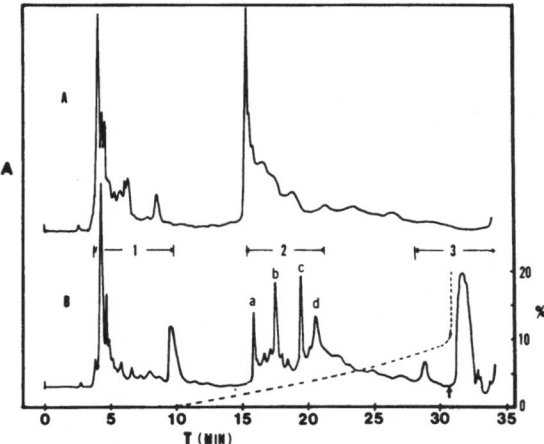

Figure 1. High pressure liquid chromatography of a G. breve
extract using (A) 4% methanol in chloroform and (B) a gradient
elution (i.e. chloroform with increasing concentrations of methanol
as indicated by the dotted line). Left ordinate: A = absorbance
at 254 nm. Right ordinate: % methanol. Abscissa: time (min).
Sample: 1 mg in 25 μl chloroform. Column: Partisil 10/50 (4.6
mm I.D.). Flow rate: 1.5 ml/min. Arrow: methanol wash.
From Kim and Padilla[12] by permission of the Publishers.

other TLC bands were mixtures of HPLC group 2 sub-fractions.
These results demonstrate the greater analytical efficiency of HPLC.

   An indication of the greater resolving power of HPLC became
evident when we compared the relative toxicity of the crude G.
breve extract with that of fractions obtained after HPLC (Figure
3[12]). As shown in this figure the crude toxin (Panel A) had an
LD_{50} of 46.6 μg/ml. Upon separation by HPLC, sub-fractions 2a
(Panel B) and 2b (Panel C) were approximately 100 times more potent
in terms of the LD_{50}. In addition, the steeper slope of the lethal-
ity vs. concentration curves (best seen when plotted as a probit
curve) for the HPLC fractions are an indication of increased homo-
geneity (i.e. purity) of the sample.

   The toxin derived from Prymnesium parvum seems also to be
composed of ichthyotoxic and haemolytic components as recently reviewed by Shilo[4]. They may be complementary to the G. breve toxins
in terms of their relative abundance and polarity. That is to say,
the major component of P. parvum toxin is a relatively polar

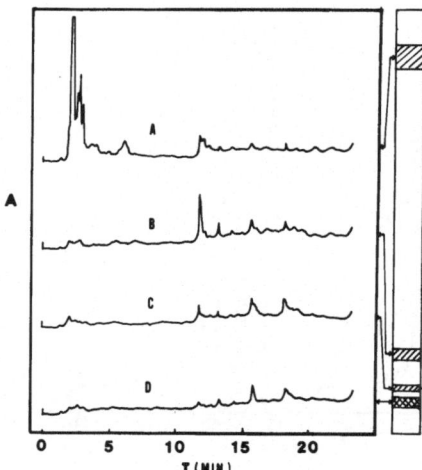

Figure 2. HPLC elution patterns of TLC fractions of partially purified G. breve toxin. Chromatography was done as in Figure 1(B). Tracings A, B, C and D correspond to TLC fractions Id, Ic, Ib and Ia as indicated on the right[5]. Ordinate: A = absorbance at 254 nm. Abscissa: T = time (min). From Kim and Padilla[12] by permission of the Publishers.

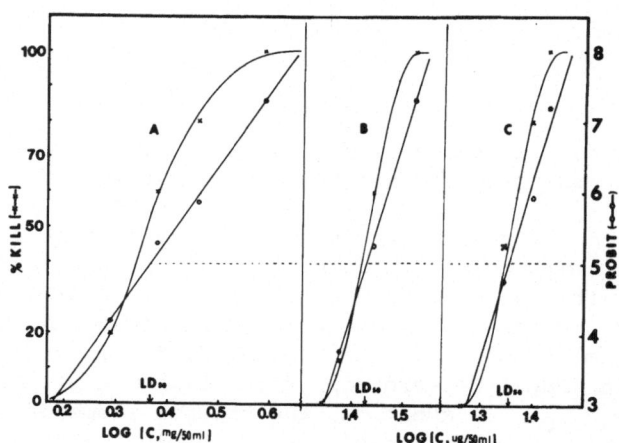

Figure 3. Ichthyotoxicity of a crude extract of G. breve and fractions 2a and 2b. (A) Crude extract, LD$_{50}$ = 46.6 ± 3.6 µg/ml. (B) Fraction 2a, LD$_{50}$ = 0.56 ± 0.04 µg/ml. (C) Fraction 2b, LD$_{50}$ = 0.45 ± 0.05 µg/ml. Left ordinate: % kill. Right ordinate: Probit units. Abscissa: logarithm of toxin concentration (mg) 50/ml for crude extract and µg/50 ml for fractions 2a and 2b). From Kim and Padilla[12] by permission of the Publishers.

haemolysin (PPTX) while in G. breve the major toxic fraction is
not haemolytic but ichthyotoxic (GBTX). It is relatively non-
polar. In addition, as shown in Figure 4, HPLC of a P. parvum
methanol extract (freed of pigments by extraction with chloroform
or acetone) was resolved into as many as seven fractions (if we do
not consider the closely eluting minor peaks in fractions 1-4 as
separate entities). Fractions 1, 2 and 7 were not haemolytic.
Fraction 5 was the major haemolytic component, which was eluted at
∿ 40% methanol. Fractions 3, 4 and 6 were only slightly haemolytic. These results are somewhat at variance with a previous
study by Ulitzur and Shilo[13], who showed that P. parvum extracts
could be resolved by TLC into six haemolytic components. However,
with HPLC the fractions are obtained directly, rather than after
a prolonged step-wise isolation procedure which Ulitzur and Shilo
employed. It should also be emphasized that this toxin is resolved
by a gradient with increasing polarity. At least 10% of the crude
toxin consisted of the haemolytic component on a per weight basis.

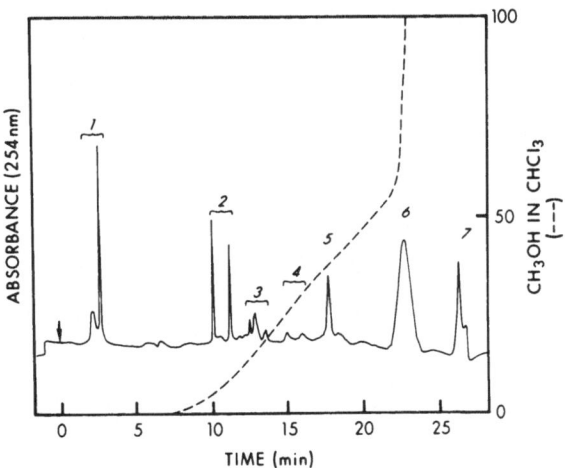

Figure 4. HPLC of a crude methanol extract of P. parvum. Ordinate,
abscissa and chromatography conditions as in Figure 1B. Column;
ODS, 10 micron particles (4.2 mm ID x 250 mm). Solvent gradient
shown by dotted line. Sample: ∿0.5 mg in 25 μl of methanol.

PHYSIOLOGICAL STUDIES

Since calcium ions have been suggested as modulators of the activity of GBTX[14], we decided to study the action of this toxin on isolated sarcoplasmic reticulum (SR) vesicles from rabbit skeletal muscle[15]. This preparation possesses a calcium-dependent ATPase which governs the in vitro transport of Ca ions into the lumen of the vesicles. This system is ideally suited to studying the action of membrane-active compounds, particularly those which might have an interaction with the Ca-transport mechanism, a disruptive effect on the membrane or an iontophoretic activity. The non-haemolytic toxin obtained from G. breve reduced the uptake of Ca ions by SR vesicles, but only at concentrations greater than 4 µg/ml and exposures longer than 10 min (Table 1). The haemolytic component from P. parvum, which has been shown to be a potent membrane-disruptive agent on a great variety of cell types, was much more effective. At 0.5 µg/ml the uptake of Ca ions was reduced to 33% of the control vesicles (Table 1). In the presence of

Table 1. Effect of GBTX and PPTX on Calcium Uptake by Sarcoplasmic Reticulum Vesicles.

Toxin	Conc. (ug/ml)	Relative Ca-Uptake
Control	0	100[a]
	0.5	99
	2.0	102
GBTX	4.0	91
	8.0	60
Control	0	100[b]
	0.5	33
	1.0	21
PPTX	2.0	15
	6.0	12

[a] Ca-uptake: $100 = 0.35$ µmoles $Ca \cdot mg^{-1}$ protein

[b] Ca-uptake: $100 = 2.2-4.4 \times 10^3$ $cmp \cdot ml^{-1}$ vesicle suspension.

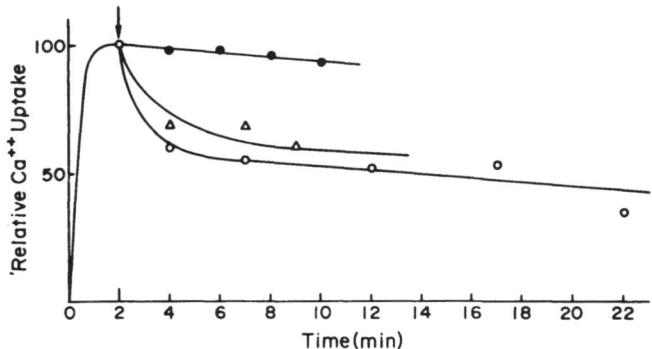

Figure 5. Efflux of calcium in sarcoplasmic reticulum vesicles induced by GBTX in the absence of oxalate. Medium: 0.1 M KCl, 5 mM $MgCl_2$, 2 mM ATP, 50 µM $CaCl_2$, pH 7.3. Vesicle concentration: 80 µg protein ml^{-1}. Toxin was added after 2 min incubation in $^{45}Ca^{++}$ at 23°. Abscissa: time (min). Ordinate: % Ca retained by the vesicles. Control (●———●). Toxin conc: extraction 7.5 µg ml^{-1} (△———△); 12 ug.ml^{-1} (○———○).

oxalate (which supposedly acts as an intraluminal Ca ion trapping agent), G. breve toxin did not induce the efflux of Ca ions except when the SR vesicles were exposed to the toxin for over 30 minutes. It was only in the absence of oxalate ions that GBTX caused a 50% efflux of Ca ions from the SR vesicles within a few minutes (Figure 5). These results suggest that GBTX did not greatly disrupt the membrane but increased the net loss of Ca ions to the surrounding medium.

In contrast, the haemolytic toxin from P. parvum induced a marked efflux of Ca ions, even in the presence of oxalate, which at this concentration (5 mM) has formed an insoluble calcium salt (Figure 6). It would thus appear that PPTX caused a substantial disruption of the SR membrane structure. This effect was probably unrelated to the calcium-dependent ATPase, since its activity was unaffected by the presence of the toxin (results not shown). GBTX, on the other hand, produced a 25% stimulation of the ATPase activity.

A series of electrophysiological voltage-clamp experiments were conducted to determine how GBTX and PPTX affected the activity of the giant axon from the squid. As reported elsewhere, it was found that GBTX induced a spontaneous train of action potentials when applied to the giant axon at ~ 10^{-9} g/ml in artificial or filtered natural sea water [16-18]. The action potentials occurred at frequencies of 200-400 per second in a regular fashion until the

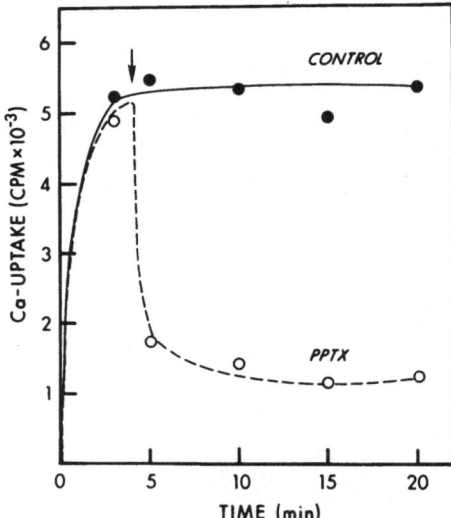

Figure 6. Efflux of calcium from SR vesicles induced by PPTX in
the presence of oxalate (5 mM). Conditions as in Figure 5.
Vesicle concentration: $\sim$ 24 μg protein.ml^{-1}. Toxin concentration:
35.6 μg.ml^{-1}. Abscissa: time (min). Ordinate: ^{45}Ca-uptake
(cpm x 10^{-3}).

nerve was rendered inexcitable. A resting potential of less than
–30 mV was detected at this time. The effect was completely
reversible if GBTX was removed within 2 min of the onset of repetitive firing.

    The toxin-induced action potentials had a shape similar to
that seen in the action potentials obtained by normal electrical
stimulation in untreated axons, except for an increase in the rate
of recovery from the afterhyperpolarization. Because of the faster
recovery induced by GBTX, the membrane potential exceeded the
resting potential and threshold, triggered another spike to produce
the sequence of repetitive action potentials. Voltage-clamp experiments using the sucrose gap technique[19] revealed that GBTX had
no effect on the normal sodium or potassium conductance changes
produced by step depolarization. In addition, consistent with the
faster recovery following an action potential, Westerfield et al.[18]
found that GBTX accelerated the recovery of the "shut-off" currents
to their steady state values following a depolarization. The
authors suggested that GBTX probably caused the repetitive
firing through the induction of a small additional inward current

which was reduced by prehyperpolarization. The toxin-induced
current was blocked by tetrodotoxin (100 mM),added in combination
with 10^{-7} g/ml of GBTX. This observation indicates that the GBTX-
induced current presumably flows through the sodium channel.

As illustrated in Figure 7, PPTX produced a similar effect on
the axonal action potential when added to the external bathing solution at a concentration of 10^{-5} g/ml. Comparison of the two toxins
shows that the potency of GBTX was approximately ten times greater
than that of PPTX on a w/v basis. Moreover, the PPTX-induced
repetitive firing was not reversed by washing the axon in filtered
sea water, even after one hour. The membrane potential finally
became steady at a depolarized value of approximately -30 mV
(after several minutes of repetitive firing). No depolarization
was observed prior to the onset of repetitive firing,as was the
case with GBTX[18]. Additionally, double axial electrode[20] revealed
that PPTX did not induce any changes in the membrane currents,as
were observed with GBTX.

It should be noted that a number of compounds, including the
amino-pyridines[21], scorpion venom[22], tetraethylammonium[23], aconi-
tine[24] and the absence of divalent cations (Westerfield, personal
communication) are known to produce repetitive firing in the giant
axon. The preliminary results presented here suggest that although
GBTX and PPTX induce the same physiological response (repetitive
firing), they may be acting through different mechanisms. GBTX

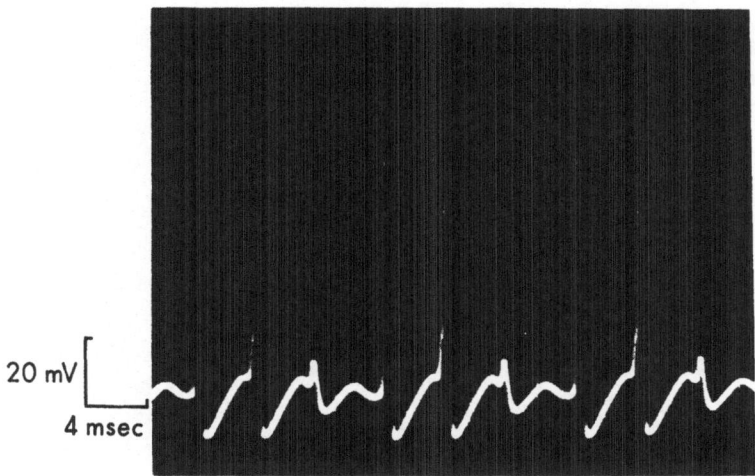

Figure 7. Repetitive firing induced by 10^{-5} g/ml PPTX. The membrane potential (20 mV per division) was recorded with an internal
axial electrode and is plotted as a function of time (5 msec per
division). No electrical stimulation was applied.

modified the "shut-off" currents following an action potential,
while PPTX did not. GBTX depolarized the axon just prior to the
onset of repetitive firing, while PPTX did not.

PARTIAL CHARACTERIZATION OF GBTX

    As noted earlier, it is one of the goals of investigators in
this field to ultimately derive the chemical structure of bioactive compounds from marine sources. We are continuing to subject GBTX to several physicochemical analyses to achieve this
goal. The major obstacle is the lack of material, inasmuch as G.
breve red tides do not occur with great regularity and in the laboratory the cultures are fastidiously slow-growing. We have performed   infrared and NMR analyses of GBTX obtained by HPLC. The
salient features are summarized in Table 2. The UV and fluores-
cence spectra were obtained in a previous study and are presented
elsewhere[12].

Table 2. Proton NMR and IR spectra of GBTX obtained by high
pressure liquid chromatography.

NMR			IR
Chemical Shift (ppm)	Coupling Constant	Possible func- tional group	Wave number (cm^{-1})
9.55 (s)	–	–CHO	1718
8.1 (s)	–	–CONH–, $-NH-\overset{O}{\overset{\|}{C}}-R$	1678
6.21 (d)	J=22	olefin	
5.76 (d)	J=3.5	olefin	} 1605
5.23 (d,m)	J=22	olefin	
4.3–3.6 (m)	–	?	
3.85 (s)	–	$-CH_2-O-$	
3.4–3.0 (m)	–	$-CH_2-$	} 2962–2853
2.6–0.7 (m)	–	$-CH_2-$	

Although it is generally impossible to completely elucidate
the structure of a complex molecule such as GBTX from only absorption spectra, several structural features of the molecule may be
identified (Table 2). The 100 MHz proton NMR immediately shows
that the toxin is primarily aliphatic in nature and contains more
than one double bond. The doublets at 6.21 and 5.23 ppm could
arise from a trans bis-allylic (-CH=CH-CH$_2$-CH=CH-) system, while
the doublet at 5.76 ppm is probably due to an isolated double bond
involved in a small (pentacyclic) ring system. The complex aliphatic region bears a resemblance to oleic acid and is probably a
long chain. The 9.55 ppm signal is characteristic of a saturated
aliphatic aldehyde (the IR and UV data also agree with this). The
8.1 ppm signal could arise from either a secondary amide or a
foramide. The UV absorption maximum of the inactive decomposition
product (fraction 2d) occurred at 234 nm[12]. It is characteristic
of a conjugated hydroperoxide:

$$\begin{array}{c} \text{O-OH} \\ | \\ \text{-C-C=C-C=C-} \end{array}$$

The kinetics of GBTX's decomposition (more rapid when a pure liquid)
also are characteristic of this type of autoxidation. Perhaps the
conversion of the bis-allylic system to a conjugated hydroperoxide
on exposure to O$_2$ is a primary mechanism of environmental detoxification.

ACKNOWLEDGEMENTS

This research was supported by the Food and Drug Administration
grant FDA 00120 (to G.M.P.) and was accomplished with the excellent
technical assistance of Ms. P. Brown. We wish to thank Drs.
D. Brent and S. Hurlburt of the Burroughs-Wellcome Research
Laboratories for the IR and NMR analyses.

REFERENCES

1. E. J. Schantz, U. E. Ghazarossian, H. K. Schnoes, F. M. Strong,
    T. P. Springer, J. O. Pezzanite, and J. Clardy, Proc. 1st
    Int. Conf. on Toxic Dinoflagellate Blooms, Boston, Mass.,
    November 1974, Massachusetts Science and Technology Foundation,
    Wakefield, Mass., 1974, p. 267.

2. T. Narahashi, Proc. 1st Int. Conf. on Toxic Dinoflagellate
    Blooms, Boston, Mass., November 1974, Massachusetts Science
    and Technology Foundation, Wakefield, Mass., 1974, p. 395.

3. G. M. Padilla and D. F. Martin, in Marine Chemistry in the
    Coastal Environment, ed., J. M. Church, Am. Chem. Soc.
    Symposium Series, No. 18, Am. Chem. Soc., Washington, D. C.,
    U.S.A., 1975, p. 596.

4. M. Shilo, in Microbial Toxins, Vol. VII, eds. S. Kakis, A.
    Ciegler and S. J. Ajl , Academic Press, New York, 1971, P. 67.

5. G. M. Padilla, Y. S. Kim, and D. F. Martin, Proc. 1st Int.
    Conf. on Toxic Dinoflagellate Blooms, Boston, Mass., November
    1974, Massachusetts Science and Technology Foundation, Wake-
    field, Mass., 1974, p. 299.

6. N. M. Trieff, V. M. S. Ramanujam, M. Alam, S. M. Ray, and
    J. E. Hudson, Proc. 1st Int. Conf. on Toxic Dinoflagellate
    Blooms, Boston, Mass., November 1975, Massachusetts Science
    and Technology Foundation, Wakefield, Mass., 1974, p. 309.

7. B. C. Abbott, A. Siger and M. Spiegelstein, Proc. 1st Int.
    Conf. on Toxic Dinoflagellate Blooms, Boston, Mass., November
    1974, Massachusetts Science and Technology Foundation,
    Wakefield, Mass., 1974, p. 355.

8. M. T. Doig, III, Ph.D. Dissertation, University of South
    Florida, Tampa, Florida, U.S.A., 1973.

9. M. Alam, J. J. Sasner, Jr., and M. Ikawa, Toxicon $\underline{11}$, 201
    (1973).

10. D. F. Martin and A. B. Chatterjee, U. S. Fish Wildlife Service,
    Fish. Bull. $\underline{68}$, 433 (1970).

11. M. Y. Spiegelstein, Z. Paster, and B. C. Abbott, Toxicon $\underline{11}$,
    85 (1973).

12. Y. S. Kim and G. M. Padilla, Toxicon, in press.

13. S. Ulitzur and M. Shilo, Biochim. Biophys. Acta $\underline{201}$, 350 (1970).

14. Z. Paster, in Animal and Plant Toxins, ed. E. Kaiser, Wilhelm
    Goldman Verlag, Munich, Austria, 1972, p. 115.

15. A. Martonosi and R. Feretos, J. Biol. Chem. $\underline{239}$, 648 (1964).

16. Y. S. Kim, L. J. Mandel, M. Westerfield, G. M. Padilla, and
    J. W. Moore, Environ. Lett. $\underline{9}$, 255 (1975).

17. M. Westerfield, Y. S. Kim, G. M. Padilla, and J. W. Moore,
    Biophys. J. $\underline{16}$, 188a (1976).

18. M. Westerfield, J. W. Moore, Y. S. Kim, and G. M. Padilla.
    Am. J. Physiol. Cell Physiol., in press.

19. J. W. Moore, J. Gen. Physiol. $\underline{48}$, 11 (1965).

20. C. M. Wang, T. Narahashi, and M. Scuka, J. Pharmacol. Exp. Ther. <u>182</u>, 442 (1972).

21. J. Z. Yeh, G. S. Oxford, C. H. Wu, and T. Narahashi, Biophys. J. <u>16</u>, 77 (1976).

22. M. D. Cahalan, J. Physiol. <u>244</u>, 511 (1975).

23. C. Bergman, W. Nonner, and R. Stampfli, Pflugers Arch. <u>302</u> 24 (1968).

24. H. Schmidt and O. Schmitt, Pflügers Arch. <u>349</u>, 133 (1974).

25. M. O. Funk, R. Issac, and N. A. Porter, Lipids <u>11</u>, 113 (1976).

* Full address: Department of Physiology and Pharmacology, Duke University Medical Center, Durham, North Carolina 27710, U.S.A.

ISOLATION AND CHARACTERIZATION OF TOXIC POLYPEPTIDES FROM

SEA ANEMONES

L. Beress and R. Beress

Institut für Meereskunde an der Universität Kiel

G. Wunderer

Institut für Klinische Chemie und Klinische Biochemie

der Universität München, West Germany

In all classes of marine animals there are species which produce toxins for the capture of prey or for defence[1,2]. These toxins belong predominantly to the following chemical groups: organic amines (histamine, serotonin); quaternary ammonium compounds (tetramine); cholinesters (murexin); steroid glycosides (holothurin); diverse heterocyclic compounds (tetrodotoxin, mytilotoxin, trigonellin); and, finally, polypeptides and proteins. Several poisonous marine animals also contain substances with antimicrobial[3,4], antiviral[5], hormonal[6], anti-inflammatory[7,8], cystostatic[9], cardiovascular[10], and cardiotonic[11,12,13] activities.

Many of these substances are destined for further pharmacological investigation as potential drugs, and the research in this field opens many possibilities to the pharmacologist, the physiologist, and the biochemist. The isolation and characterization of sea anemone toxins is an example. From the sea anemone Anemonia sulcata, three toxic polypeptides were isolated, and from Condylactis aurantiaca, four were isolated[14,15]. All these toxins paralyse crustaceans, following intramuscular injection, by acting on the neuromuscular transmission[16,17]. Toxin II of Anemonia sulcata acts on the sodium channel[18,19,20]. It has been shown that Toxins I and II of Anemonia sulcata are very potent cardiotoxins[11]; this has given rise to further pharmacological investigations[12,13].

The sea anemones are nettling animals which belong to the phylum Coelenterata. They possess tentacles which are equipped

285

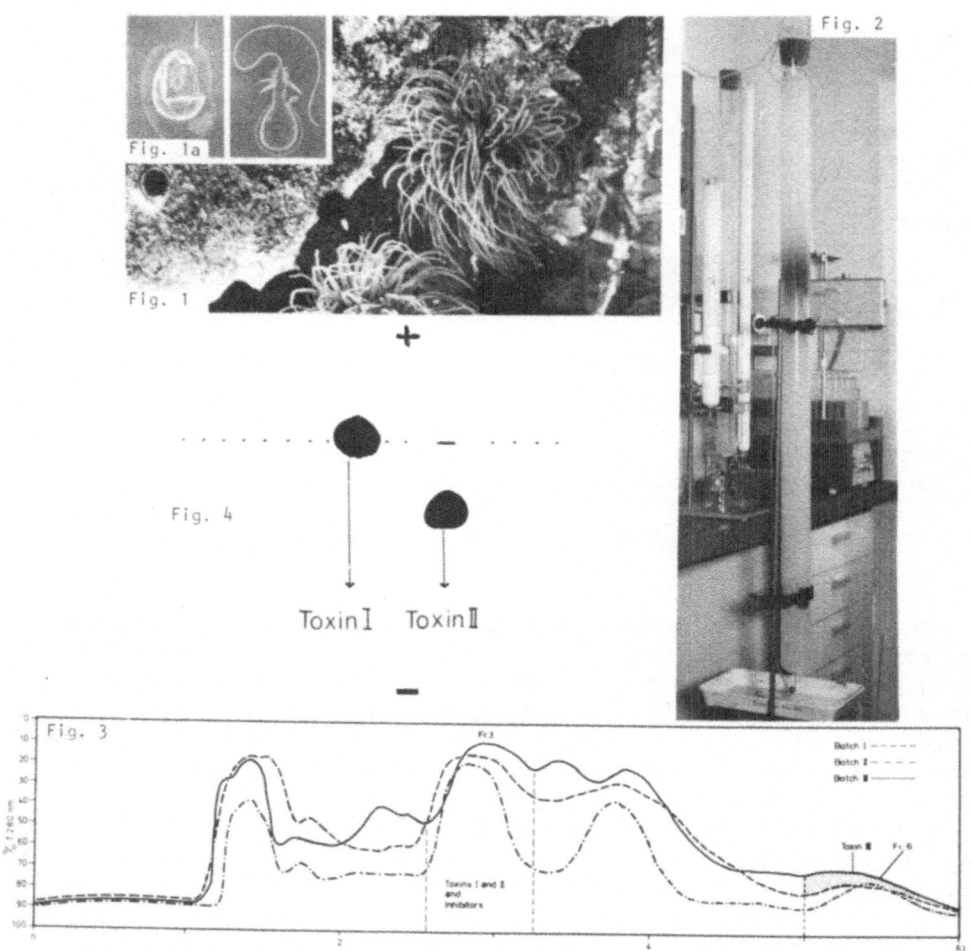

Figure 1.   Anemonia sulcata Pennant.   Figure 1a.   Nematocysts.
Figure 2.   Chromatographic column for gel filtration on Sephadex G-50.
Figure 3.   Chromatographic patterns of Anemonia sulcata peptides
            after gel filtration on Sephadex G-50.
Figure 4.   Electrophoretic pattern of·the pure toxins from Anemonia
            sulcata after electrophoresis on PAA gel, pH 8.6.

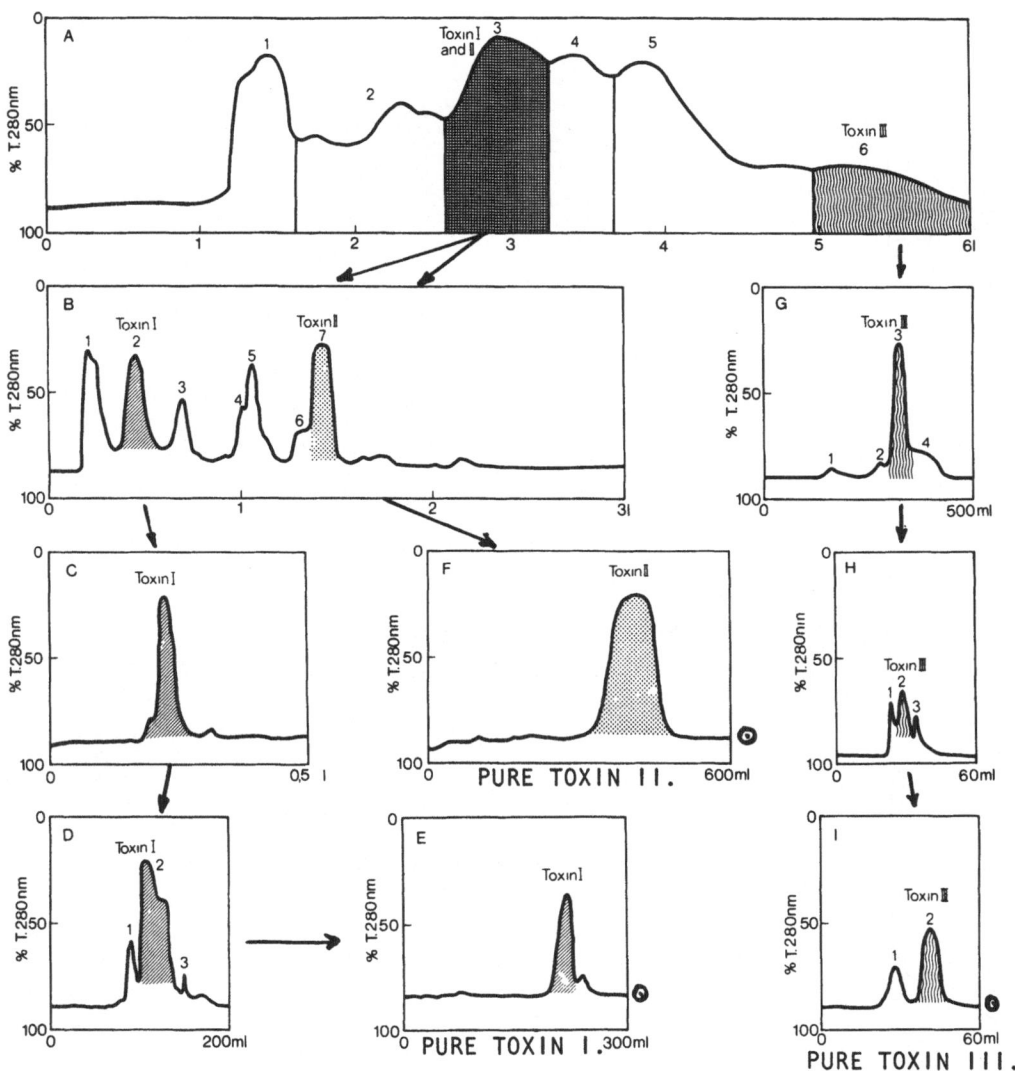

Figure 5.

A Gel filtration of <u>Anemonia</u> <u>sulcata</u> crude toxin on Sephadex G-50.
B Ion exchange chromatography of the toxin fraction on SP-sephadex C25.
C Ion exchange chromatography of Toxin I on QAE-Sephadex A-25.
D Chromatography of Toxin I on Sephadex G-25 Superfine.
E Ion exchange chromatography of Toxin I on SP-Sephadex C-25.
F Ion exchange chromatography of Toxin II on SP-Sephadex C-25.
G Gelfiltration of Toxin III on Sephadex G-25.
H Gelfiltration of Toxin III on Sephadex G-10.
I Gelfiltration of Toxin III on Biogel P-2.

Table 1. Amino acid compositions of Condylactis aurantiaca toxins and Anemonia sulcata toxins

Amino Acid	Condylactis aurantiaca				Anemonia sulcata		
	Toxin I	Toxin II	Toxin III	Toxin IV	Toxin I	Toxin II	Toxin III
Asp	6.02 (6)	7.00 (7)	4.00 (4)	2.10 (2)	5	4	1
Thr	2.80 (3)	3.10 (3)	2.72 (3)	3.98 (4)	2	2	0
Ser	5.60 (6)	7.00 (7)	3.25 (4)	3.70 (4)	4	4	2
Glu	6.06 (6)	4.80 (5)	6.08 (6)	3.93 (4)	2	1	2
Pro	1.16 (1)	0.94 (1)	2.09 (2)	3.08 (3)	2	4	2
Gly	5.00 (5)	6.05 (6)	5.30 (5)	8.00 (8)	8	8	4
Ala	0.87 (1)	0	1.95 (2)	0.53 (0-1)	3	1	0
Cys (a)	4.99 (6)	5.90 (6)	4.11 (6)	6.20 (6)	6	6	4
Val	2.07 (2)	1.94 (2)	1.15 (1)	1.07 (1)	1	1,5 ⎫	1
Met (a)	0.81 (1)	0.85 (1)	0.59 (1)	0.75 (1)	1	0 ⎬5 (b)	0
Ile	0.99 (1)	1.00 (1)	1.20 (1)	0.96 (1)	2	3,5	0
Leu	2.96 (3)	3.20 (3)	2.08 (2)	1.95 (2)	1	3	0
Tyr	0.95 (1)	1.03 (1)	2.35 (3)	4.07 (4)	1	0	2
Phe	0.97 (1)	1.08 (1)	1.20 (1)	1.86 (2)	1	0	0
Trp (c)	1.95 (2)	1.93 (2)	1.20 (1)	0.89 (1)	2	3	2
His	1.50 (1-2)	1.89 (2)	0.55 (0-1)	0.86 (1)	0	2	0
Lys	1.98 (2)	2.12 (2)	4.95 (5)	4.00 (4)	2	3	1
Arg	1.22 (1)	1.16 (1)	2.00 (2)	1.07 (1)	2	1	1
Total	49 – 50	51	49 – 50	49 – 50	45	47	24
Mol. wt.	5400-5537	5599	5444-5630	5397-5468	4702	4770	2678
N-terminus	Gly	Gly	Gly	Gly	Gly	Gly	Arg

a) Determined as cysteic acid and methionine sulfone, respectively, after performic acid oxidation.
b) Microheterogeneity Ile/Val in position 2 of the sequence (24).
c) Determined spectrophotometrically.

with thousands of nematocysts. Small fishes and crabs are quickly paralysed after coming into contact with the tentacles of the sea anemone. Anemonia sulcata is the most common sea anemone in the Mediterranean and has the highest toxin content (Figure 1).

The isolation of toxins from Anemonia sulcata was accomplished by alcoholic extraction of the homogenised sea anemones, followed by batchwise adsorption onto cation exchangers, gel filtration on Sephadex G-50 (Figures 2 and 3), and ion exchange chromatography on SP-Sephadex C-25 and QAE-Sephadex A-25 (Figure 5). After the samples were desalted with Sephadex G-25 and G-10 and Biogel P-2, pure anemone toxins were obtained (Figure 5). Their purity was ascertained by electrophoresis on polyacrylamide gel at pH 8.6 (Figure 4). The isolation of Condylactis toxins[15] and Bolocera toxins[21] was carried out in a similar fashion.

The toxins were tested on the shore crab Carcinus maenas as described earlier[22]. The lethal dose (LD$_{100}$) of the toxins ranges between 2 and 6 µg/kg Carcinus maenas[14,15]. The amino acid composition and molecular weights of the toxins from Anemonia sulcata and Condylactis aurantiaca are given in Table 1.

Detailed investigations were carried out on the structural chemistry of toxins from Anemonia sulcata; the sequence of Toxin II has been recently established[23,24,25]. Further structural studies on Toxin II of Anemonia sulcata have been carried out by Laser-Raman spectroscopy[26].

The method elaborated for the isolation of the toxins from Anemonia sulcata has proved to be a generally applicable method for the isolation of biologically-active polypeptides such as proteinase inhibitors[7,8] and some glycopeptides and glycoproteins[27]. The method opens up possibilities for the isolation of further biologically-active polypeptides of marine origin.

ACKNOWLEDGEMENTS

We thank Prof. Dr. Hans Fritz, Institut für Klinische Chemie und Klinische Biochemie der Universität München, for help and advice, and Mr. Joachim Zwick for technical assistance. This work was supported by the Deutsche Forschungsgemeinschaft. Grants Be 554/4-6 (L.B.) and SFB 51, München (G.W.).

REFERENCES

1. B. W. Halstead, Poisonous and Venomous Marine Animals of the World, U. S. Government Printing Office, Washington, D.C., 1965, 1967, 1970.

2. M. T. Doig, D. F. Martin, and G. M. Padilla, in Marine
    Pharmacognosy, eds. D. F. Martin and G. M. Padilla, Academic
    Press, New York and London, 1973.

3. M. F. Stempien, Jr., J. S. Chib, R. F. Nigrelli, and R. A.
    Mierzewa, in Food and Drugs from the Sea Proceedings, ed.
    L. R. Worthen, Marine Technology Society, Washington, D. C.,
    1972.

4. R. J. Andersen and D. J. Faulkner, in Food and Drugs from the
    Sea Proceedings, ed. L. R. Worthen, Marine Technology Society,
    Washington, D. C., 1972.

5. P. R. Burkholder, in Food and Drugs from the Sea Proceedings,
    ed. H. D. Freudenthal, Marine Technology Society, Washington,
    D. C., 1968.

6. W. P. Schneider, L. E. Rhuland, R. D. Hamilton, G. L. Bundy,
    E. G. Daniels, F. H. Lincoln, and J. E. Pike, in Food and
    Drugs from the Sea Proceedings, ed. L. R. Worthen, Marine
    Technology Society, Washington, D. C., 1972.

7. H. Fritz, B. Brey, and L. Beress, Hoppe-Seyler's Z. Physiol.
    Chem. 353, 19 (1972).

8. G. Wunderer, K. Kummer, H. Fritz, L. Beress, and W. Machleidt,
    in Bayer Symposium V: Proteinase Inhibitors, eds. H. Fritz,
    H. Tschesche, L. J. Greene, and E. Truscheit, Springer Verlag,
    Berlin, Heidelberg and New York, 1974.

9. J. Hurst, E. Premuzic, and J. Bairdi, in Food and Drugs from
    the Sea Proceedings, ed. L. R. Worthen, Marine Technology
    Society, Washington, D. C., 1972.

10. P. N. Kaul, in Food and Drugs from the Sea Proceedings, ed.
    L. R. Worthen, Marine Technology Society, Washington, D. C.,
    1972.

11. C. Alsen, Naunyn-Schmiedeberg's Arch. Pharmacol. Suppl. 287,
    105 (1975).

12. C. Alsen, L. Beress, K. Fischer, D. Proppe, T. Reinberg, and
    R. W. Sattler, Naunyn-Schmiedeberg's Arch. Pharmacol., in
    press.

13. U. Ravens, Naunyn-Schmiedeberg's Arch. Pharmacol., in press.

14. L. Beress, R. Beress, and G. Wunderer, Toxicon 13, 359 (1975).

15. R. Beress, L. Beress, and G. Wunderer, Hoppe-Seyler's Z. Physiol. Chem. 357, 409 (1976).

16. W. Rathmayer, B. Jessen, and L. Beress, Naturw. 62, 538 (1975).

17. W. Rathmayer and L. Beress, J. Comp. Physiol., in press.

18. G. Romey and M. Lazdunski, Abstract, Biophysics Congress, Copenhagen, 1975.

19. C. Bergmann, J. M. Dubois, E. Rojas, and W. Rathmayer, C. R. Acad. Sc. Paris 282, 1881 (1976).

20. G. Romey, J. P. Abita, H. Schweitz, G. Wunderer, and M. Lazdunski, Proc. Nat. Acad. Sci., in press.

21. L. Beress and J. Zwick, Mar. Chem., submitted.

22. L. Beress and R. Beress, Kieler Meeresforsch. 27, 117 (1971).

23. G. Wunderer, Dissertation, Technische Universität, München, 1975.

24. G. Wunderer, W. Machleidt, and E. Wachter, Hoppe-Seyler's Z. Physiol. Chem. 357, 239 (1976).

25. G. Wunderer, H. Fritz, E. Wachter, and W. Machleidt, Eur. J. Biochem. 68, 193 (1976).

26. B. Prescott, G. J. Thomas, Jr., L. Beress, G. Wunderer, and A. T. Tu, FEBS Lett. 64, 144 (1976).

27. H. Theede, R. Schneppenheim, and L. Beress, Mar. Biol. 36, 183 (1976).

CHEMISTRY RELATED TO THE SEARCH FOR DRUGS FROM THE SEA

F. J. Schmitz et al.*

Department of Chemistry, University of Oklahoma

Norman, Oklahoma 73019 USA

In the search for bioactive compounds from marine sources, the selection of specific organisms and classes of compounds to be studied is dictated by bioassay, rather than the conscious choice of the investigator. As a result, efforts are often not focused on a given phylum or class of compounds. Instead, animals from different phyla and compounds varying greatly in size, type, and structural complexity are dealt with. The collection of different organisms and assorted types of natural products described in this paper illustrates these points.

The two types of drug candidates for which we have elected to search are central nervous system active agents and tumor-inhibitory substances. Some results in these two areas will be discussed first; the remainder of the paper will deal with new marine natural products discovered as a spin-off of the search for drugs or in parallel basic natural products research.

CENTRAL NERVOUS SYSTEM DEPRESSANT SUBSTANCE

Bioassay of a random collection of marine organisms revealed that extracts of the sea hare Aplysia dactylomela possessed central nervous system (CNS) depressant activity, as evidenced by the potentiation of pentobarbital hypnosis and by depression in both spontaneous and locomotor activities. The extract had a consistent effect on the parameters investigated, and the activity of some fractions at a dose of 10 mg/Kg was comparable to that of a 3 mg/Kg dose of chlorpromazine. Bioassay-guided fractionation, using a sequence of solvent partitioning followed by LH-20 and adsorption chromatography, led to the isolation of an active

293

DACTYLYNE

fraction that caused strong potentiation of pentobarbital hypnosis but caused little or no depression in spontaneous and locomotor activities. The active compound was identified as dactylyne, a halogenated ether isolated earlier in an investigation of this animal for tumor inhibitory substances[1].

Pharmacological evaluation of pure dactylyne indicated that it is non-toxic up to 200 mg/Kg administered intravenously and that it potentiates pentobarbital-induced sleep-time significantly (200-600% at 25 mg/Kg; significant potentiation is still noted at 10 mg/Kg). Dactylyne does not induce hypnosis by itself, nor does it exhibit any antipsychotic (tranquilizer) or analgesic activity. Some anticonvulsant activity was observed. These results suggest that dactylyne's principal pharmacological potential is in the area of general cortical depressants. Dactylyne was assayed for antibiotic activity against fifteen different microorganisms and was found to have activity against only Sarcina lutea, a gram-positive organism. An 18-mm zone of inhibition was obtained using a 6.35-mm pad and solution concentrations of 1 mg/ml.

TUMOR-INHIBITORY SUBSTANCES

We are currently fractionating extracts of a number of marine organisms in order to isolate tumor-inhibitory compounds whose presence is indicated by NCI tests of the crude extracts. Two active materials isolated from sponges are described below. The first of these is a very simple compound, while the second is a toxic complex of high molecular weight that has resisted all our efforts to resolve it into pure components.

Fractionation of the in vivo and in vitro active fractions of the sponge Anthosigmella varians has led to the isolation of one of the antineoplastic agents from this sponge. Several other active fractions are still under investigation. The active purified substance was easily identified as p-hydroxyphenylacetamide by comparison of spectral data and mp (174-175°) with an authentic sample. This phenol shows in vitro activity in the National Cancer Institute's P388 cell culture system[2] at 0.5 μg/ml. In the in vivo PS test[2], T/C values of 140% were obtained for the active chromatographic fraction from which p-hydroxyphenylacetamide was obtained.

Efforts are underway in our laboratory to prepare analogs and derivatives of this phenol for antineoplastic testing.

Extracts of the sponge Haliclona rubens showed activity against the KB cell line[2] and were toxic in PS testing down to doses of 5 mg/Kg. At lower, non-toxic doses, no in vivo tumor inhibition was noted. Fractionation of this sponge extract was undertaken to identify the cytotoxic agent and to determine whether the cytotoxicity and toxicity to mice and fish were due to the same or different components in the sponge extract.

In an earlier investigation[3] it had been found that the crude aqueous extracts of various species of the sponge genus Haliclona were toxic to mice at doses of >275 mg/Kg wt, and the name halitoxin was coined for the crude, uncharacterized extract derived from a Caribbean species, H. viridis. We have elected to use the name halitoxin-R for the toxic factor in the extracts of H. rubens, the suffix R serving to distinguish the toxin of this species from similar toxins that may be isolated from other species.

The most expeditious method we have found for isolating halitoxin-R in fairly pure form is outlined in Scheme 1. The toxin is quite effectively extracted by 1-butanol from the defatted water-soluble fraction. The 1-butanol-soluble fraction was purified by ultrafiltration, using a series of membranes having different molecular-weight cut-off ranges. After lyophilization, our purest fractions are hygroscopic brown foams or glasses. The typical activity of crude extracts and various fractions are listed in Table 1.

Numerous other methods of purification of halitoxin-R have been attempted. Most are more involved experimentally and more time-consuming; none appear to be more effective. Little additional purification was effected by chromatography subsequent to the ultrafiltration. The alternate or supplementary means of purification consisted largely of chromatography, using a variety of the common inorganic adsorbents, ion exchange resins, Sephadex polypropylene powder, 10% acetylated cellulose, and carbowax-treated controlled pore glass beads.

All of the fractions containing <500 mol. wt. materials exhibit virtually the same nmr spectrum and show comparable levels of cytotoxicity and toxicity to mice (Table 1). The 500-1000 mol. wt. range fraction has been used in structure elucidation work for the most part, but not exclusively. Halitoxin does not possess any type of carbonyl functionality, as judged from the absence of any appropriate ir absorption. The absence of hydroxy groups was inferred from the failure of halitoxin-R to incorporate any acetyl or formyl groups under a variety of experimental conditions.

SCHEME 1

Isolation of Halitoxin-R

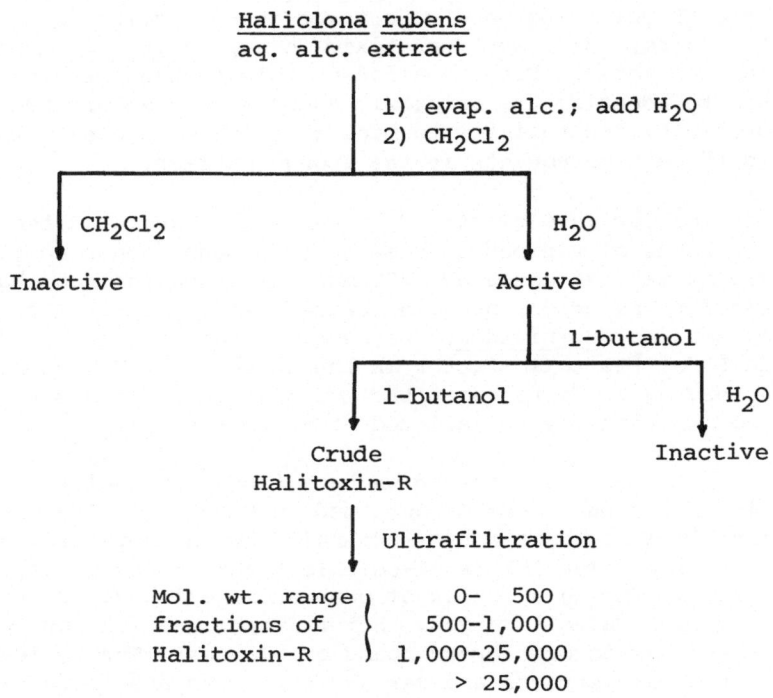

<u>Haliclona</u> <u>rubens</u>
aq. alc. extract

1) evap. alc.; add H_2O
2) CH_2Cl_2

CH_2Cl_2 H_2O

Inactive   Active

1-butanol

1-butanol H_2O

Crude   Inactive
Halitoxin-R

Ultrafiltration

Mol. wt. range ⎫   0- 500
fractions of   ⎬   500-1,000
Halitoxin-R   ⎭ 1,000-25,000
                        > 25,000

Table 1.   Cytotoxic activity[2] of <u>H</u>. <u>rubens</u> extract.

<u>H</u>. <u>rubens</u>

<u>KB</u>

Crude Extr.   7.0 (LD_{50} 5 mg/Kg)

n–BuOH Extr.   4.8

<u>M. wt. Frac.</u>

    0 -   500   >100

  500 -   1,000   5.8

1,000 - 25,000   (∿5)

      > 25,000   ∿5

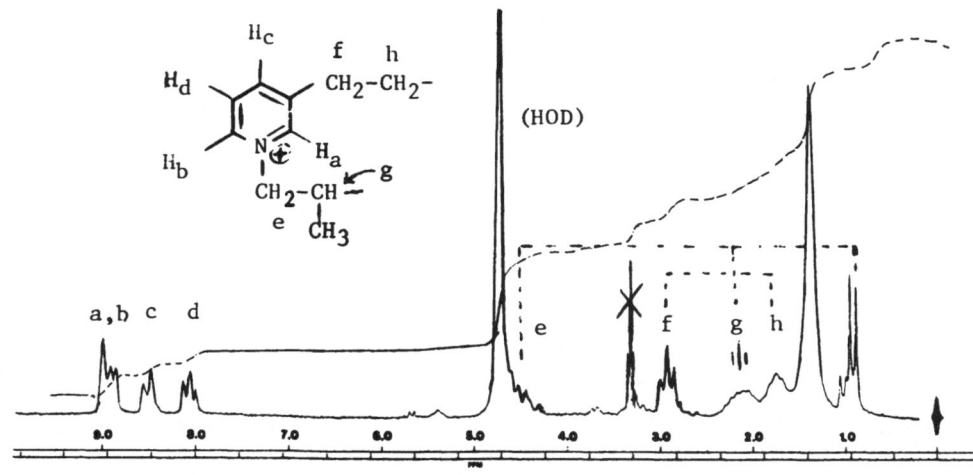

Figure 1. Nmr spectrum of halitoxin-R (100 MHz, CD₃OD).

Furthermore, the proton magnetic resonance spectrum (see Figure 1)
does not show any signals attributable to protons deshielded by
hydroxyl groups. The toxin exhibits a uv spectrum characteristic[4]
of an alkyl pyridinium salt (λ_{max} 267 nm, 273 (sh), 212).

The nmr spectrum (see Figure 1) confirms the alkyl pyridinium
nature of halitoxin-R and reveals many of its structural features.
The position of the low field signals and their multiplicity pattern
reveal that the pyridine ring is quaternized and is 3-substituted.
The absorption at δ 4.5, labeled e, is attributed to the methylene
group attached to the pyridinium nitrogen. This signal is clearly
observable as a complex multiplet after the large signal, due to
water (δ 4.8), is shifted downfield by the addition of a few
percent of pyridine-d₅ to the solution. A methyl group is located
on the aliphatic carbon β to the pyridinium nitrogen. This is
evident from the mutual coupling of the e signal and the larger
methyl doublet (δ0.9) with the g multiplet. At 360 MHz, signal g
is resolved into two broad signals, only one of which (one proton)
is coupled to both the larger methyl doublet and the e multiplet.

The first segment of the pyridine 3-alkyl substituent is shown
to consist of two contiguous methylene groups by the triplet nature
of the benzylic protons f; the f triplet is collapsed to a singlet
by irradiation at the position of the h multiplet. The remaining

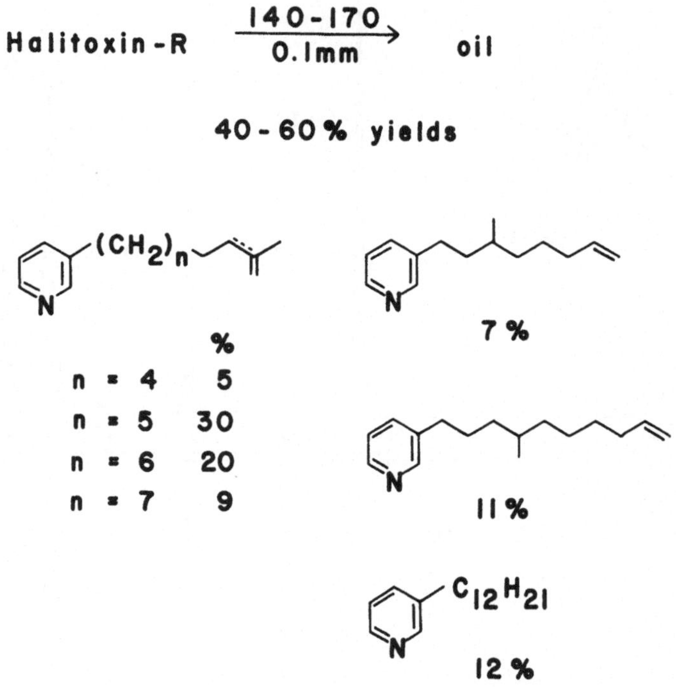

Figure 2. Pyrolysis products of halitoxin-R.

structural features evident from the nmr spectrum are a moderate
number of methylene groups and an additional secondary methyl
group (δ 1.0,d). At 360 MHz, the δ 1.0 methyl doublet is clearly
separated from the larger doublet, and the complete absence of any
n-alkyl terminal methyl groups is confirmed, a fact whose impli-
cation will be developed below.

Additional details of the N-alkyl 3-alkylpyridinium structure
suggested by the spectral data were obtained by chemical degradation. Pyrolysis of halitoxin-R at 140-160°C afforded a mixture of
3-alkenylpyridines in 40-60% yields. Attempts to break the N-alkyl
linkage in halitoxin-R by nucleophilic displacements or base-
catalyzed elimination in solution were unsuccessful. The 3-alkenyl-
pyridines obtained by pyrolysis were assigned the structures shown
in Figure 2 on the basis of nmr and ms data of fractions collected
by preparative gc. Combined gc/ms analysis, using electron impact
and chemical ionization, supports these assignments. The positions
of the secondary methyl groups are regarded as somewhat tentative.

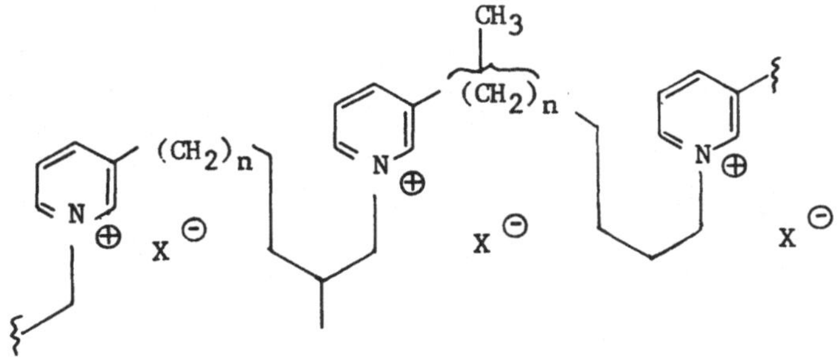

n = 3, 4, 5, 6

Halitoxin-R (Mol. Wt.: >500- ?)

Since there were no olefinic proton or vinyl methyl signals
in the nmr spectrum of halitoxin-R, the unsaturation sites in the
3-alkenylpyridines clearly mark the sites of N-alkyl links in the
toxin. This corroborates the structure deduced from nmr data.
The isopropylidene groups probably arise by acid-catalyzed isomer-
ization of the expected isopropenyl group, the acid having arisen
from the elimination reaction. This leads to the generalized
linear polymeric structure shown above for halitoxin-R. The toxin
is thus a mixture in which two types of variations are present:
(a) Within a given molecule the pyridine rings are connected by an
assortment of alkyl chains differing in length and position of the
methyl substituent; and (b) among different molecules there are
variations in the relative amounts of the different alkyl connect-
ing links.

We further suggest an overall macrocyclic structure for the
500-1000 mol. wt. range materials. This conclusion is based on
the following arguments: (1) There is no evidence for any non-
quaternized pyridine rings in the nmr spectrum, even when the

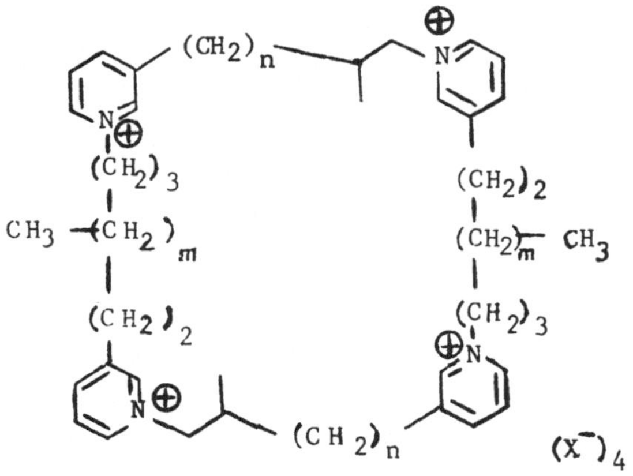

Halitoxin-R (n = 6, 7, 8, 9; m = 3, 5)

spectrum is run in the presence of base ($Na_2CO_3/D_2O/CD_3OD$).
(2) Alkyl chains terminating in methyl or isopropyl groups are
ruled out, judging from the nature of the alkenylpyridines obtained
as pyrolysis products and the nmr spectrum of the toxin itself.
In view of the apparent absence of saturated C-terminal chains and
non-quaternized pyridine rings, a macrocyclic structure is inferred.
Phosphate or sulfate links between the nitrogen and the alkyl groups
are ruled out by the elemental analyses of the toxin.

The indicated tetrameric nature of the 500-1000 mol. wt. range
fraction of halitoxin-R is a very tentative conclusion based on
sketchy FD mass spectral data of the saturated product obtained
upon catalytic hydrogenation. Very weak ions as high as m/e 630
have been observed, with no indication that the molecular ion has
been detected. Efforts to obtain more reliable and complete FD
mass spectra of the reduction product are underway.

We have detected halitoxin-like material in several different
Haliclona species (see Table 2). The presence of such material
was confirmed by nmr analysis of the butanol-soluble fractions of
the sponge extracts (cf. Scheme 1). In addition to cytotoxicity,
halitoxin-R causes haemolysis at concentrations of 1 µg/ml and also
causes irreversible neuromuscular blockade. The latter activity is
presumably due to the repeating quaternary ammonium functionality,
a feature present in other neuromuscular blocking agents such as
the curare alkaloids and decamethonium. Out of a battery of
fifteen microorganisms, halitoxin-R showed significant antibiotic
activity against only two, namely Bacillus subtillus and Strepto-
coccus pyogenes, both gram-positive organisms.

We have thus far been unable to detect in the total sponge
extract any simple alkyl pyridine derivatives that might be likely
precursors of halitoxin-R.

Table 2. Occurrence and bioactivity of halitoxin from various
Haliclona species.

	Halitoxin-R	KB	LD
H. rubens	+	7.0	5 mg/Kg
H. viridis	+	2.8	2-3
H. permallis (?)	-	100	Not toxic
Haliclona sp.	+	26	∿3

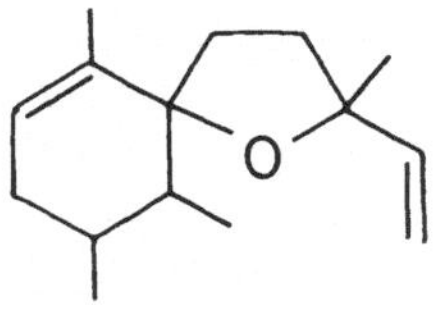

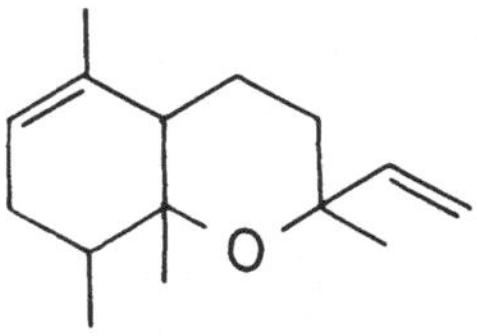

SESQUITERPENOIDS FROM THE SEA HARE <u>APLYSIA</u> <u>DACTYLOMELA</u>

Earlier work on the lipid extracts of the digestive glands of <u>Aplysia</u> <u>dactylomela</u> has resulted in the structure elucidation of sesquiterpenoids[5] and halogenated C_{15} ethers[1,6]. The structures of some sesquiterpene ethers isolated earlier are shown above.

In further work with the extracts of this sea hare, two ter- tiary sesquiterpene alcohols, dactylenol and dactylol, and also the acetate of dactylenol, have been isolated. Dactylenol, $C_{15}H_{24}O$, M^+ 220, ir: 3400, 3090, 1640 cm^{-1}, $[\alpha]_D$+ 204^0, and its acetate, M^+ 60, 202, $[\alpha]_D$ + 168^0, were interrelated by removal of the acetate by lithium aluminum hydride reduction and partial acetylation of the tertiary alcohol, using acetic anhydride in pyridine at 50^{0}C. Dactylenol, an oil, possessed spectral charac- teristics which were reminiscent of those of dactyloxene-A, -B, and -C. On the basis of its nmr spectrum[7] and its co-occurrence with dactyloxene-A, -B, and -C, the structure shown below was proposed for dactylenol. However, two other possible structures (see partial structures A and B below) could not be ruled out from

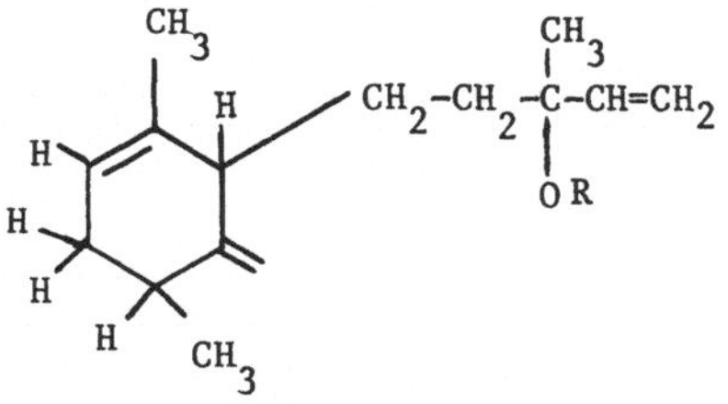

DACTYLENOL   R = H

DACTYLENOL ACETATE R = Ac

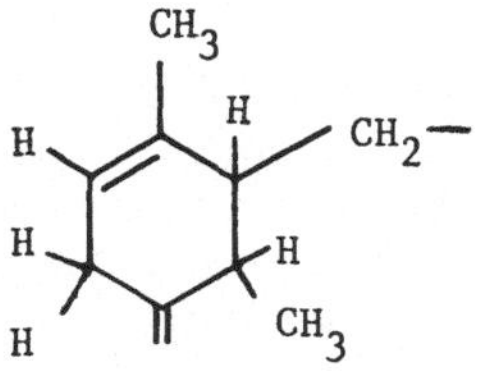

A   OR   B

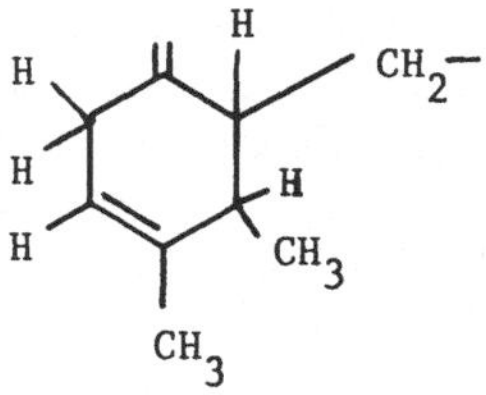

this spectrum. One feature that distinguishes the full structure
proposed for dactylenol from the other two possible partial structures, A and B, is that the methine proton on the carbon bearing
the side chain in the full structure is doubly allylic and would
be coupled in the nmr spectrum vicinally to only the side-chain
methylene group. In the partial structures A and B, this same
methine proton would also be coupled to an adjacent ring proton.
The signal for this methine proton could not be observed independently in the routine nmr spectrum of dactylenol, due to inadequate
chemical shift differences of the various allylic protons in this
compound, but in the presence of 0.45 molar shift reagent Eu(fod)$_3$
it was clearly discernible. Under these conditions, each of the
non-equivalent protons resonates at a distinctly different position,
and by decoupling it was shown that the distinguishing methine
proton (δ 4.36, br d, J = 10, 4($\frac{1}{2}$ w)) is coupled only to the side-
chain methylene group (δ 6.26, 6.82, m, 1 ea) as was expected in
the structure shown for dactylenol. This alone rules out partial
structures A and B, since, in these, the methine proton in question
would also be coupled to another vicinal proton in the ring. Additional evidence in favor of the full structure selected for
dactylenol is that in the europium-shifted spectrum the methine
proton (δ 3.66, m) of the ring carbon bearing the secondary methyl
group is definitely coupled to the allylic methylene group (δ 2.02,
2.63), an interaction that would not occur in A and B. In summary,
the complete structure proposed for dactylenol could be deduced
solely on the basis of the chemical shift and complete proton coupling information derived from the Eu(fod)$_3$-shifted spectrum.

The second alcohol isolated from A. dactylomela is a solid,
mp 50-51°, $C_{15}H_{26}O$ (M$^+$, 222), ir: 3600 cm^{-1}. The tertiary nature
of this alcohol, named dactylol, was confirmed by the off-resonance
decoupled ^{13}C nmr spectrum, singlet at δ 83.3, and the absence of
any signals attributable to -CH-OH in the pmr spectrum. The pmr
spectrum (Figure 3a) showed signals indicative of a -CH$_2$-CH=C(CH$_3$)-
unit (δ 1.8 (3H); 5.45 (1H, t, with further fine splitting)), two
quaternary methyl groups (δ 0.9, 6H, s), and one secondary methyl
(δ 0.93, only one member of which is visible in Figure 3a). The
^{13}C nmr spectrum confirmed that there were only two sp^2-hybridized
carbons (δ 135.4 and 125.5) in dactylol, and hence it must be
bicyclic.

One other significant feature in the pmr spectrum of dactylol
is the AB quartet whose members are centered at δ 2.1 and 2.35.
The chemical shift of these signals suggests that they are allylic
and possibly also α to a carbinol carbon. Thus the partial formula
containing the double bond may be expanded to -CH$_2$-CH=C(CH$_3$)-CH$_2$-C-OH.

Further structural information was obtained by decoupling of
the Eu(fod)$_3$-shifted spectrum of dactylol (Figure 3b). In this
spectrum, the AB quartet signals have been shifted to 6.6 and 8.45,
consistent with the proposed proximity of the isolated methylene
group to the tertiary hydroxyl.

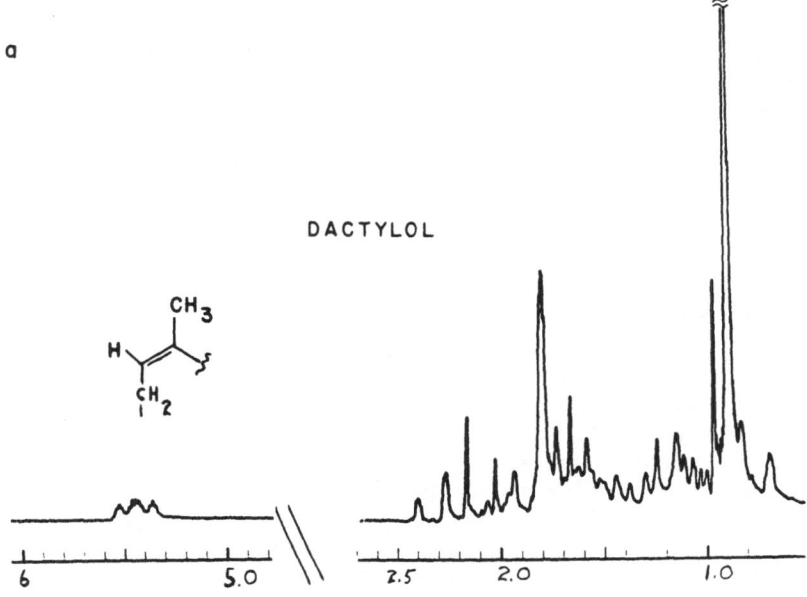

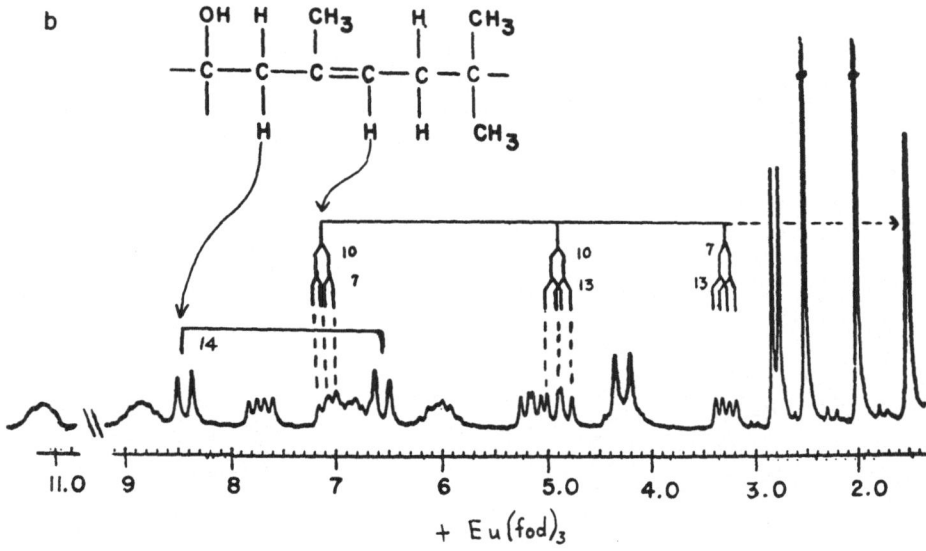

Figure 3. Nmr spectra of dactylol: (a) CCl₄; (b) CCl₄/0.45 mole ratio of Eu(fod)₃.

The olefinic proton in the shifted spectrum was easily identified (δ 7.1) by its coupling to the vinyl methyl group. From
decoupling data it is evident that those allylic methylene protons
(δ 3.3 and 4.9) that are coupled to the olefinic proton (J = 7,10)
interact further only with each other (J = 13). Hence, the carbon
adjacent to this allylic methylene group must be fully substituted,
as is shown in the partial formula in Figure 3b. The geminal
dimethyl arrangement is demanded by the ^{13}C data (0.45 mole ratio
Eu(fod)$_3$), which shows that there is only one quaternary carbon
(δ 41.65) in addition to the tertiary carbinol.

The foregoing partial structure was confirmed by a chemical
degradation (see below), which also gave further vital information
regarding the remaining bonds to the carbinol carbon. Lemieux-
von Rudloff oxidation (KMnO$_4$-KIO$_4$) of dactylol gave an hydroxy
keto acid (3600, 1710 cm^{-1}) which, upon esterification with diazomethane, gave an hydroxy keto ester (3600, 1750, 1710 cm^{-1}) which
retained all of the carbon atoms of dactylol (M$^+$ 284), thus confirming that the vinyl group was contained within a ring. Treatment
of this hydroxy keto ester with sodium carbonate in methanol at
room temperature resulted in a facile removal of acetone, in agreement with expectations for the proposed β-hydroxy ketone unit. The
carbonyl absorption (1750 cm^{-1}) of the resulting keto ester (M$^+$ 226)
indicated that the new carbonyl unmasked by the reverse aldol reaction is in a five-membered ring. This was confirmed when basic

Dactylol:

1) KMnO$_4$, KIO$_4$

2) CH$_2$N$_2$

1) Na$_2$CO$_3$
   MeOH

2) ",
   H$_2$O

hydrolysis of the keto ester gave a keto-acid-retaining cyclopentanone carbonyl absorption (1750 cm^{-1}), in addition to an acid
absorption (1710 cm^{-1}). The structure proposed for the keto ester
is supported by mass spectral data. This established that dactylol
contains a five-membered ring.

The remaining details of the structure proposed for dactylol
are based on further analysis of its europium-shifted nmr spectrum.
This analysis proceeds from two different reference points. The
first of these is the lowest field C-H signal, δ 11.1, which is
assumed to be due to a proton immediately adjacent to the tertiary
hydroxyl group. This proton is coupled to three other protons,
δ 6.85, 6.04 and 4.3, the last of which is a broad signal over
which is superimposed a sharp doublet from a different proton (see
Figure 3b and the labelled formula of dactylol (below). Thus, one
can conclude the presence of a sequence of two methylene groups
next to the tertiary carbinol carbon: -C(OH)-CH$_2$-CH$_2$-. The two
superimposed signals at δ 4.3 are more clearly resolved at higher
temperatures and in another solvent.

The second reference point in the europium-shifted spectrum
is the signal due to the methine proton (δ 8.85) on the carbon
bearing the secondary methyl group. This signal can easily be
identified by decoupling, which collapses the secondary methyl
doublet, δ 2.8. Decoupling reveals that this methine proton is
coupled to two of the protons (δ 6.04 and 4.3) which form part of
the four-proton system that begins at the tertiary carbinol center.
Thus the carbon bearing the secondary methyl is bonded on one side
to the two-carbon methylene chain that begins at the tertiary
carbinol center: -C(OH)-CH$_2$-CH$_2$-CH(CH$_3$)-.

Further analysis of the nmr spectrum established that the
remaining bond to the carbon bearing the secondary methyl group

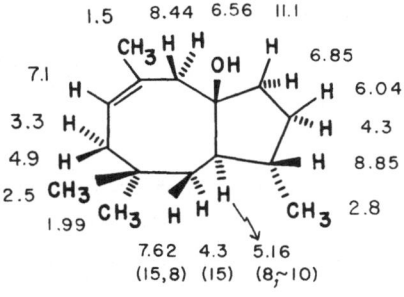

Aromadendrene

in the above partial formula comes from a tertiary carbon which
bears the proton resonating at δ 5.16 (dd, 8, 10). This proton
was shown to be coupled (10 Hz) to the methine proton at the
secondary methyl position and (8 Hz) to the proton resonating at
δ 7.62 (dd, 8, 15). The larger coupling in the δ 7.62 signal
corresponds to the 15 Hz doublet at δ 4.3 (proven by decoupling),
and its magnitude signifies a geminal relation between these two
protons. The lack of observable coupling between the δ 4.3 and
5.16 protons is ascribed to a dihedral angle close to 90°. Thus
the partial formula is expanded to $-C(OH)-CH_2-CH_2-CH(CH_3)-CH-CH_2-$.

The lack of furthur coupling for the geminal protons resonating at δ 7.62 and 4.3 reveals that this methylene group is flanked
by a fully-substituted carbon. Fulfillment of this requirement
and that of forming a five-membered ring, using the partial
formulae generated thus far, leads to the overall structure shown
above for dactylol. The numbers on the periphery summarize the
chemical shift assignments and some of the coupling constants in
the europium-shifted spectrum.

The _trans_ ring juncture is proposed on the basis of chemical
shift and coupling data. The proton at the ring juncture, being
α to the carbinol carbon, would be expected to be much further
downfield in the case of a _cis_ ring fusion (cf. other protons
adjacent to the carbinol center). The _trans_ ring fusion also best
accounts for the negligible coupling between the ring juncture
proton and one of the neighboring methylene protons (δ 4.13).
Only relative stereochemistry is implied by the formula.

The eight-membered ring in the proposed structure for dactylol
is unusual for sesquiterpenes[8]. The structure is most closely
related to aromadendrene (see above) and related compounds[8]. Opening of the cyclopropane ring in the skeleton of aromadendrene
would give the ring system of dactylol. The proposed structure
for dactylol is non-isoprenoid, the vinyl methyl group being
situated one carbon to the left, relative to the isoprenoid skeleton
of aromadendrene, as depicted above.

SOFT CORAL CHEMISTRY

In earlier work we reported[9] the structures of two cembrene
derivatives, nephthenol and epoxynephthenol acetate, from a Pacific
soft coral, _Nephthea_ sp. From the soft coral _Xenia elongata_, collected near Heron Island, Australia, we have now isolated a new
oxygenated diterpene, xenicin, $C_{28}H_{38}O_9$, mp 141.5-142.3°, $[\alpha]_D^{23.5}$
- 36.7°(CHCl$_3$)[10]. Interestingly, specimens of _X. elongata_ from
Picnic Bay, Magnetic Island, Australia, and from the Fiji Islands
did not contain any of the new diterpenoid. The infrared spectrum
of xenicin lacked hydroxyl absorption but displayed a strong, broad

Xenicin

carbonyl band centered at 1735 cm^{-1} (acetate) with a shoulder at
1700 cm^{-1}. The nmr spectrum (see Table 3) contained signals for
four acetates and three vinyl methyl groups, in addition to
downfield multiplets corresponding to nine protons, from which
some partial structural information could be gleaned but from
which it was not possible to deduce a complete structure. Preliminary hydrolytic and catalytic reduction experiments did not yield
encouraging results.

The structure of xenicin, including absolute configuration,
was elucidated by single crystal X-ray diffraction and is shown
above. The crystals of xenicin are monoclinic, space group C2,
four molecules per unit cell, with dimensions a=17.704(3),b=9.061(2),
c=18.656(4) Å; β=113.33 (2)$^{\circ}$. The intensity data (3015) were collected on an automatic diffractiometer. The structure, determined
by direct methods, was refined by least-square methods. The final
R was 0.046 for all the data.

Xenicin possesses a dihydropyran ring <u>trans</u>-fused to a nine-
member carbocyclic ring. The nine-membered ring contains a <u>trans</u>
double bond with a torsion angle considerably less than 180° and
an exocyclic double bond. Bond lengths and torsion angles indicate
that the nine-membered ring is slightly strained. No short inter-
molecular distances were found in the crystal structure.

The nmr chemical shift assignments and proton couplings for
xenicin, confirmed by double irradiation experiments at 100 and
220 MHz, are shown in Table 3. The enol ether proton H$_3$ exhibits
allylic coupling to H$_{4a}$ but not to the conformationally mobil H$_{12}$.
At 220 MHz (benzene-d$_6$) the H$_9$ signal is clearly visible as a
broadened triplet, J=7, indicating a coupling of nearly 7 Hz with
one of the C$_{10}$ protons and only a very small coupling to the other,
as would be suggested by the conformation of crystalline xenicin.

Table 3. Nmr chemical shift and multiplicity data for xenicin.

Proton	Solvent		Multiplicity, J (coupling proton(s))
	C_6D_6	$CDCl_3$	
H_1	6.10	5.87	d, 2 (H_{11a})
H_3	6.65	6.58	d, 2 (H_{4a})
H_{4a}	2.26	2.19	
H_8	5.42	5.27	br d, 8-9 (H_9)
H_9	5.75	5.70	br t, 8-9 (H_8, one H_{10})
H_{11a}	1.85	1.88	
H_{12}	5.62	5.38	d, 9-10 (H_{13})
H_{13}	6.10	5.82	t, 9-10 (H_{12}, H_{14})
H_{14}	5.08	5.08	br d, ∿10 (H_{13}; $H_{16,17}$)
H_{16}; H_{17}	(1.53, 1.84)	(1.74, 1.84)	
H_{18}	1.48	1.74	
H_{19}	4.91	4.82, 4.96	
OAc	1.60(3)	2.04(3)	
	1.74(9)	2.06(6)	
		2.08(3)	

The biosynthesis of xenicin may involve cyclization of geranyl geraniol in a manner alalogous to that proposed[11] for caryophyllene and related compounds, followed by oxidative cleavage of the result- ing cyclobutane ring and eventual closure of the dihydropyran ring. However, a more direct formation of the nine-membered ring can be envisioned as occurring _via_ oxidative cyclization of geranyl linalool, as outlined above. We are searching for compounds related to xenicin which might provide some insight into this question.

ACKNOWLEDGEMENTS

The work described above was carried out with various collab- orators. Drs. P. Kaul and S. Kulkarni conducted the pharmacological studies on dactylyne. Dr. K. Hollenbeak conducted the bioassay- guided isolation of dactylyne and p-hydroxyphenylacetamide. He also isolated quantities of halitoxin-R and dactylol and carried out the degradative studies on these compounds. Dr. D. C. Campbell first isolated halitoxin-R and proposed a structure for it based on spectral analysis. The isolation and most of the structural analysis of dactylenol and dactylenol acetate were completed by

Dr. D. J. Vanderah, who also first isolated dactylol and made preliminary structural evaluations based on spectral analyses.
Dr. Ciereszko and Mr. P. Steudler first discovered xenicin;
complete purifications and detailed spectral analysis of it were
performed by Dr. Vanderah, who deduced some partial structures for
it. Dr. D. van der Helm and Mr. J. D. Ekstrand carried out the
X-ray structure determination of xenicin.

This work was supported by NCI Grants CA-17256-01 and -02,
Commerce Department Sea Grant 5-158-63, and NIH Grant GM 20250-02.

REFERENCES

1. F. J. McDonald, D. C. Campbell, D. J. Vanderah, F. J. Schmitz,
    D. M. Washecheck, J. E. Burks, and D. van der Helm, J. Org.
    Chem. 40, 665 (1975).

2. R. I. Gueran, N. H. Greenberg, M. M. MacDonald, A. M. Schumacher
    and B. J. Abbott, Cancer Chemother. Rep. Part 3, 3, No. 2
    (Sept. 1972). P388 = cell culture of lymphocytic leukemia;
    in vitro activity (ED$_{50}$) is measured as the dose in μg/ml
    that causes 50% inhibition of growth; ED$_{50}$'s of less than 10
    are considered active. PS = lymphocytic leukemia assay in
    mice; T/C % = (survival time of treated animals)/(control's
    survival time) X 100; values >125 are considered active.
    KB = cell culture of human epidermoid carcinoma of the naso-
    pharynx; activity is measured as for P388.

3. M. H. Baslow, Marine Pharmacology, The Williams and Wilkins
    Co., Baltimore, 1969, p. 86.

4. H. L. Bradlow and C. A. Vanderwerf, J. Org. Chem. 16, 1143
    (1951).

5. (a) F. J. Schmitz and F. J. McDonald, Tetrahedron Lett., 2541
    (1974); (b) Unpublished results.

6. D. J. Vanderah and F. J. Schmitz, J. Org. Chem., in press.

7. 100 MHz (CDCl$_3$) 1.07(d,3,2^0-Me), 1.24 (s, 3, 3^0-Me), 1.68
    (br s, 3, vinyl Me), 4.73 (m, 2, = CH$_2$), 5.36 (br m, 1,
    -CH=C(CH$_3$)-), 5.94 (ABq, 1, J = 12, 16, -CH=CH), 5.06 (dd, 1,
    J = 12, 2, -CH=CH$_2$), 5.20 (dd, 1, J = 16, 2,-CH=CH$_2$).

8. T. K. Devon and A. I. Scott, Handbook of Naturally Occurring
    Compounds, Vol. II, Academic Press, New York and London,
    p. 153.

9.   F. J. Schmitz, D. J. Vanderah, and L. S. Ciereszko, Chem.
     Comm., 407 (1974).

10. Calcd for $C_{28}H_{38}O_9$: C, 64.86; H, 7.38; M^+ 518.2518; Found:
     C, 64.90; H, 7.32, M^+ 518.24888.

11. D. J. Vanderah, P. A. Steudler, L. S. Ciereszko, F. J.
     Schmitz, J. D. Ekstrand, and D. van der Helm, J. Am. Chem.
     Soc., submitted.

12. J. B. Hendrickson, Tetrahedron 7, 82 (1959).

*   Dr. Schmitz has requested that the authorship of this paper
     be quoted as follows: F. J. Schmitz, D. C. Campbell,
     K. Hollenbeak, D. J. Vanderah, L. S. Ciereszko, P. Steudler,
     J. D. Ekstrand, D. van der Helm, P. Kaul, and S. Kulkarni.

ANTIBODY-LIKE SUBSTANCES IN MARINE ORGANISMS

D.J. Rogers

School of Pharmacy, Portsmouth Polytechnic

King Henry I Street, Portsmouth, PO1 2DZ, U.K.

INTRODUCTION

It has been known for many years that substances which, like antibodies, have the power to agglutinate erythrocytes, may be extracted from plants[1]. However, it was not until 1948 that Renkonen[2] demonstrated that some extracts may have the ability to preferentially agglutinate the erythrocytes of certain human blood-groups. Boyd[3], later, proposed that extracts showing specificity should be called lectins, derived from the Latin legere, to pick out or choose. This term has now come into common use.

The search for human blood-group specific substances was later extended to many other life-forms including molluscs[4,5] and fish ova[6]. The term protectin[7] has been used to describe the animal extracts since it was considered that they may provide some primitive, fixed specificity, immunological protection system for the organism. Gold and Balding[8], however, prefer the term "receptor specific protein" (RSP) to describe both lectins and protectins since the few which have been examined biochemically have been found to be composed, principally, of protein and because a considerable amount of uncertainty exists regarding their biological role in the organisms[9,10].

In comparison with land and freshwater species, marine organisms have been investigated less extensively.

The work by Gold and Balding[8] must now be regarded as the principal reference source for workers with RSPs and they appear to make no mention of the presence of RSPs in marine viruses or bacteria. Among the Cyanophyta only one species has been

examined[11],Lyngbya majuscula. The cell sap had the unusual human
blood-group specificity anti A + H.

Two groups of workers have investigated marine algae for RSPs.
Boyd et al[11] examined the cell sap from 24 tropical species. They
demonstrated that the cell sap from Codium isthmocladum agglutinated human A, B and O erythrocytes to an equal degree. Six of
the brown algae they examined: Dictyota bartayresii, D. cervicornis,
D.divaricata, D. deliculata, Sargassum rigidulum and Padina
vickersiae reacted preferentially with human A and O erythrocytes
and were, therefore, considered to show A + H specificity.
Sargassum natans and Turbinaria turbinata agglutinated A, B and O
erythrocytes. From the Rhodophyta they found that Spyridia
filamentosa sap possessed human blood-group A specificity. Haemagglutination inhibition experiments were not performed to determine
the specificity more fully.

A more complete study was conducted by Blunden, Rogers and
Farnham[12] on British marine algae. Extracts from 105 species and
subspecies were examined by haemagglutination tests in 0.9% sodium
chloride medium and 24 were found to contain agglutinins directed
against human blood-group A, B and O erythrocytes equally. Only
one species, Ptilota plumosa showed human blood-group specificity,
reacting preferentially with erythrocytes carrying the B antigen.

No marine protozoa appear to have been investigated for
haemagglutinic activity.

There is a similar absence of an extensive systematic search
for RSPs amongst the Porifera. Those species which have been investigated and found to be positive were examined as part of more
basic biological studies of the factors influencing sponge cell
aggregation[8]. Sponge cell aggregation factors are glycoprotein in
nature and therefore similar chemically to RSPs. They react with
specific carbohydrate residues on neighbouring cells specifically
promoting adhesion. MacLennan and Dodd[13] demonstrated that the
aggregation factor from Axinella sp. had heterohaemagglutinating
properties reacting with a variety of fowl and mammalian erythrocytes. They also described a haemagglutinin from Cliona celata.
Extracts from both species reacted with human blood-group A, B,
AB and O erythrocytes failing to show any specificity within this
antigenic system. Gold et al[14] extended the investigations with
the Axinella agglutinin to other types of cell and demonstrated
that human lymphocytes, fibroblasts, HeLa cells and Burkitt
lymphoma cells could agglutinate in this extract. Bretting[15] has
shown that Aaptos papillata produced an agglutinin which was non-
specific for human blood-group ABH antigens.

Other marine invertebrate groups have been investigated more
extensively but considerable care is needed in the interpretation

of the results of research which was largely concerned with phylogenetic considerations. Investigations have been carried out with
whole body extracts, haemolymph and albumen gland extracts.

The first invertebrate heterohaemagglutinins described were
demonstrated in the haemolymph of two crustaceans Limulus
polyphemus and Homarus americanus[16]. The Limulus polyphemus
agglutinin has found a variety of uses, including the investigation of turtle blood-groups[17], and the isolation of leucocytes and
the detection of tumour cells in mammalian peripheral blood[18].
Brown et al[19] demonstrated a weak haemagglutinin with the human
blood-group specificity anti H in the haemolymph of Dardanus
venosus, Mithrax sculptus and Panulirus (Palinurus) anceps and a
weak anti A in Callinectes danae.

Molluscs are important sources of RSPs. Since many react
with bacterial cells in addition to erythrocytes a biological role
has been suggested for RSPs as a primitive, fixed specificity protection system. Since the RSP concentration remains unchanged by
specific antigenic challenge, on this criterion alone they cannot
be considered to be antibodies. Megathura crenulata[20] haemolymph
contains a powerful non-specific agglutinin and the blood of
Crassostrea virginica[21] reacts similarly. It has been suggested
that this last agglutinin may play a role in the phagocytosis of
substances injected into the organism[21]. Johnson[4] reported the
first human blood-group specific agglutinin from molluscs in whole
body extracts of the clam Saxidomus giganteus, which reacted preferentially with erythrocytes possessing the A antigen. Most
invertebrates have been investigated for RSPs using native
(untreated) erythrocytes. Uhlenbruck et al[22] have demonstrated
the limitations of this by demonstrating a haemagglutinin in
Buccinum undulatum with pronase-treated erythrocytes, which was
unreactive with untreated erythrocytes. Within the Aplysia genus
variation seems to exist with regard to haemagglutinin presence
since McKay et al[23] found the Aplysia sp. they examined to be positive, but Brown et al[19] found none in A.protea. Pauley et al[24]
demonstrated a bacterial and erythrocyte agglutinin in serum from
A. californica.

Haemagglutinins with human blood-group Anti-A activity appear
to be more unusual from marine than from terrestrial sources. It
is, therefore, interesting to note that Bretting and Renwrantz[25]
have recorded specific anti-A agglutinins in extracts from Venus
verrucosa and the soft coral Alcyonium palmatum.

Tyler[26] found agglutinins in the body and seminal fluids of
the marine annelid worm Chaetopterus variopedatus, which agglutinated invertebrate spermatoxoa and vertebrate erythrocytes. A
non-specific human blood-group agglutinin was demonstrated in the
body fluid of Sabellastarte magnifica by Brown et al[19].

The Echinodermata have been moderately extensively investigated. Brown et al[19] demonstrated that extracts from Strongylocentrotus dropachiensis may react with human blood-group O and B
erythrocytes, but not with A and O_h (Bombay) erythrocytes, suggesting H specificity. Extracts from Echinus esculentus agglutinate
human blood-group B, native, erythrocytes and human blood-group O
erythrocytes treated with pronase[22]. It is unfortunate that this
report does not mention the use of human blood-group A erythrocytes and without this information no conclusions regarding human
blood-group specificity can be drawn. In view of the results
reported in this paper concerning human blood-group B specific
agglutinins from marine sources, this question has particular
relevance.

Extracts from the ascidian, Ciona intestinalis, agglutinate
group A human erythrocytes[22] in addition to pigeon and pronase-
treated rabbit erythrocytes. However, no mention is made of the
reactivity of this agglutinin with human O or B erythrocytes. Two
separable agglutinins have been demonstrated by Fuke et al[27] in the
coelomic fluid of Styela plicata. Both showed human blood-group
anti H activity.

Heterohaemagglutinins have been demonstrated in the serum of
a wide variety of marine vertebrates. Since many of the agglutinins in the sera are true immunoglobulins they will not be considered here. Ova from marine vertebrates have been investigated
only superficially. From these reports, the more notable observations include the recognition of an agglutinin with human blood-
group anti B activity in the ova of Salmo salar[6] and another of
similar specificity in the ova of Clupea harengus[28]. The ova from
Solea solea and Pleuronectes platessa have also been examined and
reported to possess no significant activity against human blood-
group B erythrocytes[29]. No mention is made of the activity of
these last two extracts against erythrocytes from other human
blood-groups. The Solea solea extract did agglutinate pronase-
treated rabbit erythrocytes.

THE USES OF RECEPTOR SPECIFIC PROTEINS

RSPs have found a variety of uses. If a good anti-B were
more readily available, it would be possible to perform human ABO
blood-grouping entirely with RSPs and a considerable amount of
research effort has been in this direction[30]. There are, of course,
many problems involved in the substitution of reliable human
immunoglobulins with unrelated natural products. These include
the reactivity of the RSP with erythrocytes of rare or modified
antigenic structure and the availability of the RSP. However,
RSPs are used for the determination of human M and N blood-groups,
investigation of the human blood-group H antigen, the detection

and elucidation of variants in the A and B antigens and studies of
polyagglutinable states. They have also been used in anthropolo-
gical research and in basic scientific investigations of the
structure of erythrocyte antigens and cell membranes. Gold and
Balding[8] provide an extensive review of this work.

RSPs have also been used to rapidly sediment erythrocytes to
obtain leucocyte suspensions from whole blood and as sources of
mitotic agents used in the culture of lymphocytes for karyotyping
purposes[31]. More recently, the response of lymphocytes to RSP-
associated mitotic agents has been used as a measure of their
immunological competence[32]. It should be made clear, however,
that the mitotic agent and the agglutinin are not necessarily the
same molecule[33].

Perhaps the most exciting use of RSPs has been the observation that certain cancer (malignant) cells may be agglutinated by
RSPs and that the same cells in an untransformed (non malignant)
state are not. It has been suggested that the unmasking, or
changes in distribution of RSP receptor sites on neoplastic cell
membranes might reflect a return to the embryonic state with
resultant decreased cell adhesion, contact inhibition and increased
cell mobility[34,35]. This is consistent with recent clinical
studies showing the appearance of foetal cell membrane factors in
the serum of individuals with gastro-intestinal neoplasms[36].
Certain RSPs have also been shown to be toxic to tumour cells
in vitro[33].

Most work on RSPs has followed a similar pathway. First, they
are demonstrated in extracts from organisms using a cell carrying
antigenic receptors. Erythrocytes have been most commonly used
for two reasons. First, much research has been directed to dis-
covering new sources of human blood-group specific reagents and
second, the erythrocyte membrane is well endowed with representative antigens. However, there is no guarantee that if other
cells, for example, bacteria, were used instead, many organisms
which do not apparently contain RSPs might react positively.
Similarly, RSPs which react non-specifically with human erythrocytes, with respect to blood-group, might have some undetermined
specificity for antigens carried by the surfaces of other types
of cell.

Initial discovery is followed by determination of antigen
specificity and optimal reaction conditions. Immunochemical specificity may only follow after the RSP has come into empirical use,
although this is now uncommon. Finally molecular structure of the
RSP may be determined. Relatively few, unfortunately, have been
investigated, so far, at this level.

This paper describes a search among marine organisms for RSPs directed against human erythrocytes and their subsequent investigation.

RESULTS

In view of the demonstration by Rogers[37] that haemagglutinins from marine algae may be detected more readily in bovine albumen medium or by pre-treatment of the erythrocytes with proteolytic enzymes, 96 species and subspecies of British marine algae were examined for haemagglutinic activity using these more sensitive methods. Table 1 shows the 36 species which were found to have haemagglutinic activity with human A_1 rr, B rr and O rr erythrocytes. The concentration of the agglutinins varied between species from those where agglutination occurred in undiluted extract only, to those such as Codium fragile, which gave titration values of 1:512. Most species found positive appeared to be non-specific for human blood-group antigens, reacting equally with A_1 rr, B rr, O rr and O R^1R^2 erythrocytes. Exceptions to this were: Chylocladia verticillata extract, which appeared to have the unusual blood-group specificity anti A + H, Ptilota plumosa extract, which reacted preferentially with blood-group B erythrocytes and Gymnogongrus norvegicus extract, which also contained an agglutinin with B specificity.

The detailed results from these three species are shown in Table 2. Complete information concerning this survey will be published elsewhere.

The anti B agglutinin from Ptilota plumosa was further investigated by collecting material from 5 different sites around the coast of North Wales, The Isle of Man, Western Scotland, and the Shetland Isles. On each occasion, the extract prepared from fresh, undried alga reacted preferentially with papain-treated B erythrocytes. If the material were dried at room temperature for 48 hrs or at 50°C for 24 hrs the extract prepared usually, but not always, lost its B specificity.

P.plumosa extract was also tested using erythrocytes which had been treated with the proteolytic enzymes trypsin, ficin, bromelin, pronase and N-acetyl neuraminidase. The results are shown in Table 3. Pronase treatment produced the greatest enhancement of the titre against B cells, raising it from 1:8 in 0.9% sodium chloride solution to 1:1000. The enzymes bromelin, neuraminidase, papain and ficin were equally effective in enhancing the reaction, while trypsinisation of the cells produced only a two-fold increase in titre. None of the enzymes induced significant non-specific agglutination in the extracts, although this was seen in the titration performed in the presence of 30%

Table 1. Seaweed species showing haemagglutinic activity.

Chaetomorpha capillaris	Stictyosiphon tortilis
Clodophora rupestris	Callithamnion corymbosum
Codium fragile sub sp. atlanticum	Plumaria elegans
C. fragile sub sp. tomentosoides	Ptilota plumosa
Ulva lactuca	Chylocladia verticillata
Cystoseira baccata	Delesseria sanguinea
C. foeniculacea	Membranoptera alata
C. tamariscifolia	Phycodrys rubens
Halidrys siliquosa	Polyneura gmelini
Desmerestia aculeata	P. hilliae
Spongonema tomentosum	Gracillaria bursa-pastoris
Ascophyllum nodosum	G. verrucosa
Fucus ceranoides	Gymnogungrus norvegicus
F. serratus	Phyllophora pseudoceranoides
F. spiralis	Polyides rotundus
F. vesiculosus	Chondria coerulescens
Pelvetia canaliculata	Cystoclonium purpureum
Himanthalia elongata	Rhodymenia (Palmaria) palmata

Table 2. Titration values of seaweed extracts under different reaction conditions.

Species	Titres of extract with various blood-groups											
	0.9% sodium chloride solution				Albumen addition				Papain pre-treatment			
	A_1 rr	B rr	O rr	O R_1R_2	A_1 rr	B rr	O rr	O R_1R_2	A_1 rr	B rr	O rr	O R_1R_2
Ptilota plumosa	-	1:4	-	-	1:8	1:256	1:2	1:2	-	1:128	-	-
Chylocladia verticillata	-	-	NT	-	1:64	1:4	NT	1:32	1:8	1:4	NT	1:16
Gymnogungrus norvegicus	-	-	NT	-	-	1:4	NT	-	1:4	1:64	NT	1:2

- = titration value of < 1:2

NT = not tested

Table 3. Titration values of Ptilota plumosa extract with
enzyme-treated erythrocytes (P. plumosa collected from Oban,
Scotland, September, 1974).

Reaction conditions or enzyme used	Blood-group of erythrocytes			
	A_1 rr	B rr	O rr	O R^1R^2
0.9% Sodium chloride solution	-	1:8	-	-
30% Bovine albumen	1:8	1:64	1:8	1:8
Trypsin	1:2	1:16	1:4	1:2
Ficin	1:4	1:128	1:4	1:2
Pronase	1:2	1:1000	1:2	1:4
Neuraminidase	1:2	1:64	1:2	1:4
Bromelin	1:2	1:256	1:4	1:4
Papain	1:4	1:128	1:2	1:4

- = titre $< 1:2$

bovine albumen. With other batches of P. plumosa extract, more
non-specific agglutination was induced following enzyme treatment
of the erythrocytes. On no occasion, however, did the non-specific
agglutination exceed one third of the titration value obtained
with B erythrocytes, providing the extract had been prepared from
undried alga.

Haemagglutination inhibition experiments were performed using
a variety of simple sugars and related substances. The results of
these are shown in Table 4. The most inhibitory of the substances
tested was p-nitrophenyl- α -D-galactose.

Other marine life-forms have also been examined for haemagglutinic activity and the results of this investigation are shown in
Table 5. Both species from the Chondrichthyes tested were found
to be negative but three out of the eight species from the Teleostei
produced agglutination. The agglutinin found in the ova of the

Table 4. Haemagglutination inhibition results of the reaction
between Ptilota plumosa extract and papainised, group B, human
erythrocytes.

Sugar	Haemagglutination inhibition titre
p-Nitrophenyl-α-D-galactose	1:64
Salicin	1:16
Maltose	1:16
L-Glucose	1:16
Trehalose	1:8
Cellobiose	1:8
D-Arabinose	1:8
Mellibiose	1:8
Sucrose	1:4
Methyl-α-D-galactose	1:4
D-Galactose	1:2
D-Glucose	1:2
L-Arabinose	1:2
L-Rhamnose	1:1
N-Acetyl-D-glucosamine	1:1
Fructose	1:1
D-Mannose	Non-inhibitory
N-Acetyl-D-galactosamine	Non-inhibitory

bass, Dicentrarchus labrax, was markedly enhanced by the addition
of bovine albumen and produced the highest titres of all against
papainised erythrocytes, but did not appear to be specific for any
human ABH blood-group antigen. The agglutinin from sole, Solea
solea, ova was of lower concentration than that found in the bass,
but shared the similarity of reacting weakly in saline media and
at a higher titre following albumen addition or erythrocyte papainisation. The sole agglutinin reacted non-specifically. Ova from
the black sea-bream, Spondyliosoma cantharus, contain an agglutinin against group B and O erythrocytes in 0.9% sodium chloride
solution, albumen and papain and more weakly against papainised A
erythrocytes. Six different collections of S. cantharus have been
made and each gave broadly similar results. In all collections,
extracts have reacted more strongly with B erythrocytes than with
O erythrocytes and not at all with A erythrocytes when the reaction is performed in 0.9% sodium chloride solution.

The nudibranch, Aeolidia papillosa, was the only mollusc to
give a positive reaction, the extract from the whole body reacting
with papainised group B erythrocytes only. The eggs from this
species reacted preferentially with B erythrocytes and produced
a titre of 1:256 when the B erythrocytes were papain treated. Non-
specific agglutination was produced following albumen addition and
papainisation of erythrocytes, but the titre of this (1:32 in
papain) was well below that obtained with B erythrocytes.

DISCUSSION

The results from this survey of British marine algae clearly
demonstrate the importance of examining extracts for haemagglutinic
activity with albumen addition tests and enzyme-treated cells since
a higher proportion of species gave positive results when these
reaction conditions were included. A further improvement to a
screen of this nature would be to use erythrocytes treated with
additional enzymes and indicator cells from other species. Further
investigations of this nature could reveal undetected specificities
in the species in Table 1 or enable agglutinins to be detected in
those species showing negative results.

Presumably the presence of agglutinins in organisms reflects
the genetic capability of that organism and one would therefore
expect to be able to use this information taxonomically. However,
the limitation to the use of results in this way is the fact that
the presence of haemagglutinins seems to vary in some species with
geographical position and also with the season. One consistent
finding which stands out is that all species of Fucaceae tested
gave positive, but non-specific, results.

Table 5. Examination of extracts from marine organisms for human erythrocyte agglutinins.

Species	Extract prepared from	Titres of extract with erythrocytes of various blood-groups											
		0.9% sodium chloride sol.				Albumen addition				Papain treatment			
		A_1	B	O rr	$O\ R^1R^2$	A_1	B	O rr	$O\ R^1R^2$	A_1	B	O rr	$O\ R^1R^2$
Scyliorhinus caniculus	Ova	–	–	–	–	–	–	–	–	–	–	–	–
Squalus acanthias	Ova	–	–	–	–	–	–	–	–	–	–	–	–
Dicentrarchus labrax	Ova	–	–	–	1:2	1:16	1:16	1:32	1:32	1:128	1:16	1:64	1:256
Spondyliosoma cantharus	Ova	–	1:16	1:4	–	–	1:64	1:64	1:32	1:8	1:128	1:128	1:128
Solea solea	Ova	–	–	1:2	–	1:2	1:4	1:8	1:8	1:8	1:8	1:16	1:16
Scomber scombrus	Ova	–	–	–	–	–	–	–	–	–	–	–	–
Trachurus trachurus	Ova	–	–	–	–	–	–	–	–	–	–	–	–
Merlangus merlangus	Ova	–	–	–	–	–	–	–	–	–	–	–	–
Pollachius pollachius	Ova	–	–	–	–	–	–	–	–	–	–	–	–
Pleuronectes platessa	Ova	–	–	–	–	–	–	–	–	–	–	–	–
Archidoris tuberculata	Whole body	–	–	NT	–	–	–	NT	–	–	–	–	–
Archidoris tuberculata	Eggs	–	–	NT	–	–	–	NT	–	–	–	–	–

Table 5 continued.

Organism	Tissue												
Aeolidia papillosa	Whole body	-	NT	1:8	-	-	NT	-	-	-	NT	-	-
Aeolidia papillosa	Eggs	1:32	NT	1:256	1:32	1:8	NT	1:32	1:8	-	NT	1:2	-
Crepidula fornicata	Eggs	-	NT	-	-	-	NT	-	-	-	NT	-	-
Patella vulgata	Whole body	-	NT	-	-	-	NT	-	-	-	NT	-	-
Sepia officinalis	Eggs	-	-	-	-	-	-	-	-	-	-	-	-
Crangon vulgaris	Whole body	-	NT	-	-	-	NT	-	-	-	NT	-	-
Alcyonidium hirsutum	Whole body	NT	NT	NT	NT	-	-	-	-	-	-	-	-
Membranipera membranacea	Whole body	NT	NT	NT	NT	-	-	-	-	-	-	-	-
Ascidiella scabra	Whole body	NT	NT	NT	NT	-	-	-	-	-	-	-	-

\- = titre $< 1:2$

NT = not tested

The apparent anti A + H specificity found in the extract from
Chylocladia verticillata is worthy of further investigation. This
observation is based upon a single collection,and further samples
from different sites are currently under investigation. Anti A + H
specificity is unusual,and haemagglutination inhibition experiments
and absorption experiments may show that this extract, in fact, contains two separable agglutinins. Interestingly, Boyd et al.[11]
demonstrated anti A + H activity in the cell sap of a blue-green
alga, Lyngbya majuscula.

Extracts from Ptilota plumosa prepared from fresh, undried
alga provide sources of a specific anti B agglutinin which per-
forms particularly well against enzyme-treated cells. The use of
this material on automated blood-grouping equipment is currently
under evaluation. Sufficient quantities of the species appear to
be available for this kind of use[38] but are only found in areas
geographically distant from the possible users. This is a dis-
advantage,since it is important that freshly collected seaweed is
extracted. The B specificity of P. plumosa extracts has been confirmed with collections from five different sites.

It is now established that the B antigen on human erythrocytes is determined by the presence of an α-linked galactose
residue in the terminal position on an oligosaccharide chain
associated with the external surface of the erythrocyte membrane.
It is, therefore, not surprising that p-nitrophenyl-α -D-galactose
was able to attach to the antigen-combining sites on the P. plumosa
agglutinin and inhibit the binding of the agglutinin to B antigens
on erythrocyte surfaces. Presumably the difference in inhibitory
capability demonstrated between methyl-α -D-galactose and p-
nitrophenyl- α-D-galactose is due to the fact that the second
substance fits the receptor site of the agglutinin more precisely
than the first.

The anti B activity demonstrated in extracts from
Gymnogongrus norvegicus against papainised erythrocytes is based
upon one collection only. This species is not common around the
coast of the British Isles,and this will certainly hamper the
intended confirmation and further investigation.

It is possible that the ova from Spondyliosoma cantharus contains two agglutinins, an anti B which reacts in 0.9% sodium
chloride solution, bovine albumen and with enzyme-treated human
erythrocytes and an anti H agglutinin reacting best with enzyme-
treated human erythrocytes. The human blood-group H antigen is
present on all human erythrocytes,with rare exceptions. It is
not, however, present on all erythrocytes to the same degree. Its
strongest expression is on erythrocytes of blood group O, to a
lesser extent on erythrocytes from group B individuals,and weakest
on group A_1 erythrocytes. Consequently, the results obtained with

the S. cantharus agglutinin against papainised erythrocytes could
be the summation of anti B + H activity. This agglutinin has been
further investigated, and results will be published in due course.

It will be interesting to discover if the anti B agglutinin
from the whole-body extract of the mollusc Aeolidia papillosa
originates from the albumen gland of the organism. This has been
found to be the case in many other molluscs and is supported by
the demonstration of higher titre anti B agglutinins in the eggs
of A. papillosa. This agglutinin is certainly worthy of a more
thorough examination.

REFERENCES

1. H. Stillmark, Über Ricin, ein giftiges Ferment aus den Samen
   von Ricinus communis L. und einigen anderen Euphorbiaceen,1888,
   Thesis, Dorpat, quoted by Oppenheimer, 1906.

2. K. O. Renkonen, Ann. Med. Exp. Fenn. 26, 66 (1948).

3. W. C. Boyd and E. Shapleigh, Science 119, 419 (1954).

4. H. M. Johnson, Science 146, 548 (1964).

5. O. Prokop, A. Rackwitz, and D. Schlesinger, J. Forens. Med.
   S. Africa, 12, 108 (1965).

6. O. Prokop, D. Schlesinger, and G. Geserick, Z. Immun. Forsch.
   132, 491 (1967).

7. O. Prokop, G. Uhlenbruck, and W. Kohler, Dtsch. Gesundh.-Wes.
   23, 318 (1968).

8. E. R. Gold and P. Balding, Receptor Specific Proteins; Plant
   and Animal Lectins, Excerpta Medica, Amsterdam, 1975.

9. K. Günterberg, M. D. Thesis, Humboldt University, Berlin, 1973.

10. G. Draxler, H. Kothbauer, and H. Schenkel-Brunner, Immun. Inform.
    3(7), 5 (1973).

11. W. C. Boyd, L. R. Almodovar, and L. G. Boyd, Transfusion 6,
    82 (1966).

12. G. Blunden, D. J. Rogers, and W. Farnham, Lloydia 38, 162
    (1975).

13. A. P. MacLennan and R. Y. Dodd, J. Embryol. Exp. Morph. 17,
    473 (1967).

14. E. R. Gold, C. F. Phelps, S. Khalap, and P. Balding, Ann. N. Y. Acad. Sci. <u>234</u>, 122 (1974).

15. H. Bretting, Z. Immun.-Forsch. <u>146</u>, 239 (1973).

16. H. Noguchi, Univ. Penn. Med. Bull. <u>15</u>, 295 (1902).

17. W. Frair, Science <u>140</u>, 1412 (1963).

18. A. L. Watne and E. Cohen, Proc. Amer. Ass. Cancer Res. <u>5</u>, 261 (1964).

19. R. Brown, L. R. Almodovar, H. M. Bhatia, and W. C. Boyd, J. Immunol. <u>100</u>, 214 (1968).

20. M. P. Rowley and D. Rowley, Experientia <u>24</u>, 1056 (1968).

21. M. R. Tripp, J. Invertebr. Path. <u>8</u>, 478 (1966).

22. G. Uhlenbruck, U. Reifenberg, and M. Heggen, Z. Immun.-Forsch. <u>139</u>, 486 (1970).

23. D. McKay, C. R. Jenkin, and D. Rowley, Aust. J. Exp. Biol. Med. Sci. <u>47</u>, 125 (1969).

24. G. B. Pauley, G. A. Granger, and S. M. Krassner, J. Invertebr. Path. <u>18</u>, 207 (1971).

25. H. Bretting and L. Renwrantz, Z. Immun.-Forsch. <u>145</u>, 242 (1973).

26. A. Tyler, Biol. Bull. <u>90</u>, 213 (1946).

27. M. T. Fuke and T. Sugai, Biol. Bull. <u>143</u>, 140 (1972).

28. O. Vetter, L. Acta Biol. Med. Germ. <u>22</u>, 427 (1969).

29. G. I. Pardoe, G. Uhlenbruck, D. J. Anstee, and U. Reifenberg, Z. Immun.-Forsch. <u>139</u>, 468 (1970).

30. P. Rees, R. Cotton, P. D. J. Holt, and D. J. Anstee, Med. Lab. Sciences <u>33</u>, 13 (1976).

31. D. A. Hungerford, A. J. Donnelly, P. C. Nowell, and S. Beck, Amer. J. Hum. Genet. <u>11</u>, 215 (1959).

32. M. G. Fitzgerald, J. Clin. Path. <u>25</u>, 163 (1972).

33. N. Sharon and H. Lis, Ann. Rev. Biochem. <u>42</u>, 541 (1973).

34. H. Ben-Bassat, M. Inbar, and L. Sachs, J. Membrane Biol. <u>6</u>, 183 (1971).

35. M. Inbar, H. Ben-Bassat, and L. Sachs, Nature New Biol. <u>236</u>, 3, (1972).

36. R. N. Lausch and F. Rapp, Progr. Exp. Tumor Res. <u>19</u>, 45 (1974).

37. D. J. Rogers, M. Phil. Thesis (CNAA), Portsmouth Polytechnic, 1974.

38. G. Blunden and D. J. Rogers, Food-Drugs from the Sea Proceedings, 1976, in press.

39. Dr. D. Smith and Mr. F. Allison of the Wessex Regional Transfusion Centre, Southampton, U. K., kindly supplied the human erythrocytes used in this study.

40. Dr. G. Blunden, Dr. C. Barwell, Mr. W. Farnham, and Mr. N. Jephson of Portsmouth Polytechnic helped in obtaining and identifying marine organisms.

CHEMICAL CONDITIONING OF SEAWATER BY ALGAL GROWTH AND DEVELOPMENT

Arne Jensen

Institute of Marine Biochemistry

N-7034 Trondheim-NTH, Norway

Sea water is generally looked upon as a growth medium of constant chemical composition. The environmental conditions in the
sea are also believed to be conservative, contrary to those of
terrestrial environments. This is quite correct for many of the
major parameters. The chemical composition of sea water is quite
constant. Water is always available. The temperature ranges are
rather narrow, $\sim 15^{\circ}C \pm 10^{\circ}C$, and, some few meters below the surface,
currents and wave action are not very strong.

This picture changes radically, however, when we look more
closely at the marine environment. Nutrient levels for the plants
in the sea are extremely variable, and the concentrations of available micronutrients, such as trace metals and vitamins, are even
more variable. Food availability for heterotrophic organisms is,
of course, also continually changing, going from starving conditions to overnourishment.

Many, and probably most, of these strong variations are of
seasonal or successional character. In temperate waters, limiting
light conditions allow high nitrate concentrations to build up
during winter, and this nutrient is rapidly consumed in spring,
when sufficient light becomes available. This is a typical seasonal
event. In areas with upwelling of deep water, the high content of
metal ions inhibits growth of a number of algae until the input of
various natural chelators resulting from biological activity or
decay in surrounding waters have detoxified the metal ions. The
latter example belongs to the successional type.

In both groups of cases, the quality of the sea water and its
ability to support and maintain life in the ocean are influenced

in a decisive way by the biological activity in the water masses. In the following, some examples of this type of biochemical conditioning of sea water will be discussed. This discussion will be limited mainly to effects caused by the growth and decay of marine algae.

CONDITIONING PHENOMENA OF TYPICAL SEASONAL CHARACTER

Among the simplest examples of sea water conditioning by algal growth are the light-limited processes in northern and southern temperate seas which lead to deprivation of nutrients. The well known removal of nitrate from sea water during spring blooms of planktonic algae is the classic example of a seasonal conditioning event. Less well recognized is another phenomenon based on seasonally regulated uptake. This is the uptake of iodine accomplished by the huge populations of brown algae, especially the Laminaria species, which cover large parts of the coastal areas of the temperate zone. This uptake will deprive the sea water of iodine (or at least lower considerably its content of this mineral) in kilometer-wide zones along the coasts of Norway, Canada, Peru, Chile, northern Japan, Tasmania, and many other countries. An important feature is that the uptake is tied to the growth of the algae and is thus limited to the spring and summer seasons. The consequences of this lowering of the iodine content of sea water are completely unknown. It is known, however, that the Laminaria species require iodine for their growth[1].

In cases where analysis reveals high selectivity in plentiful organisms for rare metals, such as the selectivity for vanadium in Ciona intestinalis, a common ascidian, one must expect seasonal deprivation of this metal in the surrounding sea water. Apart from the obvious effects caused by nutrient removal and seasonal deprivation of some few trace elements, almost nothing is known about the effects of the many specific uptake processes which are tied to seasonally dependent growth of marine algae.

Better recognized and somewhat more studied are certain release processes connected with seasonal growth of both seaweeds and planktonic algae. In spring, several of the brown seaweeds belonging to the Fucaceae exude phenolic compounds, together with carbohydrates and some nitrogenous materials, during periods of strong illumination[2]. No phenolic material, or very little, seems to be exuded in late summer and autumn. These phenols have attracted some interest because of their physiological effects on a number of marine organisms[3,4] and because they are very potent chelators for heavy metals[5] and influence the growth of marine phytoplankton[6,7]. The structures of some of the brown algal phenols have been revealed recently by Glombitza's[8,9,10] and Craigie's[11] groups. They seem to be based on a phloroglucinol pattern.

A nice example of the effects of these exudation processes and the adaptations they lead to in the environment has been given by Seshadri and Sieburth[12], who studied the fungal flora on the surface of brown seaweeds. In spring, the fungal population on the thallus of the alga was adapted to growth on algal carbohydrates and phenols, while the autumn population of fungi were inhibited by the same phenols and could utilize only the carbohydrates.

Less well defined but well timed are certain release processes which are connected with the life cycles of the algae. A massive release of organic material (including ascorbic acid) into the sea takes place every year in Norway towards the end of May, when many hundred thousand tons of Ascophyllum nodosum are shedding their fruit bodies, which account for 30-50% of the plant and which contain ca. 150 g of ascorbic acid per ton, dry weight[13]. Another seasonal phenomenon of much larger scale is the release of 30-50% of the total biomass of the dominating species of Laminaria in Norwegian waters, namely Laminaria hyperborea. The standing stock of this alga amounts to several million tons, and nearly half of this is shed and broken down rapidly in the sea water during the early summer every year. This brings up two important aspects, namely that even trace compounds present in the alga may occur in the sea water in physiologically important concentrations and that the presence of the compounds released is regulated by a complicated mechanism. The regulation is obviously seasonal, but it is not tied to the calendar directly and solely. It is a result of the total development of the alga and represents an integration of many factors, such as the nutritional state of the plant as a result of the conditions during the preceding year, the light and temperature conditions during the spring, the nutrient level in the sea, light and temperature during early summer, etc. The release of ascorbic acid from Ascophyllum nodosum thus constitutes a report of the previous environmental conditions and the resulting status of the plant, a signal which may be of great value as a trigger for other biological events.

CONDITIONING PHENOMENA OF SUCCESSIONAL CHARACTER

The literature abounds with indications of interspecies interactions in the marine environment, and this has frequently been used to explain successions in the occurrence of the various species involved. There can be little doubt that a chemical language is involved in the establishment and maintenance of succession in the sea. The isolation and structural identification of sirenin, the male-gamete-attracting pheromone of Ectocarpus siliculosus, by Jaenicke et al.[14] represents a major contribution to marine allelochemistry. There is in general, however, very little knowledge of the chemicals involved.

Among the many well documented examples of the influence of
algal growth on the quality of sea water, the conditioning of the
upwelling water in Peru is now regarded as a typical case of heavy
metal detoxification caused by the activity of living and dead
organisms in the surrounding water[15].

Studies of the exudation of organic compounds from planktonic
algae have revealed that vitamins[16], amino acids[17], carbohydrates[18,19], various chelators[15], and a number of secondary
metabolites are released by these organisms and that this exudation
often has a marked influence on the growth and development of other
organisms in the surrounding water. Carlucci and Bowes[20] showed
very clearly how the diatom <u>Pheaodactylum</u> <u>tricornutum</u> released
B-vitamins into a medium deficient in this factor, which then
became acceptable as a growth medium for the thiamin- and B_{12}-
requiring diatom, <u>Skeletonema</u> <u>costatum</u>. Pedersen and Fridborg[21]
have demonstrated cytokinin-like activity of sea water from the
<u>Fucus-Ascophyllum</u> zone. These are examples of positive influence.
Negative influences or antagonisms have also been reported frequently. The <u>Olisthodiscus-Skeletonema</u> antagonism forms a well-
known example of mutual exclusion, likely to be caused by
exudation products[22]. The diatom mentioned above, <u>P</u>. <u>tricornutum</u>,
seemed to release a principle which killed another alga, the
coccolithophoride <u>Coccolithus</u> <u>huxleyi</u>, and a very recent report by
Murphy <u>et</u> <u>al</u>.[23] describes how several blue-green algae deprive.
competing species of algae of iron by excreting hydroxamate chelators into the growth medium. This may be a major reason for
the much feared and frequently experienced dominance of blue-green
algae in eutrophic waters. Another aspect of these common exudation processes is that they provide nutrients for bacteria and other
microorganisms which also will show successional fluctuations in
occurrence and population density.

SEARCH FOR WATER CONDITIONING PHENOMENA AND INTERSPECIES INFLUENCE

Since conditioning of the sea water is a major tool in
shaping and timing in marine ecosystems, we need a better understanding of the compounds involved and the processes which occur.
This is necessary both for the protection of the environment and
for a successful use of it in well planned mariculture work. The
natural products chemists continuously demonstrate the presence of
unusual compounds in marine organisms, such as the many halogenated
compounds (for a review, see reference 30), polysulfides[24], the
unusual and still unknown selenium- and arsenic-containing compounds of Lunde[25], and many, many other strange molecules. The
fact that these compounds are quite widely distributed and are
frequently released into the sea water makes studies of their
physiological effects very important. They are likely, in many
cases, to have a decisive influence on organisms present in the
environment.

Establishment of interspecies interaction is a necessary prerequisite in the search for allelochemicals, and efficient methods
for the detection of chemical interactions between species have
been developed. Carlucci and Bowes[20] used mixed cultures, which
often produce problems when one species outgrows the other. The
dialysis or caged-culture technique[26,27] offers certain advantages.
It involves growing the organisms to be tested on each side of a
diaphragm which retains the organisms but allows free exchange of
all low-molecular-weight and most high-molecular-weight chemicals
between the populations (see Figure 1). Growth and development
of the two cultures can be followed separately by conventional
methods, with no difficulty, and compared with the growth in control
cultures. Any positive or negative influence on growth is easily
detected. We have used the system to demonstrate the effect of
stirring on the growth of marine diatoms[28] by comparing a stirred
and an unstirred population of the same species in the machine.
In principle, the dialysis chambers may contain any sort of marine
microorganism or small organism. One type of apparatus with four
chambers is already commercially available.

As is pointed out above, screening for physiological effects
of the many strange natural products isolated from marine organisms is a necessary step in the detection of species interaction.
This screening is also required in order to detect general physiological effects of the compounds, and methods for rapid screening
are therefore in great demand. By the use of modern electronic
equipment, it is possible to detect very small variations in
optical density. This has allowed us to develop a very sensitive
turbidostat which has been applied to a growth chamber fitted with
filters of well-defined pore sizes. The growth medium, i.e., sea
water, can be rapidly pumped through the chamber, which will
retain the microorganism to be tested, and changes in population

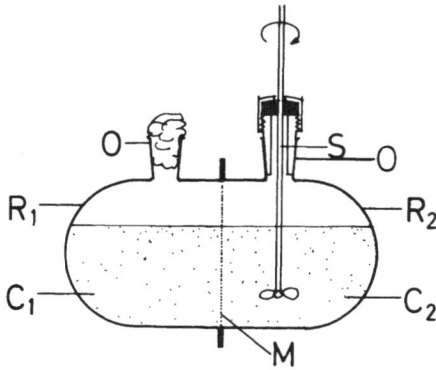

Figure 1. Culture chamber. (For explanation of symbols, see
reference 27.)

density can be registered rapidly. In these caged cultures we
can detect a 10% increase in cell number, which means that the
growth rate of a diatom dividing every ten hours can be measured
in one hour. In the conventional bioassay, determination of growth
rate takes three to four days for this type of organism. This
rapid means of detecting and measuring species interaction may be
used in a systematic search for such phenomena and the chemicals
involved.

As mentioned above, exudation and production of extracellular
compounds are frequently encountered in the marine environment.
One can wonder why some planktonic algae exude into the sea up to
50-80% of the carbon they fix by photosynthesis. In some cases
this question may have a simple answer, and a useful explanation
may be found. In many other instances the interactions with other
organisms in the environment may be very complex, and the benefit
of the exudation process to the alga may look very dubious to us.
Because of the present state of knowledge, it is probably wise to
take the attitude referred to by Wangersky (1965) and look upon
the organisms in the sea as parts of a very large and very sloppily
organised tree.

REFERENCES

1. M. Pedersen, Physiol. Plant. <u>22</u>, 680 (1969).

2. J. McN. Sieburth and A. Jensen, J. Exp. Mar. Biol. Ecol. <u>3</u>,
   275 (1969).

3. J. T. Conover and J. McN. Sieburth, Proc. Fifth Int. Seaweed
   Symp., Pergamon Press, Oxford, 1966, p. 99.

4. J. McN. Sieburth and J. T. Conover, Nature <u>208</u>, 52 (1965).

5. M. A. Rashid, Soil Sci. <u>111</u>, 298 (1971).

6. A. Prakash and M. A. Rashid, Limnol. Oceanogr. <u>13</u>, 598 (1969).

7. A. Prakash, M. A. Rashid, A. Jensen, and D. V. Subbarao,
   Limnol. Oceanogr. <u>18</u>, 516 (1973).

8. K.-W. Glombitza and E. Sattler, Tetrahedron Lett., 4277 (1973).

9. K.-W. Glombitza and H. U. Rösener, Phytochem. <u>13</u>, 1245 (1974).

10. K.-W. Glombitza, H.-W. Rauwald, and G. Eckhardt, Phytochem.
    <u>14</u>, 1403 (1975).

11. M. Ragan and I. S. Craigie, Can. J. Biochem. 54, 66 (1976).

12. R. Seshadri and J. McN. Sieburth, Appl. Microbiol. 22, 507 (1971).

13. A. Jensen, Proc. Fourth Int. Seaweed Symp., Pergamon Press, Oxford, 1963, p. 319.

14. L. Janicke, T. Donike, Akintobi, and D. G. Müller, Science, 171, 815 (1971).

15. R. T. Barber, R. C. Dugdale, I. I. MacIsaak, and R. L. Smith, Investigacion Pesquera 35, 171 (1971).

16. L. Provasoli and A. F. Carlucci, in Algal Physiology and Biochemistry Botanical Monographs, Blackwell Science Publ., Oxford, 1974.

17. J. A. Hellebust, Limnol. Oceanogr. 10, 192 (1965).

18. R. R. L. Guillard and P. J. Wangersky, Limnol. Oceanogr. 3, 449 (1958).

19. A. Myklestad, A. Haug, and B. Larsen, J. Exp. Mar. Biol. Ecol. 9, 137 (1972).

20. A. F. Carlucci and P. M. Bowes, J. Phycol. 6, 393 (1970).

21. M. Pedersen and G. Fridberg, Experientia 28, 111 (1972).

22. D. M. Pratt, Limnol. Oceanogr. 11, 447 (1966).

23. T. P. Murphy, D. R. S. Lean, and C. Nalewajko, Science 192, 900 (1976).

24. S. J. Wratten and D. J. Faulkner, J. Org. Chem. 41, 2465 (1976).

25. (a) G. Lunde, Biochim. Biophys. Acta 304, 76 (1972); (b) G. Lunde, Reports on Technological Research Concerning Norwegian Fish Industry, Vol. 5, Directorate of Fisheries, 1972, p. 1; (c) G. Lunde, J. Am. Oil Chemist's Soc. 50, 26 (1973).

26. J. S. Schultz and P. Gerhardt, Bact. Rev. 33, 1 (1969).

27. A. Jensen, B. Rystad, and L. Skoglund, J. Exp. Mar. Biol. Ecol. 8, 241 (1972).

28. L. Skoglund and A. Jensen, J. Exp. Mar. Biol. Ecol. $\underline{21}$, 169
     (1976).

29. P. J. Wangersky, Amer. Sci. $\underline{53}$, 358 (1965).

30. W. Fenical, J. Phycol. $\underline{11}$, 245 (1975).

CYTOKININ ACTIVITY OF SEAWEED EXTRACTS

G. Blunden

School of Pharmacy, Portsmouth Polytechnic

King Henry I Street, Portsmouth, PO1 2DZ, U.K.

Seaweed extracts for use in agriculture and horticulture have been commercially available for many years. These extracts are prepared from a number of different seaweeds, but Ascophyllum nodosum is the species most commonly utilised. Other species that are used include Laminaria digitata, L. hyperborea, Fucus serrata and Sargassum species. It is probable that the choice of seaweeds has been governed solely by commercial availability, rather than by determination of the suitability of a large number of species and selection of the best ones. The solvent most commonly used for extraction of the seaweeds is water, although sodium carbonate solution is used to prepare at least one commercially available product. The extracts are used, diluted with water, as sprays, both to the foliage and to the soil. Some companies recommend that seaweed sprays should be applied only to the foliage, but others also recommend application to the soil.

A wide range of different beneficial effects have been reported from the use of seaweed extracts. Although of considerable interest, the experimental plan and degree of replication reported in many of the trials makes it difficult to determine the validity of the results. Also, many of the results reported have not been analysed statistically, and so it is impossible to determine whether the differences recorded have any significance. Beneficial results from the use of seaweed extracts reported in the literature have included increased crop yields[1,2], improved seed germination[3,4], increased resistance of plants to frost[1] and to fungal and insect attack[5], increased uptake of inorganic constituents from the soil[1], and reduction in storage losses of fruit[2,6-9].

It is frequently claimed that the effects produced by seaweed extracts can be explained by their content of trace elements. However, the quantities of dissolved solids in seaweed extracts that are applied annually to one hectare are very small. For example, one major supplier recommends an application rate of 12:1 of extract, containing about 16% dissolved solids. This results in less than 2 kg of solids being applied annually to one hectare, of which only 600-800 g is inorganic. Another well-known supplier recommends variable application rates, but the highest is 72:1 of extract to one hectare annually[10]. This extract contains about 9% dissolved solids, which results in about 6.5 kg of solids being applied, of which about 40% is inorganic. Using an analysis published by the company[10] and the maximum recommended application rate, the annual quantities applied per hectare of certain important inorganic elements have been calculated, along with conservative estimations of the annual requirements of these elements for a crop such as hay[11] (Table 1). The quantities supplied by seaweed extracts form an insignificant proportion of the annual requirements.

Because of the small amount of material applied to a hectare, the substances in seaweed extracts responsible for the agriculturally beneficial results must be capable of having an effect in very low concentrations. As a result, plant hormones were considered. The presence of gibberellins was recorded by Williams et al.[12] in various freshly manufactured, commercially-available seaweed extracts. However, it was found that the gibberellin activity fell rapidly and that after storage at room temperature for 3 months, the activity was negligible. Even in freshly manufactured extracts, the levels of activities recorded were too low to produce any noticeable effect when applied at the recommended rates of application to a crop. Weak auxin activity was found by Mowat[13] in one commercially-available seaweed extract, but Williams et al.[12] did not obtain any significant auxin activity in any of the three seaweed extracts tested. The presence of cytokinins in seaweed extracts was suggested by Booth[14], who stated that many of the results recorded with the use of seaweed extracts might be explained by cytokinins. Brain et al.[15] demonstrated for the first time that one commercially-available seaweed extract had a high cytokinin activity. This study was extended and it was shown that the three products tested all had high levels of cytokinin activity, although the level varied considerably from batch to batch of the same product[12]. It was suggested that the quantities of cytokinin present in the seaweed extracts were sufficient to produce biological effects when applied to plants, even at the low rates of application used in the field.

After demonstrating the cytokinin activity of seaweed extracts, field trials were conducted using seaweed extracts of known cytokinin activities and a reference cytokinin, kinetin

Table 1. Quantities of certain inorganic elements supplied in one
year to one hectare by seaweed extract and the estimated annual
requirements of these elements.

Element	Wt. (g) per hectare from seaweed extract application	Estimation of possible annual requirement per hectare (g)
Iron	22	280
Manganese	0.3	140
Zinc	0.7	140
Copper	0.3	140
Boron	0.006	56
Molybdenum	0.07	1.4
Cobalt	0.03	1.4

(6-furfuryl-amino-purine). The seaweed extracts were assayed
using the radish leaf expansion bioassay with kinetin as the
reference compound[16,17].

POTATO TRIAL

In a trial conducted in 1974, the use,as a foliar spray, of a
commercial seaweed extract (S.M.3, manufactured by Chase Organics
Ltd.) at 1 gallon/acre (11.22 l/hectare), which resulted in a
cytokinin application rate equivalent to 0.57 g/acre (1.4 g/
hectare) kinetin, produced a significant increase in the yield of
potatoes of the variety King Edward. Application of an aqueous
solution of kinetin, also at a rate equivalent to 0.57 g/acre,re-
sulted in a significant increase in the weight of potatoes obtained,
there being no significant difference in the results from the
kinetin application and the results from the application of the
seaweed extract (Table 2). This close correlation suggests that
the beneficial result from the use of the seaweed extract may be
due to its cytokinin content. A full account of this trial has
been published by Blunden and Wildgoose[18].

Table 2. Yields from potato plants of the variety King Edward
treated with either aqueous seaweed extract or kinetin solution.

Treatment	Cytokinin application rate, g (kinetin equivalents) per acre	Tuber yield (% of control)
Seaweed extract	0.57	112.9*
Kinetin	0.57	110.8*

*   P = 0.05

GRASS TRIAL

A trial was conducted in 1975 to study the effects of the
applications, as foliar sprays, of a commercial aqueous seaweed
extract (Algistim, manufactured by Glenorganic Ltd.) and kinetin
solution to grass. The seaweed extract was applied at the rate of
1 gallon/acre (11.22 l/hectare), which was equivalent to 0.82 g/
acre kinetin. Kinetin was applied at the same application rate,
and each control block was sprayed with a volume of water equal to
that applied to each of the test blocks. No significant difference
was recorded in either the fresh or dry weights of grass obtained,
but the crude protein content (protein plus ammonium[19]) was significantly higher in both the seaweed extract treated and kinetin
treated grass (Table 3). Moreover, there was no significant difference in the results from the kinetin and seaweed extract treated
samples, which gave strength to the idea that the seaweed extract
owed its beneficial effect to its cytokinin content. The higher
protein contents of the grass treated with either seaweed extract
or kinetin solution was to be expected, as cytokinins are known to
stabilise leaf proteins and to retard their degradation[20].

LIME TRIAL

The limes used in the trial were imported from South Africa
and harvested approximately 7 days prior to use. Samples of
fruit, each sample containing 10 limes, were completely immersed
for 1 hr in either aqueous dilutions of seaweed extract having
final cytokinin activities equivalent to 30, 15 and 7.5 p.p.m.
kinetin, or in equivalent concentrations of aqueous kinetin solutions. Two commercial seaweed extracts were used, which were
Marinure, supplied by Wilfred Smith (Horticultural) Ltd. and
S.M.3. As a control, one sample of 10 limes was immersed in
water for the same period. After immersion, surface liquid was

Table 3. Mean crude protein contents of grass treated with either
aqueous seaweed extract or kinetin solution.

Treatment	Cytokinin application rate, g (kinetin equivalents) per acre	Crude protein content (%) of grass*
Seaweed extract	0.82	9.11**
Kinetin	0.82	9.06**
Control (water)	0	8.41

*   mean of 8 replicates

** P = 0.05

removed from the fruits, which were then stored in a controlled-
environment growth cabinet at 10° and at 85% relative humidity, in
subdued lighting. Examined daily for colour changes from green to
yellow, the percentage areas of different colours were recorded
until the limes were 95-100% yellow. Full details of this trial
have been given by Blunden, Jones and Passom[21].

A pronounced retarding of the rate of "de-greening" was produced by immersion of the limes in either the diluted seaweed
extracts or kinetin solutions. The best results were obtained
with both kinetin and the seaweed extracts at a concentration
equivalent to 30 p.p.m. kinetin (Table 4). The effects produced
by both the seaweed extracts and the kinetin solutions were similar, which suggests that the result from the use of the seaweed
extract was due to its cytokinin content. Stabilisation of
chlorophylls is a well-known property of cytokinins[20].

STANDARDISATION OF SEAWEED EXTRACTS

The effects produced as a result of application of seaweed
extracts to plant tissues vary considerably with the amount
applied. It is therefore necessary to know the cytokinin activities of the seaweed extracts, if reproducible results are to be
expected. Erratic results have been obtained from the use of
commercially-available seaweed extracts,and these may be due, at
least partly, to the large variations in the cytokinin activities
of different batches of the same product. The need for standard-
isation of their products with respect to cytokinin activity
has been recognised by some seaweed extract manufacturers.

Table 4. Mean time taken after immersion in either diluted
seaweed extracts or in kinetin solutions for limes to become yellow.

Immersion solution	Mean time (days) for limes to become yellow			
	Cytokinin concentration (p.p.m.), calculated as kinetin			
	30	15	7.5	0
S.M.3	75.1	67.9	61.2	
Marinure	71.7	50.8	57.7	
Kinetin	83.6	58.4	47.9	
Water				55.6

Although the need for supplying seaweed extracts of known
cytokinin activity is recognised, a reliable and simple procedure
for the routine assay of the extracts is a major problem. The
extracts contain substantial amounts of growth inhibitory materials
which interfere with many of the cytokinin bio-assays. It is possible to remove many of the inhibitors by extracting with organic
solvents, and the purified material can then be succussfully assayed
by a variety of different methods, such as the one utilising
Amaranthus seedlings[22]. However, the results obtained after
purification may not be applicable to the unpurified extracts
which are used in agriculture and horticulture. It is possible
that a more reliable result is obtained by assaying the unpurified
extracts, but the applicability of results obtained from biological
assays carried out in the laboratory to actual effects produced
in the field is uncertain. The tissue culture bioassays, such as
the tobacco pith callus, tobacco stem pith, soybean callus and
carrot root tissue, are at present the most reliable methods for
testing crude extracts (Letham[23]). However, these procedures uti-
lise sophisticated techniques and are very lengthy, which reduce
their suitability for the routine analysis of seaweed extracts by
small manufacturers. The radish leaf expansion bioassay[16,17],
although somewhat insensitive and prone to giving erratic results,
has been found useful for the routine analysis of the unpurified
seaweed extracts. Nevertheless the development of a more suitable
assay method which is rapid and reliable is necessary.

ACKNOWLEDGEMENTS

I wish to thank Mr. B. A. Plunkett, Chemistry Department, Portsmouth Polytechnic, for analysing the grass samples for their crude protein contents.

REFERENCES

1. T. L. Senn, J. A. Martin, J. H. Crawford, and C. W. Derting, South Carolina Agr. Exptl. Sta., Res. Ser. No. 23, 1961.

2. G. Blunden, Proc. Intern. Seaweed Symp. 7, 584 (1972).

3. E. F. Button and C. F. Noyes, Agron. J. 56, 444 (1964).

4. J. B. Aitken and T. L. Senn, Botan. Mar. 8, 144 (1965).

5. W. M. Stephenson, Proc. Intern. Seaweed Symp. 5, 405 (1966).

6. B. J. Skelton and T. L. Senn, Proc. Intern. Seaweed Symp. 6, 723 (1969).

7. M. Povolný, Rostlinná Vyroba 12, 335 (1966).

8. M. Povolný, Proc. Intern. Seaweed Symp. 6, 703 (1969).

9. M. Povolný, Rostlinná Vyroba 18, 703 (1972).

10. Maxicrop for Grass and Arable Crops, Maxicrop Ltd., Holdenby, Northants, U. K., 1975, 5 pp.

11. T. Wallace, The Diagnosis of Mineral Deficiencies in Plants, H. M. S. O., London, 1951.

12. D. C. Williams, K. R. Brain, G. Blunden, P. B. Wildgoose, and K. Jewers, Proc. Intern. Seaweed Symp., 1976, in press.

13. J. A. Mowat, Proc. Intern. Seaweed Symp. 4, 352 (1961).

14. E. Booth, Proc. Intern. Seaweed Symp. 5, 349 (1966).

15. K. R. Brain, M. C. Chalopin, T. D. Turner, G. Blunden, and P. B. Wildgoose, Plant Sci. Lett. 1, 241 (1973).

16. S. Kuraishi and F. S. Okumura, Bot. Mag. Tokyo 69, 817 (1956).

17. J. A. Bentley-Mowat and S. M. Reid, Ann. Bot. (N.S.) 32, 23 (1968).

18. G. Blunden and P. B. Wildgoose, J. Sci. Fd. Agric., in press.

19. S. E. Allen, H. M. Grimshaw, J. A. Parkinson, and C. Quarmby,
     Chemical Analysis of Ecological Materials, Blackwell, Oxford,
     1974, p. 186.

20. A. E. Richmond and A. Lang, Science 125, 650 (1957).

21. G. Blunden, E. Jones, and H. Passom, Trop. Sci., in press.

22. N. L. Biddington and T. H. Thomas, Planta 111, 183 (1973).

23. D. S. Letham, Ann. Rev. Plant Physiol. 18, 349 (1967).

ENHANCEMENT OF HERBICIDAL EFFECT BY SEAWEED EXTRACTS

Keith R. Brain, D.S. Lines, M. Booth and G. Ansell

Welsh School of Pharmacy

UWIST, Cardiff, CF1 3NU, U.K.

INTRODUCTION

Seaweed extracts have a number of applications in agriculture and horticulture and these are reviewed by Blunden[1] elsewhere in this volume. During a laboratory investigation on the influence of environmental conditions on the herbicidal activity of the auxin-type compound CMPP (mecoprop), it was noted that there appeared to be a beneficial interaction, in terms of weed control, between this herbicide and a seaweed extract applied as a foliar feed. This paper briefly describes some investigations on this phenomenon.

Since, in practice, increased herbicidal activity is usually of value only in the absence of detrimental effects to crop plants, experiments were carried out to assess crop vigour as well as her-bicidal effect. Finally some work on the possible mechanism of action is reported.

RESULTS

Herbicidal Activity

A small scale method which we have developed for the assessment of herbicidal activity in the laboratory was used which is based on the measurement of chlorophyll loss in treated plants over a ten-day period. We have found this method very reliable in comparative work on this type of herbicide.

Application of a range of doses of mecoprop to small plants of Stellaria media and Ranunculus scleratus resulted in death of the

345

plants in all cases, although the time taken to reach this condition
increased with decreasing concentration. In the field, the time for
death is important in relation to the possibility of recovery.
Addition of seaweed extract, marketed as foliar feed, to the herbicide spray solution potentiated activity by a factor of approximately
two. Figure 1 gives an example of the results obtained with
Stellaria media, and these were closely paralleled by those with
Ranunculus scleratus. The enhancement effect was reproduced repeatedly over a period of three years, was also observed when un-
favourable spraying conditions were simulated by spraying the plants
with water 30 minutes after application, when unformulated CMPP was
used in place of a commercial product, and when an alternative seaweed extract, of rather different composition, was tested.

     As the seaweed extracts were known to contain cytokinins (1-3)
which can affect leaf chlorophyll, particularly in senescing tissues, the effects of the extracts alone on the weed species were
checked. As anticipated, all levels of extract used showed an
enhancement of chlorophyll content and the effect was dose-
dependent. However, it must be emphasised that since this increase
in chlorophyll level was converse to the decrease found after herbicidal treatment, any effect on herbicidal assessment would be a
tendency to underestimate the potentiation on combination.

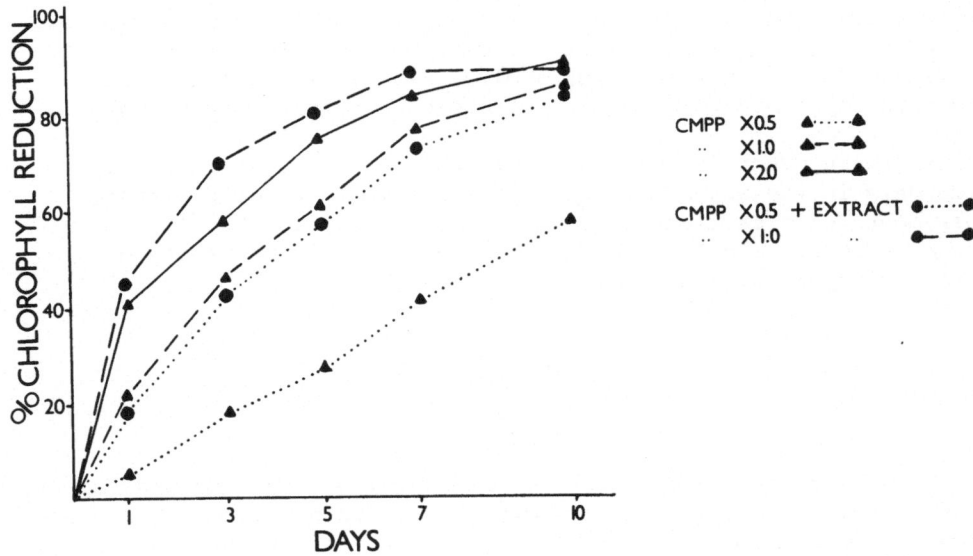

Figure 1. Influence of seaweed extract on herbicidal action of
CMPP (mecoprop) against Stellaria media.

Crop Vigour

Growth effects on a test crop were assessed by sequential height measurements over 14 days on young barley plants. Mecoprop used alone gave a reduction in height, compared to untreated control plants, of 11% at the normal recommended dose (2.4 lg/acre in an equivalent dose volume of 20 gallons/acre), and of 6% at half this dose. On the other hand, the use of the seaweed extract alone at the normal recommended dose (0.1 gallons/20 gallons/acre) increased crop height by 4%, while at one quarter of this dose an increase of 23% was obtained. Combination of half or one quarter of the normal dose of mecoprop with the normal rate of seaweed extract had no significant effect on growth, while combination of the full dose of mecoprop with the seaweed extract had a more detrimental effect (reduction of 15%) than this dose of mecoprop used alone. It can thus be seen that seaweed extract may have a protecting effect against herbicidal activity to the crop under certain conditions, although when higher levels of mecoprop were used this effect was reversed. It is possible that, since the doubling of herbicidal action against the weed means that the dose of herbicide to be applied for equivalent action can be halved, both increased herbicidal action against the weed and some protection of the crop may be achieved at the same time.

Mode of Action

It was obvious that there could be a number of causes for the observed potentiation effects on herbicidal activity, and a number of these were considered. The simplest explanation would be a chemical or physicochemical interaction between the two products, resulting in improved spray distribution and/or retention and/or uptake. However, as no significant differences were found between the pH or surface tension values of the various test solutions, or between the interaction effects with pure or commercially formulated CMPP, this seemed unlikely. An alternative proposition was that the beneficial effects could be due to adhesive or humectant properties of algal polysaccharides which could improve uptake. Comparisons were therefore made with solutions of alginate at three concentrations (50, 100 and 200 mg/l). At all dose levels tested, alginate used alone enhanced growth of the weed quite markedly, but in combination with mecoprop it increased herbicidal activity by a factor of more than two at the higher alginate levels. On the other hand, alginate used alone reduced crop height considerably and in combination with mecoprop this adverse effect was increased. In view of the known presence of plant growth regulators in the seaweed extracts (1-3), it was thought likely that these could interact with the applied herbicide, either to facilitate absorption or to further disrupt metabolism within the plant. Since cytokinins appear to be the most important compounds in these extracts, comparisons were

made with the synthetic cytokinin, kinetin, at three dose levels.
Used alone kinetin always enhanced growth of the weed. In combination with mecoprop it gave enhancement of herbicidal activity which
was equal to or greater than that found with the seaweed extract in
combination with mecoprop. Used alone, kinetin enhanced crop growth
but in combination with mecoprop it had a detrimental effect. Since
both polysaccharide and cytokinin clearly enhanced the effect of
mecoprop, combinations of these materials were tested. Spraying of
a combination of alginate and kinetin without herbicide gave marked
stimulation of growth of the weed, as was observed with either
alginate or kinetin alone, but there was no additive effect. When
mecoprop was included there was enhancement of herbicidal activity,
but this was rather less than that obtained when only alginate was
added and about equal to that obtained with kinetin alone. The combination of alginate and kinetin had no significant effect on crop
growth, but in the presence of mecoprop detrimental effects were
observed.

In an effort to determine whether the interaction was occurring
at the plant surface, absorption studies were undertaken with
Stellaria media. Mecoprop was applied alone or in combination
with seaweed extract, polysaccharide, or kinetin and the amounts
remaining on the surface and recoverable from within the tissue
determined by extraction and HPLC over a period of four hours
(Table 1). When mecoprop was used alone, none could be recovered
from the surface three hours after application. Addition of seaweed extract gave more repid uptake in the first two hours, but a
small residue of mecoprop remained at the end of four hours.
Similarly, both alginate and methylcellulose increased initial
uptake but left residues. On the other hand, addition of kinetin
noticeably slowed the uptake rate and considerable amounts remained
after four hours. These uptake figures are of interest, but since,
in all cases, uptake was essentially complete within a few hours,
it is doubtful that absorption is a rate-limiting stage which is
affected by the seaweed extract. In fact, only when mecoprop was
used alone was complete uptake observed. In addition, there is no
correlation between uptake and the time taken to kill the plants.
Mecoprop used alone required 13 days, but this was almost halved
by addition of seaweed extract or kinetin, while the carbohydrate
additives showed smaller reductions in the death time.

                              DISCUSSION

The results indicate that a combination of seaweed extract with
a herbicide can have beneficial effects, in terms of both weed kill
and protection of the crop from damage by the herbicide, and it is
possible that this may have commercial significance, although it
is not known whether the phenomenon is restricted to auxin-type
herbicides or even to CMPP, alone. Our investigations on the

Table 1. Uptake of mecroprop by Stellaria media

Hours After Application

	1		2		3		4		Death Time (Days)
	surface	internal	surface	internal	surface	internal	surface	internal	
Mecoprop alone	51	37	36	53	--	82	--	84	13
+ Seaweed Extract	42	47	24	67	5	89	--	87	7
+ Kinetin	66	21	41	49	14	76	10	79	8
+ Methylcellulose	37	60	19	69	3	83	--	91	10
+ Alginate	33	65	19	70	2	81	7	80	10

possible mechanism of the interaction suggest that it is most likely that this is occurring within the plant and that the plant growth regulators, particularly the cytokinins, are probably the compounds responsible. Although increased herbicidal activity could be produced by incorporation of polysaccharide material, the amounts of such material normally present in seaweed extracts are so low that they are unlikely candidates.

REFERENCES

1.   G. Blunden, preceeding paper.

2.   K. R. Brain, M. C. Chalopin, T. D. Turner, G. Blunden, and P. B. Wildgoose, Plant Sci. Lett. $\underline{1}$, 241 (1973).

3.   D. C. Williams, K. R. Brain, G. Blunden, P. B. Wildgoose, and K. Jewers, Proc. Intern. Seaweed Symp. $\underline{8}$, in press.

CHEMICAL BASIS OF SEXUAL APPROACH IN MARINE BROWN ALGAE

Dieter G. Müller

Fachbereich Biologie der Universität Konstanz

D-7750 Konstanz, Federal Republic of Germany

Fusion of gametes is an integral part of the various events summarized under the heading of "sexual reproduction," which is, in turn, one of the major driving forces of evolution. Because of this crucial significance, biologists have been interested for a very long time in learning more about fertilization. As early as 1854, the French pioneer Thuret[1] observed masses of sperm cells aggregating around eggs of Fucus, and a look at the older literature reveals that apparently some marine brown seaweeds are especially suitable for observing fertilization.

The enormous numbers of sperm cells aggregating around a Fucus egg demonstrate immediately and beyond any doubt that there is something that makes the male gametes approach the egg and remain excited in its vicinity until the zygote is formed.

This dramatic happening, which can be easily studied on the North Atlantic coasts, stimulated a number of workers to try to find out what it is that makes the sperm approach the egg. Although all these efforts eventually failed, the British workers Cook et al.[2] came closest to the solution by demonstrating in 1947 that the eggs secrete a highly volatile compound with male-attracting properties.

Unfortunately, further progress was prevented due to the limitations set by the tools of analytical chemistry at that time. No significant new results were reported until, in 1968, the American group of L. Machlis and co-workers[3] identified the first sex attractant in plants. They used interspecific crosses of the aquatic phycomycete Allomyces and identified a sesquiterpene, which they named sirenin, as the signal transmitter between female and

351

male gametes. At about the same time, I started to rediscover the
fine organisms studied by classical marine botanists in the last
century. By then, analytical methods had reached a degree of
sophistication that gave some hope of successfully handling the
minute quantities of substances involved in sexual chemotaxis.
Three sex attractants of marine brown algae have now been identified through the cooperative efforts of a group of biochemists and
myself. I will now briefly describe the three projects and then
give you some information about the current state of our work.

    Ectocarpus siliculosus is a filamentous brown alga growing
in the coastal zones of the North Atlantic and the Mediterranean.
There are female and male plants which release motile unicellular
gametes into the surrounding sea water. Female cells settle on a
substrate and then start to attract male gametes in the vicinity,
which accumulate until finally cell fusion takes place. The
supernumerary cells lose interest, once the zygote has been formed
(Figure 1).

    Fortunately, the species can be cultivated in the laboratory.
Clonal isolates originating at Naples, Italy, were grown in mass
cultures. Female gametes were observed to produce a very characteristic scent, reminiscent of juniper. Suspensions of female
gametes were placed in bottles, and purified air was used to flush
the volatile substances into a cold trap at -78°C. There, the condensed material could be recovered with solvents and examined by

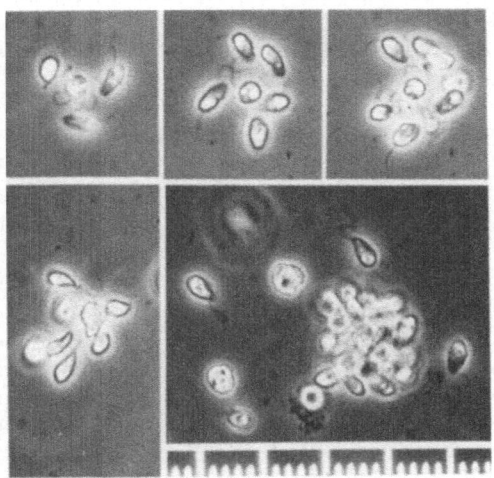

Figure 1. Mating reaction in Ectocarpus siliculosus: motile
gametes surrounding single stationary female cells. Lower left:
cellular fusion. Lower right: violent reaction of many male cells
around one female gamete, which is obscured. Phase contrast,
scale unit 2.5 µ.

gas chromatography, which revealed a single compound. Two and a
half years of continuous intense mass culture yielded 92 mg of
the material. This turned out to be far more than necessary for
the identification of the molecule. One crucial problem I have
neglected so far--the everpresent questions: Is the substance
recovered and handled by the chemist the one we are looking for,
and is it still active? In other words, the examination of the
biological activity has to accompany the chemical work. In
Ectocarpus, there are several possible bio-assays. One is to dis-
solve trace amounts of the compound in mineral oil and offer such
droplets to male gametes swimming in sea water. When the oil
droplet contains active material, the male cells will aggregate
around it.

Finally, in 1971, the compound was identified and named
ectocarpen[4]. The structural formula is given in Figure 4.

Fucus serratus is one of the coarser seaweeds of the Northern
Atlantic. Eggs and sperm are produced on separate plants. Fertilization takes place between a giant egg, visible to the unaided
eye, and minute, highly-motile sperm cells (Figure 2). During two
winter seasons, large masses of live eggs were collected at the
Marine Station at Roscoff, France. Volatiles were extracted from
the suspension and condensed in a cold trap at dry-ice temperature.
Gas chromatographic examination of the condensate revealed a single
compound. This time the bio-assay consisted of tiny droplets of
vasaline offered to male gametes. When they contained active
material, the sperm always mistook these bodies for eggs.

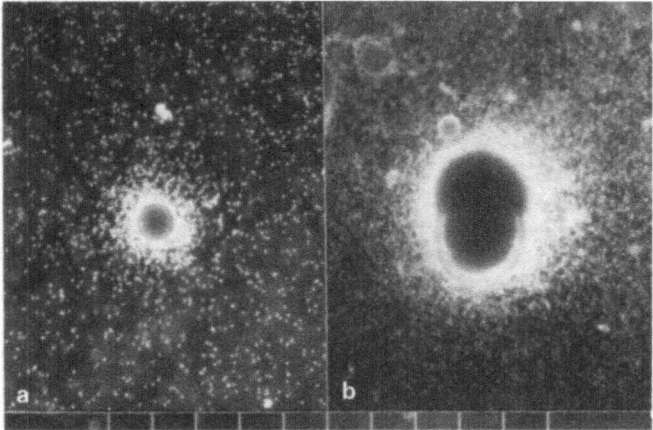

Figure 2. Fucus serratus: spermatozoids attracted by an egg (a)
and an oogonium containing eight eggs before discharge (b). Dark
field, scale unit 0.1 mm.

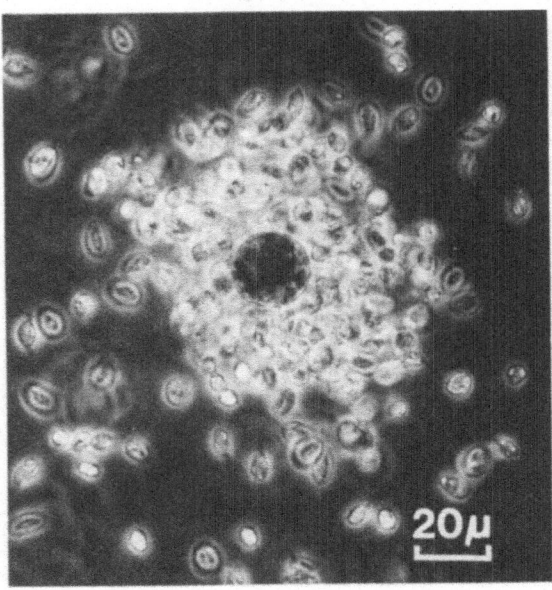

Figure 3. <u>Cutleria</u> <u>multifida</u>: female gamete (centre) attracting
numerous male cells.

This bio-assay confirmed that the recovered compound was, indeed,
the attractant. In total, 500 micrograms were obtained, which
were just about sufficient to determine its structure[5]. The substance was named fucoserraten (Figure 4).

   <u>Cutleria</u> <u>multifida</u> is one of the famous textbook algae, known
for its alternation of heteromorphic generations. Sexual plants,
again male and female as separate individuals, are present in
springtime at some places on the Mediterranean coast. Quite
unexpectedly, this alga proved to be very easily cultured in the
laboratory. Fertilization takes place between large female gametes
and tiny sperm-like male gametes (Figure 3). The sexual differentiation of this species is intermediate between the equality in size
of the gametes in <u>Ectocarpus</u> and the extreme inequality in <u>Fucus</u>.

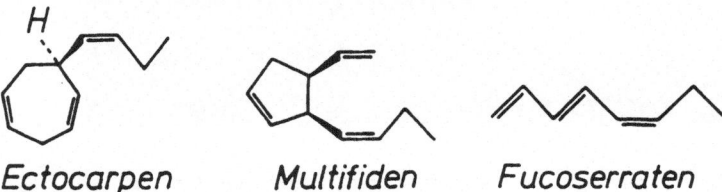

Figure 4. Chemical structure of the three attractants discussed.

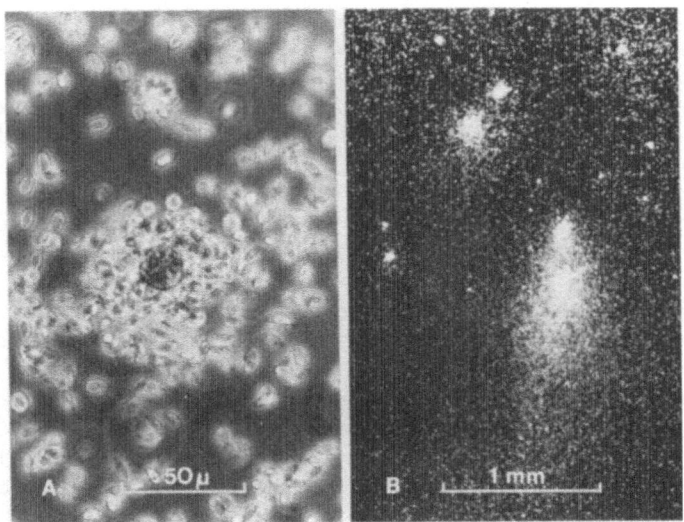

Figure 5. "Sexual" cross-reactions. A: female gamete of <u>Cutleria</u>
attracting male gametes of <u>Ectocarpus</u>. B: male gametes of <u>Cutleria</u>
accumulating at places where several female gametes of <u>Ectocarpus</u>
have settled. A, phase contrast; B, dark field.

Half a year of mass culture and extraction of female gametes
yielded 3.7 mg of material. Gas chromatography revealed a mixture
of three main fractions present in a ratio of 10:4:1. The components were separated by preparative gas chromatography and their
biological activity determined. The bio-assay this time consisted
of microbeads of porous glass placed in a suspension of male
gametes. Untreated particles were ignored, whereas particles which
had been previously allowed to absorb an active substance exhibited
attractive properties. Using this technique, the most prominent
fraction in the chromatogram (Figure 5) was identified as the
attractant. The chemists determined the structures of all three
compounds. The attractant was named multifiden, the second fraction was named aucanten, and the third was found to consist of the
already well-known ectocarpen[6].

By 1973, the attractants in the three marine brown algae discussed here were identified. All of them are unsaturated hydrocarbons, in two cases carrying rather unusual five- or seven-
membered rings. All three compounds are extremely hydrophobic and
practically insoluble in water. They probably move out of the
water into the air almost immediately upon formation. These
properties certainly ensure that there is no noticeable build-up
of the attractant in the vicinity of female cells and thus
guarantee steep concentration gradients.

So far, we have been concerned with female gametes announcing their presence by emitting a chemical signal that is received by male gametes. However, it can be easily demonstrated that male gametes of all three species mentioned here respond to a large number of volatile hydrophobic chemicals in addition to their native attractants.

Ectocarpus males, for instance, have been found to be stimulated by various compounds, such as aldehydes, esters, ketones, and hydrocarbons[7]. Very clearly, this would seem to indicate a devaluation of the species specific interaction because there is no doubt that the habitat of the plants contains many organic substances which might possibly disturb the sexual approach[8]. A simple but nevertheless impressive example of such an "abnormal" interaction is seen when male gametes of Cutleria are confronted with female gametes of Ectocarpus and vice versa: the foreign males are definitely attracted, although, to the experimenters' comfort, no fusion occurs in these cases (Figure 6).

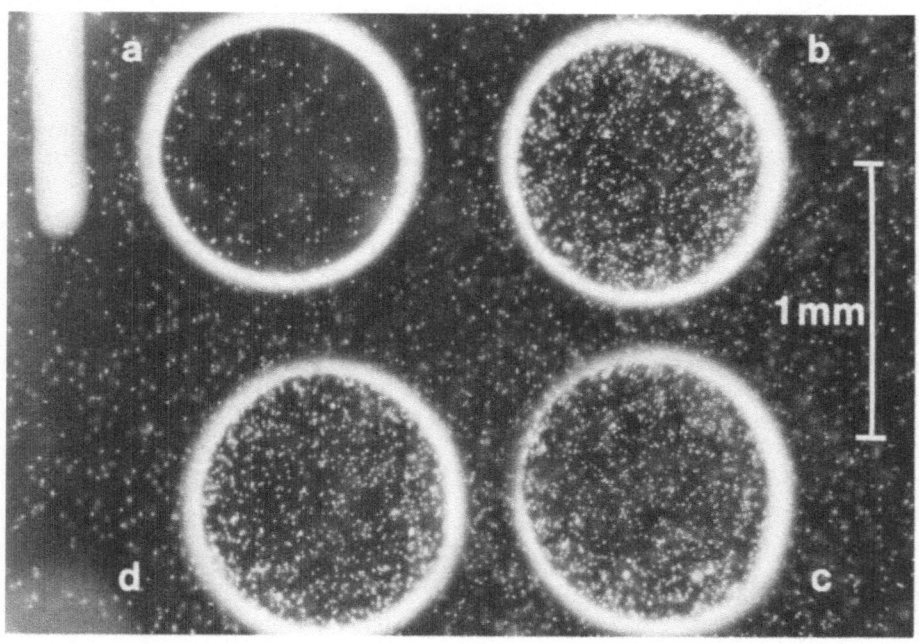

Figure 6. Bio-assay for synthetic attractants. Four drops of the fluorochemical FC-78 in a suspension with male gametes of Cutleria. Drop a (identified by a scratch mark on the outside of the dish) contains pure solvent. Drops b, c, and d contain 3.1×10^5 M multifiden. Dark field.

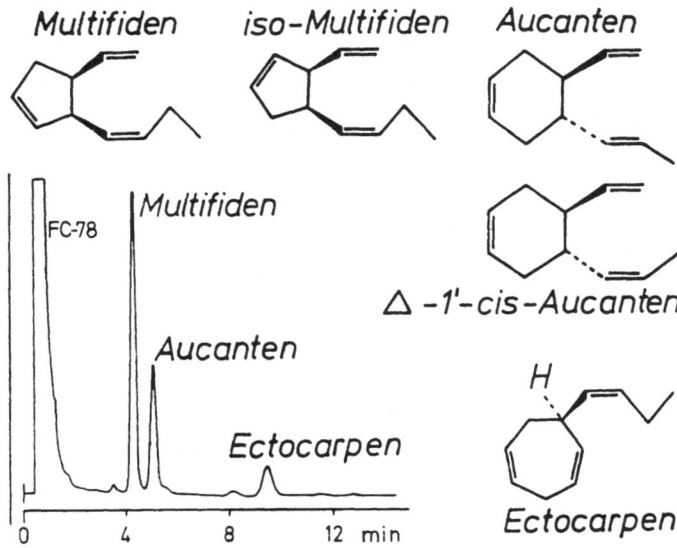

Figure 7. Gas chromatographic analysis of the eluate from female gametes of <u>Cutleria</u> and structural formulae of the synthetic compounds used for assaying their attractive properties. Due to separation difficulties, the synthetic samples of multifiden and aucanten consisted of 1:1 mixtures of the isomers shown.

This called for an attempt to find out if, and to what degree, the male gametes are able to differentiate between various chemicals. For these studies, known amounts of synthetic attractants were dissolved in a biologically-inert solvent and offered as flat microdroplets to a suspension of male gametes. A drop of pure solvent was placed in the same preparation as an internal standard. After documenting the experiment with a flashlight microphotograph, the degree of preference for the experimental concentrations by the male gametes can be measured (Figure 7). With proper statistical treatment of the data, it is possible to determine the minimum concentration of attractant which gives a significant response of the males[9]. Figure 7 gives the compounds which were assayed with male gametes of <u>Ectocarpus</u> and <u>Cutleria</u>. The results are presented in Table 1.

The most sensitive response of the male gametes is directed toward their species-specific attractant. Responses toward other substances are at least ten-fold less sensitive. Aucanten, which is formed by <u>Cutleria</u> macrogametes, is clearly not acting as a significant attractant; its function is not understood. Apart from the fact that the male gametes respond most sensitively to the females of their own species, which is not surprising, some other questions remain unsettled at the moment. One of these questions

Table 1. Data showing the behaviour of male gametes toward droplets with various concentrations of synthetic attractants (Figure 6). Figures represent arithmetic means of relative cell numbers in fields b, c, and d, when cell number in reference field a is set at 1.00. Numbers of experiments are given in brackets. Significant responses (p < 0.5%) are indicated.

Attractant Concentration M

Male Gametes	0	10^{-7}	3.1×10^{-7}	10^{-6}	3.1×10^{-6}	10^{-5}	3.1×10^{-5}	10^{-4}	3.1×10^{-4}	10^{-3}	3.1×10^{-3}	10^{-2}
ectocarpen												
Ec. Na-164	0.98 (27)	1.01 (15)	1.10 (41)	1.00 (30)	1.54 (22)	1.68 (19)	1.60 (14)	2.10 (19)	2.26 (6)	3.41 (9)		
Cutleria	1.05 (111)	1.12 (34)	1.03 (56)	0.99 (29)	1.01 (17)	1.21 (23)	1.67 (17)	3.79 (20)	9.87 (12)	15.82 (6)		
multifiden												
Ec. Na-164	0.98 (27)	0.97 (12)	0.97 (18)	0.95 (45)	0.96 (9)	0.96 (36)	0.99 (18)	1.04 (27)	1.06 (30)	1.38 (9)	1.23 (6)	1.25 (6)
Cutleria	1.05 (111)	0.97 (9)	1.01 (21)	1.10 (21)	1.31 (23)	2.41 (12)	6.26 (18)	9.75 (9)	14.44 (10)	12.67 (12)		
aucanten												
Ec. Na-164	0.98 (27)			0.96 (12)	0.98 (27)	0.95 (12)	0.97 (12)	1.49 (18)	1.52 (21)	1.28 (12)	1.04 (27)	0.96 (21)
Cutleria	1.05 (111)			1.07 (15)	1.14 (21)	1.18 (23)	1.19 (15)	1.52 (12)	1.91 (12)	2.37 (12)	1.85 (12)	1.22 (18)
n-hexane												
Ec. Na-164	0.98 (27)			1.02 (9)	1.04 (9)	1.03 (9)	1.00 (54)	1.02 (63)	1.07 (24)	1.22 (12)	4.02 (9)	3.58 (12)
Cutleria	1.05 (111)	1.01 (9)	1.01 (15)	1.03 (24)	1.07 (24)	1.05 (48)	1.07 (30)	1.08 (34)	0.97 (38)	1.45 (27)	1.79 (9)	2.44 (12)

concerns the cross-attraction between Cutleria and Ectocarpus
gametes, which can be very clearly demonstrated. Is it of any
biological significance in the natural habitat? At certain places
in the Mediterranean, both species occur at the same time of the
year at the same habitat. If there should be something like
species competition on the gametic level by chemical means, both
Cutleria and Ectocarpus have the ability to survive without
fertilization. Male and female gametes of Ectocarpus, as well as
female gametes of Cutleria, can develop parthenogenetically. The
situation is somewhat different in Fucus. Fucus eggs have to be
fertilized by a sperm cell in order to start development, and
there is no parthenogenetic reproduction possible. In this case,
the presence of hydrocarbons with attractive power toward spermatozoids might indeed endanger the success of fertilization and thus
the survival of the species. Studies on the specificity of sexual
chemotaxis in Fucus have not yet been done, since the work has to
be carried out at the coast, but plans for such studies are being
made.

     Another line of current interest is the analysis of the loco-
motory behaviour of male gametes, which are finally guided to the
surface of the female cells when stimulated by a gradient of the
attractant. After watching carefully the locomotory behaviour of
male gametes, we can make the following statements: When swimming
freely in the medium, they move in straight lines. Coming into
contact with the surface of the slide or cover-glass, they tend
to stay in close contact with the surface. In this case, they
swim in wide counter-clockwise circular paths. This is obviously
due to the asymmetrical construction of the cells. When the male
gametes are stimulated by an attractant, these circles become

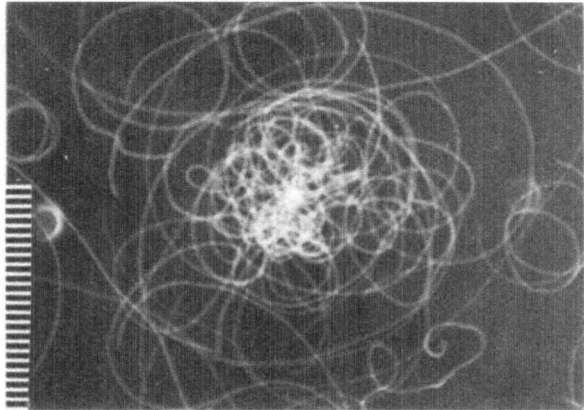

Figure 8. Light tracks of male gametes of European Ectocarpus
around one American female cell (center). Dark field, exposure
5 seconds, scale unit 10µ.

narrower, which almost inevitably brings the male cell closer to the source and keeps it in its vicinity. This mechanism seems to be at least part of the procedure of sexual approach in Ectocarpus. Figure 8 shows a long-time exposure microphotograph of an encounter of some European males with one American female.

This combination of European and American gametes seems to be a system especially favourable for the study of cellular chemotaxis, since the American female cell serves as a point source of ectocarpen, and the reaction is kept going undisturbed for a long time because cellular fusion is not possible in this combination[10]. If the reaction proceeds, in compatible matings the male cell anchors itself with the tip of its front flagellum on the surface of the female cell.

This experimental system may also justify a marine botanist's desire to visit various coasts. I did not anticipate the incompatibility combined with intact sexual attraction in Italo-American matings when I visited Woods Hole in 1975 to pick up Ectocarpus. I just felt that some time this visit might pay.

REFERENCES

1. G. Thuret, Ann. Sci. Nat. Bot., Ser. IV, 2, 197 (1854).

2. A. H. Cook, J. A. Elvidge, and I. Heilbron, Proc. Roy. Soc. B. 135, 293 (1947/48).

3. L. Machlis, W. H. Nutting, and H. Rapoport, J. Amer. Chem. Soc. 90, 1674 (1968).

4. D. G. Müller, L. Jaenicke, M. Donike, and T. Akintobi, Science 171, 815 (1971).

5. D. G. Müller and L. Jaenicke, FEBS Lett. 30, 137 (1973).

6. L. Jaenicke, D. G. Müller, and R. E. Moore, J. Amer. Chem. Soc. 96, 3324 (1974).

7. D. G. Müller, Planta 81, 160 (1968).

8. W. M. Sackett and J. M. Brooks, Amer. Chem. Soc. Symp., Ser. 18, 211 (1975).

9. D. G. Müller, Z. Pflanzenphysiol. 80, 120 (1976).

10. D. G. Müller, J. Phycol. 12, 252 (1976).

BIOLOGICALLY-ACTIVE LIPIDS FROM THE ANDROGENIC GLAND OF THE CRAB

CARCINUS MAENAS

J.-P. Ferezou, J. Berreur-Bonnenfant, A. Tekitek,

M. Rojas, and M. Barbier, CNRS, Gif-sur-Yvette, France[*]

M. Suchý and H. K. Wipf, SOCAR, Ltd., Switzerland[*]

J. J. Meusy, Université Paris, France[*]

The existence of sexual hormones in crustaceans is now well established. The discovery of the androgenic gland by Charniaux-Cotton in 1954 has led to great progress in the comprehension of the mechanisms[1]. Androgenic glands, first studied in the amphipod Orchestia gammarellus, have since been found in a variety of crustaceans[2]. Every individual possesses a pair of androgenic glands, but they develop only in males. Grafts of androgenic glands to females of Orchestia gammarellus have demonstrated their role in the determination of sex in these animals[1,3]. Several teams have tried to isolate the compounds responsible for this activity. According to Sarojini, the hormone is known to be lipidic because of its behaviour during the isolation[4]. In 1963 Sarojini[5] reported the masculinisation of crabs, using testosterone propionate. More recently, Katakura et al.[6] obtained from the androgenic gland of the isopod Armadillidium vulgare a protenic fraction of molecular weight 15,000 which induced the development of male sexual characteristics.

A series of biological tests has been reported which allows the isolation of biologically-active compounds to be followed. The development of male sexual characteristics is rapidly observed in Talitrus saltator, since grafts of an androgenic gland produce an orange colour in the antennae due to the accumulation of astaxanthin[7]. The induction of spermatogenesis is another conse-quence of the action of the androgenic gland. The masculinisation is also observed in the ovaries of females, since grafts of andro-genic glands inhibit the biosynthesis of proteins in the vitellus. This last observation has permitted the development of an in vitro

bioassay which has been employed in this work[8,9]. Ovaries of
Orchestia gammarellus are kept as organ subcultures in a media
containing ^{3}H-leucine. Addition of extracts from the androgenic
glands inhibits the incorporation of leucine. This test can also
be performed in vivo. When using this test, non-specificity is
noticed among crustaceans, leading to the idea that the compounds
responsible for the activity could be the same in all species.

We have found two of the known activities of the androgenic
glands in a lipid extract, the inhibition of the biosynthesis of
proteins in the ovaries and the accumulation of astaxanthine in
the antennae of Talitrus saltator. However, we have not yet
succeeded in inducing spermatogenesis in females with the isolated
products. In the present communication we report the identification of two biologically-active components from the lipid extract
of the androgenic glands of the male crab Carcinus maenas[10].

Using the inhibition of incorporation of ^{3}H-leucine in ovary
subcultures of Orchestia gammarellus and the accumulation of
astaxanthine in the antennae of Talitrus saltator as bioassays, a
biologically-active lipid fraction was obtained from both the
hemolymph and the androgenic gland of the male crab Carcinus
maenas[9]. The activity was present in a lipid fraction extracted
by ethanol-ether. Using silica gel thin-layer chromatography, and
eluting with 9:1 benzene-ether, it was found in a zone of R_f 0.55-
0.75 but could not be visualised due to the very small amount of
material. A maximum inhibition of 50% was observed in the fraction
isolated from the preparative TLC. A problem occurred due to the
presence of phthalates in the extracts, but it was later found that
following butyl phthalate as an internal standard to localise the
interesting fraction on TLC was a useful technique. The next step
of the purification was filtration on a column of Sephadex-LH20,
using chloroform as solvent. Different standards were used to
check the column. A good relationship between the elution order
and the molecular weights was observed with cholesterol propionate
(442), cholestanone (366), butyl phthalate (222), and 2-pentyl-
cyclopentanone (154). Each fraction from the column was checked
by TLC, and the products of R_f 0.55-0.75 submitted for bioassay.
The fractions eluted from the Sephadex-LH20 column between butyl
phthalate and ethyl phthalate were found to be active. It was
thus possible to assess an approximate molecular weight of 250.
From the behaviour of the fraction on TLC, polar compounds such as
alcohols could be excluded and the functionality was restricted to
esters, ketones or similar groups. On GLC, using a DEGA column
(8%), the biologically-active compounds were eluted at 150°C, with
an RRT of 1.69, using methyl stearate as reference. Using preparative GLC under these conditions, a final product was collected,
representing a purification of ca. 10[5].

(1)

(2)

(3)

After the Sephadex-LH20 column, the product was analysed by coupled GLC-MS. A 20 m capillary column, coated with OV-1, was interfaced with a Varian MAT CH5 mass spectrometer. The column temperature was programmed from 30°C to 180°C and then kept under isothermal conditions. A relatively abundant component was identified as 6,10,14-trimethylpentadecan-2-one (hexahydrofarnesyl-acetone (1)[10]. The molecular ion in the MS was not present. However, the following ions were found: m/e 250 (M-18)$^+$ 4%, 225 (M-43)$^+$ 1%, 210 (M-58)$^+$ 2%, 43 (79%), 58 $(C_3H_6O)^+$ in agreement with a saturated methyl ketone (100%), 71 (59%). The mass spectrum was identical to that of the synthetic product and also corresponded to the published spectrum[11]. The synthetic product was prepared by H_2/Pt reduction of farnesyl acetone (2) and purified by preparative column chromatography. Its retention time on GLC was identical to that of the natural compound by co-injection. 6,10,14-Trimethyl-5-trans,9-trans,13-pentadecatrien-2-one (t,t-farnesyl acetone (2)) was also present (ca. 10% of (1)), as indicated by the coupled GLC-MS and also by co-injection of an authentic sample. Most of the other peaks observed in GLC were identified as impurities consisting mainly of phthalates or fatty acid esters ranging from C_{14} to C_{18}.

In vitro bioassays of (1) have shown an inhibition of the incorporation of ^{3}H-leucine in ovary subcultures at a concentration of 10 nanograms per ovary. No difference was observed using the pure 6R,10R compound (3), prepared from phytol by oxidation with chromium trioxide in acetic acid. In vitro bioassay of (2) indicated a higher activity, with inhibition being observed with as little as 250 picograms per ovary.

The C_{18}-isoprenoid ketone (1) has previously been found in recent marine sediments[11], and one of its possible sources is phytol, originally derived from chlorophyll. It is very difficult to assess the amount of the two compounds present in a single androgenic gland or to make comparisons between the biological activity of the total lipid extract and that of the identified compounds. However, several arguments lead to the hypothesis that other biologically-active components may be present. They are probably hidden by impurities in the GLC-MS analysis, and the active mixture should be re-investigated with this in mind. The secondary male characteristics, such as the pink colouration of antennae due to accumulation of astaxanthin[7], which were produced by injection of the purified lipid extract, were not obtained using (1) or (2). The inhibitory effect on the incorporation of ^{3}H-leucine into the ovaries in the subcultures was lowered when the lipid extract of the androgenic glands was saponified, but full activity was recovered after treatment with an ether solution of diazomethane. This indicated the existence of some biologically-active esters. In fact, the existence of methyl esters could be demonstrated by adding ^{14}C-methionine to cultures of androgenic glands. The saponification of the isolated lipids has led to labelled methanol, isolated as its 3,5-dinitrobenzoate. Lipid extracts from the androgenic glands inhibit biological transmethyl-ations of E. coli B tRNA in vitro (using the transmethylases of crab testes and ^{14}C-S-adenosylmethionine as methyl donor). A 50% inhibition was observed, with about 10 μg of the purified lipid extract corresponding to five androgenic glands[12]. Repetition with the methylases from crab ovaries showed an inhibition with the extract of four androgenic glands, the difference being due to the lower level of such enzymes in the ovaries. The inhibition of biological methylation of tRNA, using an enzyme preparation from crab testes, reached a maximum of 48%, with 330 nanograms/ml of farnesyl acetone (2) and 26% with 50 μg/ml of hexahydrofarnesyl acetone (1); no synergism was observed with the two products. (For details, see references 12 and 13.) Since a maximum of 95% inhibition of the biological methylation of E. coli tRNA can be obtained with the lipid extract from sixty androgenic glands, it seems that a more active component than (1) or (2) was present in the mixture. The corresponding in vivo inhibition has not yet been demonstrated. Such a process, if established, would give an idea of the mechanism by which androgenic secretions could control the biosynthesis of protein in vitellus. (For a review on tRNA methylations, see reference 14).

The inhibition of E. coli tRNA methylations by androgenic lipid extracts or by farnesyl acetone also occurred when methylases from rat liver were used[15].

For compounds (1) or (2), the biological activities are restricted to narrow concentration ranges, a phenomenon which is not observed with the extracts from androgenic glands. For example, the inhibitory effect on methylases does not exist above or below the indicated concentrations.

CONCLUSIONS

A purified lipid fraction was obtained from the androgenic glands of the male crab Carcinus maenas. It possessed two of the known biological activities of the glands; inhibition of the incorporation of ^{3}H-leucine into ovary subcultures and pink colouration of the antennae, a secondary sexual characteristic of the male. Masculinisation, that is, induction of spermatogenesis, has never been obtained using this preparation. The possibility of water-soluble hormones, proposed by other workers, cannot be excluded. From this lipid fraction, hexahydrofarnesylacetone (1) and farnesylacetone (2) have been identified, (2) being more active in the in vitro bioassays. The two products have also shown interesting properties as inhibitors of biological transmethylation in an artificial in vitro system. In vivo repetitions of these tests are under investigation. Several arguments can be used to propose the existence of other biologically-active components in our extract. At least one of them could be a methyl ester.

ACKNOWLEDGEMENTS

We wish to thank Professors H. Charniaux-Cotton and E. Lederer for their continued interest in this work. We also wish to thank Mrs. M. C. Fried-Montaufier and M. C. Carre-Lecuyer for technical assistance.

REFERENCES

1. H. Charniaux-Cotton, Ann. Sc. Nat. Zool. Biol. Animale 19, 411 (1957); J. Berreur-Bonnenfant and H. Charniaux-Cotton, Ann. Biol. 9, 187 (1970).

2. H. Charniaux-Cotton, in Organogenesis, eds. R. De Haan and H. Ursprung, Rinehart and Wilson, New York, 1965.

3. T. Ginsburger-Vogel, Ann. Biol. 9, 441 (1970).

4. S. Sarojini, Current Science 2, 55 (1964).

5. S. Sarojini, Current Science 9, 411 (1963).

6. Y. Katakura, Y. Fujinaki, and K. Unno, Annotationes Zoologicae Japonenses 48, 203 (1975).

7. M. Barbier, H. Charniaux-Cotton, and M. C. Fried-Montaufier, Comptes-rendus Acad. Sc. Paris 263, Ser. D, 1508 (1966).

8. J. Berreur-Bonnenfant and J. J. Meusy, Comptes-rendus Acad. Sc. Paris 272, Ser. D, 1641 (1972).

9. J. Berreur-Bonnenfant, J. J. Meusy, J. P. Ferezou, M. Devys, A. Quesneau-Thierry, and M. Barbier, Comptes-rendus Acad. Sc. Paris 277, Ser. D, 971 (1973).

10. J. P. Ferezou, J. Berreur-Bonnenfant, J. J. Meusy, M. Barbier, M. Suchy, and H. K. Wipf, Experientia, in press.

11. R. Ikan, M. J. Baedeker, and I. R. Kaplan, Nature 244, 154 (1973).

12. A. Tekitek, J. Berreur-Bonnenfant, J. P. Ferezou, J. J. Meusy, M. Barbier, and E. Lederer, Biochimie, in press.

13. A. Tekitek, J. Berreur-Bonnenfant, M. Barbier, and E. Lederer, in preparation.

14. F. Nau, Biochimie 58, 629 (1976).

15. M. Rojas, A. Tekitek, M. Barbier, and E. Lederer, in preparation.

*J.-P. Ferezou, A. Tekitek, M. Rojas, and M. Barbier, Institut de Chimie des Substances Naturelles, CNRS, 91190 Gif-sur-Yvette, France

J. Berreur-Bonnenfant, Laboratoire de Biologie et de Génétique Evolutive, CNRS, 91190 Gif-sur-Yvette, France

M. Suchy and H. K. Wipf, SOCAR, Ltd., Agrochemical Research and Development, 8600-Dübendorf, Switzerland

J. J. Meusy, Laboratoi e de Sexualité et Reproduction des Inverté-brés, Université Paris-VI, Tour 32,4 Place Jussieu, 75230 Paris Cedex 05, France

DETECTION AND IDENTIFICATION OF MOULTING HORMONE (ECDYSONES) IN

THE BARNACLE BALANUS BALANOIDES

P.M. Bebbington[*], E.D. Morgan and C.F. Poole

Department of Chemistry, Keele University,

Keele, Staffordshire, U.K.

INTRODUCTION

The accumulation of barnacles, weed, and other fouling organisms on ships' bottoms has been a problem for centuries. This is
because it increases the skin frictional resistance of the hull,
which in turn increases fuel consumption. Attempts have been made
to control the fouling problem by applying various coatings to
ships' hulls below the water-line, but present treatments give
only about a two-year out-of-dock period, and dry docking is ex-
pensive[1-5].

In an attempt to discover a more effective anti-fouling
method, we have made some initial studies on barnacles whose life-
cycle, like that of other arthropods, is under hormonal control.

There are three points at which it might be possible to
interfere with the development of barnacles and effect some kind
of control. These are: the moulting process, the cementation of
cyprids to substrate, and the calcification of the outer shell.
Such interference might be brought about by hormone mimics, which
could cause mis-timing of developmental stages (Figure 1). Metabolic blocking agents might be found which would prevent proper
development, or enzymes might be used to disrupt biochemical
function.

We embarked on a project to extract and identify the moulting
hormones from the barnacle Balanus balanoides. In arthropods,
steroidal hormones of the ecdysone type (see Figure 2) have been
shown to control moulting, and ecdysone was the first hormone to
be isolated and identified from insects[6,7]. 20-Hydroxyecdysone has

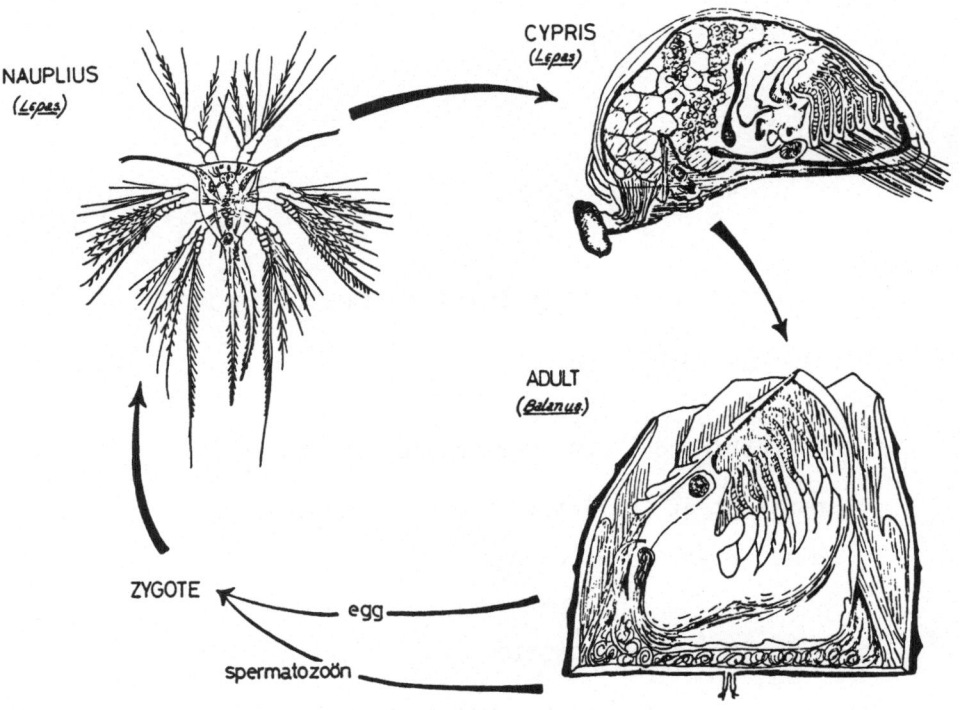

Figure 1. The life cycle of the barnacle.

been shown to be the major moulting hormone in insects, and three
other ecdysones, 20,26-dihydroxyecdysone, 26-hydroxyecdysone, and
makisterone A, have also been found in insects[9-11]. The first
crustacean moulting hormone to be identified was 20-hydroxyecdysone
(crustecdysone, ecdysterone, β-ecdysone) and it was isolated from
the crayfish Jasus lalandei[12]. Three other crustacean moulting
hormones have also been identified: 2-deoxy-20-hydroxyecdysone
from J. lalandei[12], inokosterone and makisterone A from the crab
Callinectes sapidus[13].

It was reasonable to suppose that the moulting hormones of
B. balanoides would be of the ecdysone type. This view was further
supported by the work of Tighe-Ford et al. who caused increased
moulting in B. balanoides by injections of 20-hydroxyecdysone[14],
and also by Cheung et al. who caused metamorphosis of the cyprids
of Balanus eburneus by treatment with 20-hydroxyecdysone[15].

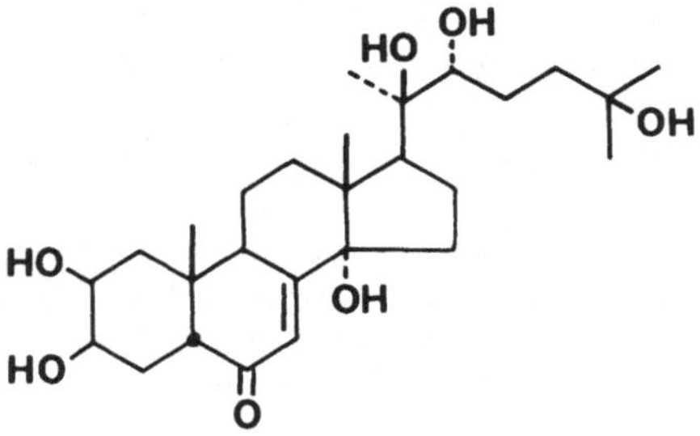

20-HYDROXYECDYSONE

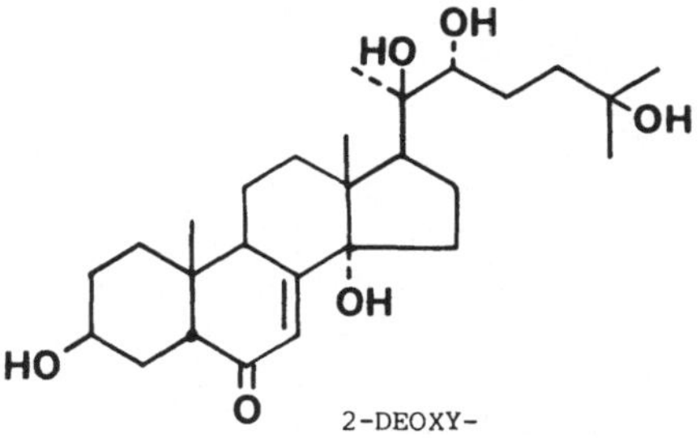

2-DEOXY-
20-HYDROXYECDYSONE

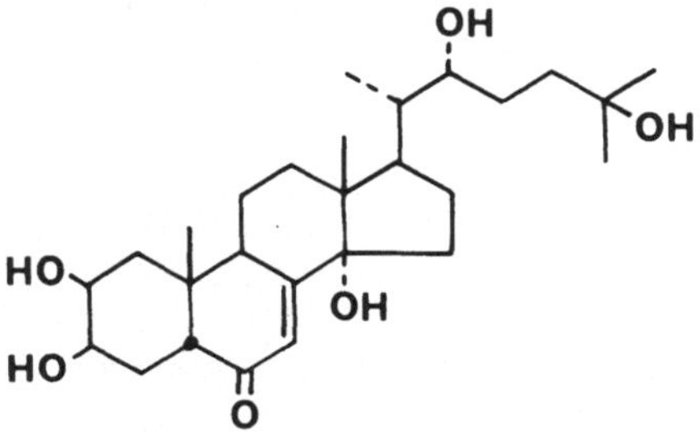

ECDYSONE

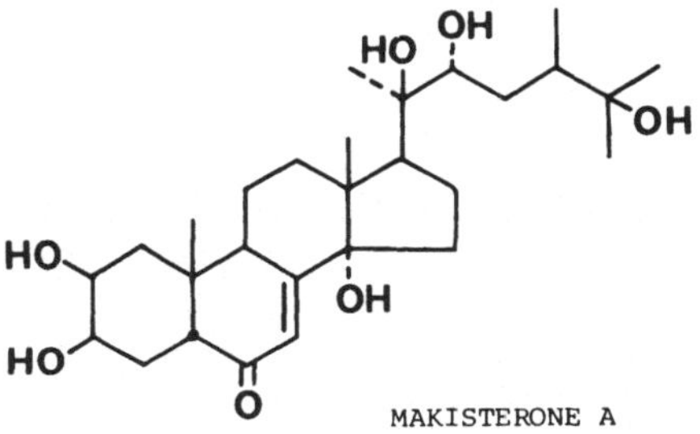

MAKISTERONE A

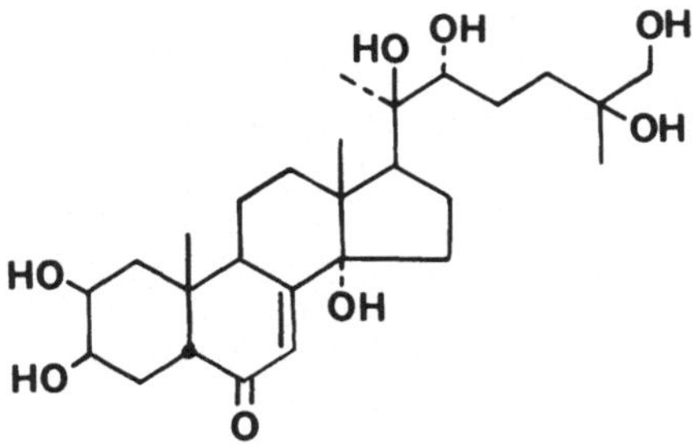

20-26-DIHYDROXYECDYSONE

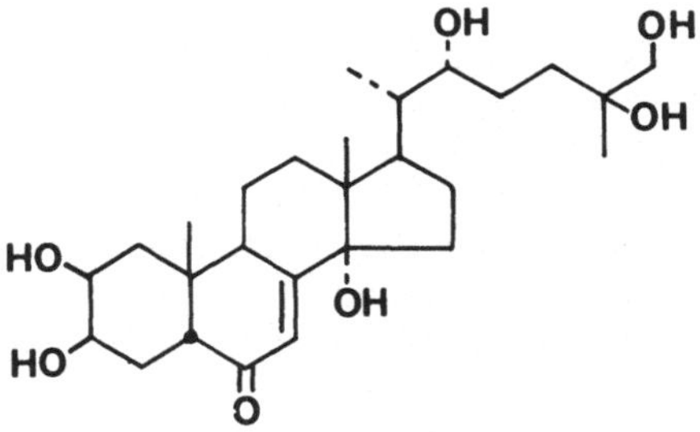

26-HYDROXYECDYSONE

Figure 2. Ecdysone and related compounds.

ANALYSIS OF ECDYSONES

Before the mass extraction of barnacles could be undertaken, it was necessary to have a very sensitive method for detecting, and if possible quantifying, the moulting hormone during each stage of the isolation. The principal methods generally used for the analysis of ecdysones are: bioassay[16], which can detect ecdysone at levels as low as 1-100 x 10^{-9} g; radio-immune assay (RIA), which can detect ecdysone at levels as low as 8 x 10^{-11} g; and gas liquid chromatography (GLC). The bioassay and the RIA lack specificity and require both considerable facilities and experience for effective routine use. GLC, on the other hand, can distinguish between individual ecdysones, and this method of analysis is used in most laboratories. We, along with another group[17-19], found that ecdysones were unexpectedly sensitive to the electron capture detector (ECD) so that, after conversion to their trimethylsilyl (TMS) ethers, they can be directly determined down to the picogram level.

A variety of methods for the formation of TMS ethers of ecdysones has been reported. We investigated the method of Ikekawa et al.[20,21] who claimed that 20-hydroxyecdysone could be quantitatively silylated with trimethylsilylimidazole (TMSI) at 100° for one hour. Ikekawa et al. maintained that the C-20 hydroxy group proved most difficult to silylate. Other workers have produced results in agreement with this. In an attempt to repeat Ikekawa's work, we heated 20-hydroxyecdysone at 100° for one hour in TMSI. On GLC analysis of the product (3ft 1% OV-101 on CQ, 266°, 60 ml min⁻¹), two peaks were produced with retention times of 5.3 and 4.6 min. If the reaction time and temperature were increased, the peak at 5.3 min predominated with a trace of a peak at 4.1 min. After reaction for 20 hours at 140°, the peak at 4.1 min became the sole peak. This sequence of events was repeated for other ecdysones. In an effort to explain this single peak at 4.1 min, obtained under conditions somewhat more severe than those reported by Ikekawa, several possibilities were considered. It was thought that the peak could be due to the formation of a TMS-enol ether at the 6-position, but this was discounted when UV analysis of the product, formed from 22,25-dideoxyecdysone (Figure 3), showed the 7-ene-6-one system to be intact. Epimerisation at the 5-position was discounted when both the 5α and 5β isomers of 22,25-dideoxyecdysone·(2β,3β,14α-trihydroxycholest-7-ene-6-one) produced peaks at shorter retention time after reaction. Epimerisation at, say, C-9, was shown to be unlikely when the keto-methoxime derivative of 22,25-dideoxyecdysone was reacted at 140° for 20 hours and gave a product having a shorter retention time than the low temperature product. After comparing the reactivity of 2β,3β-dihydroxy-5α-cholestane (Figure 3) at room temperature with 22,25-dideoxyecdysone for one hour at 100° with TMSI, it was found that the 14α-hydroxy group reacted slowly. This was con-

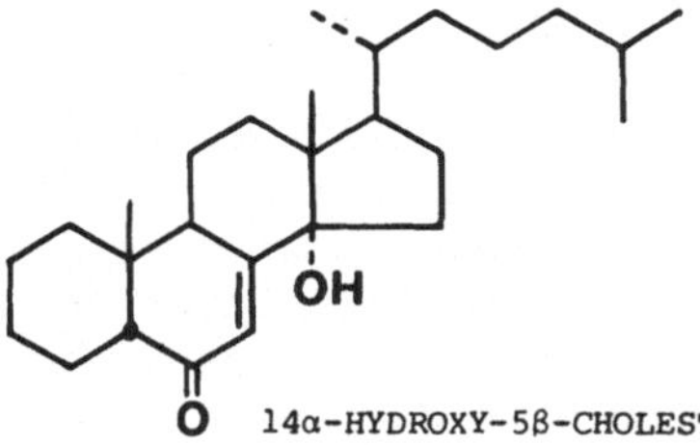

14α-HYDROXY-5β-CHOLEST-7-ENE-6-ONE

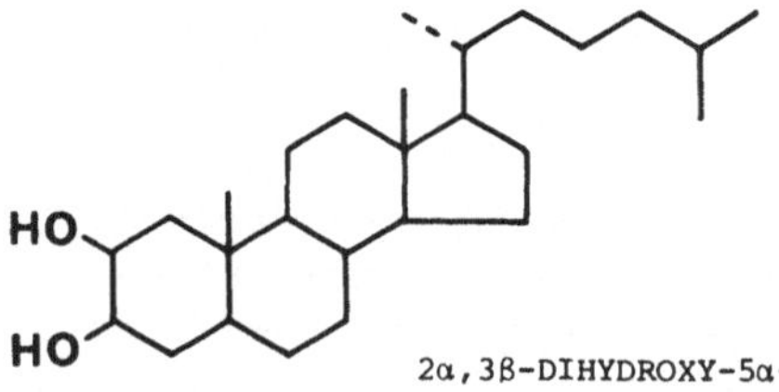

2α,3β-DIHYDROXY-5α-CHOLESTANE

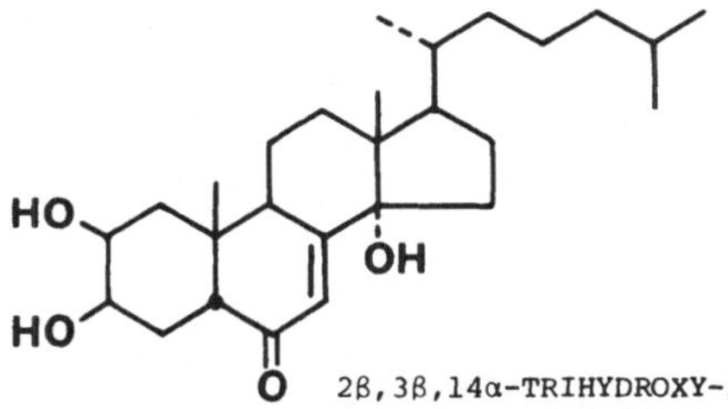

2β,3β,14α-TRIHYDROXY-5β-CHOLEST-7-ENE-6-ONE
(22,25-DIDEOXYECDYSONE)

Figure 3

firmed by reacting 14 α-hydroxy-5 β-cholest-7-ene-6-one (Figure 3)
with TMSI at 100° for one hour, when reaction was slow, and at
140° when 12 hours was required for complete reaction. The rate of
silylation of the 14 α-hydroxy group was speeded up by trimethyl-
chlorosilane (TMCS). However, too high a concentration led to
enolization, which was moderated to some extent by a 14 α-oxy
substituent. Pyridine tended to slow down the enolization reaction.

As a result of our work, it was decided to use mild silylation conditions for the analysis of ecdysones in arthropod extracts, leaving the 14 α -hydroxy group unreacted[22]. Dried arthropod extracts were heated at 100° for six   hours in a mixture of pyridine and TMSI in the ratio of 5:2. Excess reagents were removed under vacuum, and the residue was refined by thin-layer chromatography on silica gel plates, developed with toluene-ethyl acetate (7:3), and the required band was eluted with diethyl ether for GLC analysis, using an EC detector. The sensitivity of the TMS-ethers of ecdysone to ECD was in the region of 10^{-12} g.

EXTRACTION AND ISOLATION OF ECDYSONES

The concentration of ecdysones in arthropods is very low, and much lower in crustaceans than in insects (See Table 1). Hence, in order to isolate enough material for identification purposes, large amounts of starting material have to be used and lengthy isolation procedures undertaken.

Karlson et al.[23] extracted 3 tons of the shrimp Crangon vulgaris without identifying a single moulting hormone, and Horn et al. obtained 2 mg of 20-hydroxyecdysone and 200 ug of 2-deoxy-20-hydroxy-ecdysone from 1 ton of the marine crayfish Jasus lalandei[12]. The method we used for the isolation of the hormones was based on that used in our laboratory for the desert locust[27] and on that used by Horn et al. for J. lalandei[12]. It is shown in the flow diagram (Figure 4)[24].

After a preliminary extraction of 150 Kg of fresh barnacles, it was apparent that a larger sample would be required, together with special large-scale equipment to handle it. Therefore, 1,500 Kg of barnacles, consisting of about 90% B. balanoides and 10% Eliminium modestus, was extracted. Attempts were made to assay the activity during purification, but the initial fractions had too low an activity to be measured on barnacles[25], and they were too toxic for the locust abdomen assay[26,27]. Therefore, solvent partition systems were used, for which the partition ratios for ecdysones were known and the phase containing activity could then be predicted.

After the first chromatographic step, the activity was sufficiently concentrated to be detectable using the   ECD-GLC method previously described. Using our extraction procedure, two ecdysones were detected and quantified in B. balanoides. These values must be approximate because of the possibility of loss in the large-scale extraction steps and because barnacles must, of necessity, be collected at all stages of the moult-cycle. These concentrations must therefore represent averages. However, it is the first time that ecdysone has been found in a crustacean mass extraction.

Table 1. Concentration of ecdysones in arthropods.

Species	Stage	'Ecdysone' Isolated	Weight Extracted Kg.	Weight 'Ecdysone' Isolated mg.	Conc. ug.Kg^{-1}
Insecta					
Antheraea pernyl	pupae	20-hydroxyecdysone	31	0.2	6.0
Schistocerca gregaria	5th instar	20-hydroxyecdysone	32	1.9	60.0
Bombyx mori	pupae	ecdysone	3000	206	68.0
		20-hydroxyecdysone		48	16.0
Crustacea					
Homarus americanus	postmoult	20-hydroxyecdysone	5	0.09	6.0
Jasus lalandei	intermoult	20-hydroxyecdysone	1000	2.0	2.0
		2-deoxy-20-hydroxyecdysone	3000	0.2	0.07
Callinectes sapidus	premoult 'green'	inokosterone	25	0.125	5.0
	premoult 'peeler'	inokosterone	25	0.5	20.0
		20-hydroxyecdysone		0.1	4.0
	postmoult 'soft shell'	20-hydroxyecdysone	25	7.0	280.0
		makisterone A		0.6	24.0
Balanus balanoides	variable	ecdysone	1500		0.006
		20-hydroxyecdysone			1.0
Crangon vulgaris	intermoult	not isolated	3000		

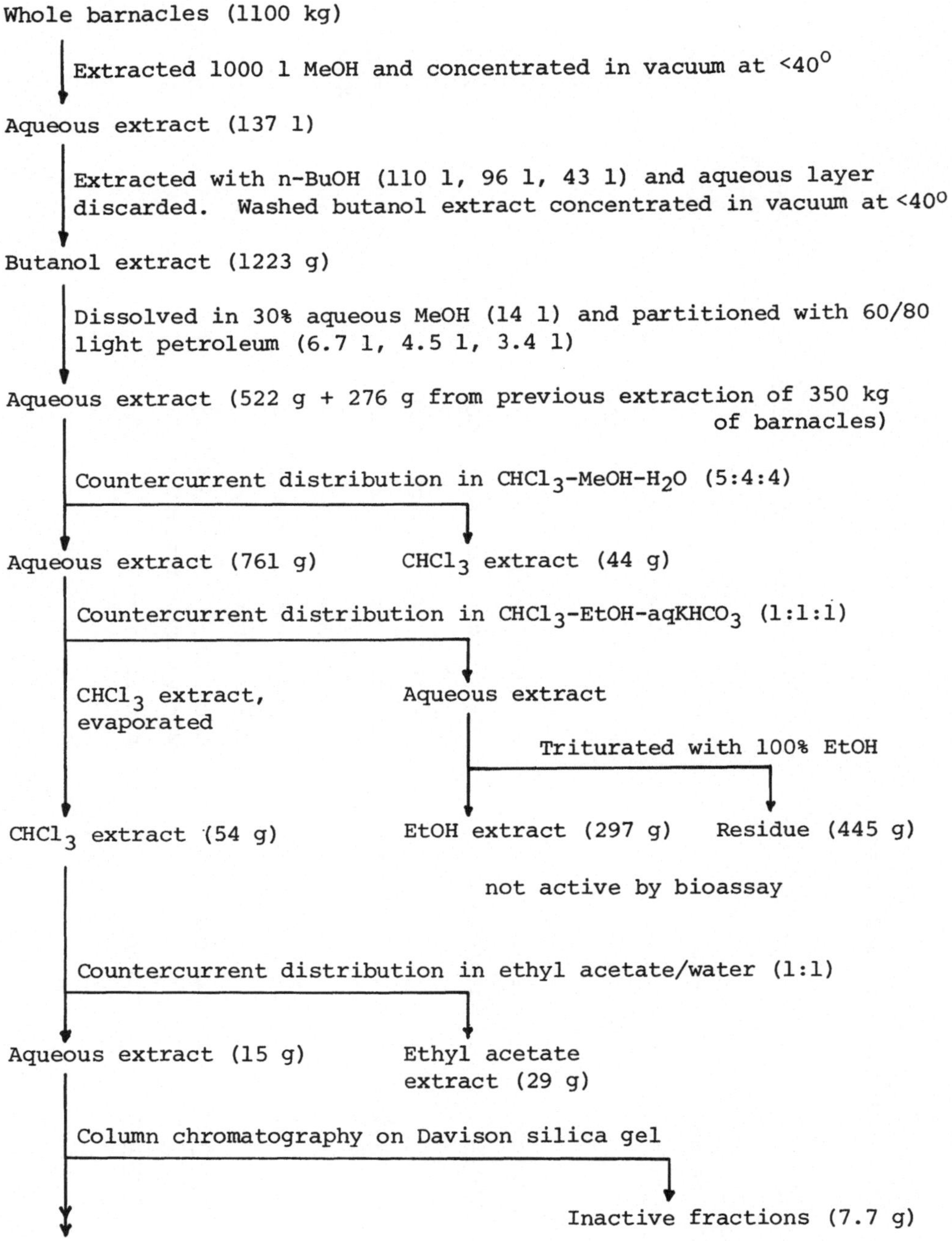

Whole barnacles (1100 kg)

| Extracted 1000 l MeOH and concentrated in vacuum at <40^0

Aqueous extract (137 l)

| Extracted with n-BuOH (110 l, 96 l, 43 l) and aqueous layer
| discarded. Washed butanol extract concentrated in vacuum at <40^0

Butanol extract (1223 g)

| Dissolved in 30% aqueous MeOH (14 l) and partitioned with 60/80
| light petroleum (6.7 l, 4.5 l, 3.4 l)

Aqueous extract (522 g + 276 g from previous extraction of 350 kg
                                                of barnacles)

    Countercurrent distribution in $CHCl_3$-MeOH-H_2O (5:4:4)

Aqueous extract (761 g) $CHCl_3$ extract (44 g)

    Countercurrent distribution in $CHCl_3$-EtOH-aqKHCO$_3$ (1:1:1)

    $CHCl_3$ extract,   Aqueous extract
    evaporated

                                    Triturated with 100% EtOH

$CHCl_3$ extract (54 g)   EtOH extract (297 g)   Residue (445 g)

                                    not active by bioassay

    Countercurrent distribution in ethyl acetate/water (1:1)

Aqueous extract (15 g)   Ethyl acetate
                              extract (29 g)

    Column chromatography on Davison silica gel

                                        Inactive fractions (7.7 g)

Figure 4. Flow diagram of the attempted isolation of moulting
hormones from Balanus balanoides.

Figure 4. (continued)

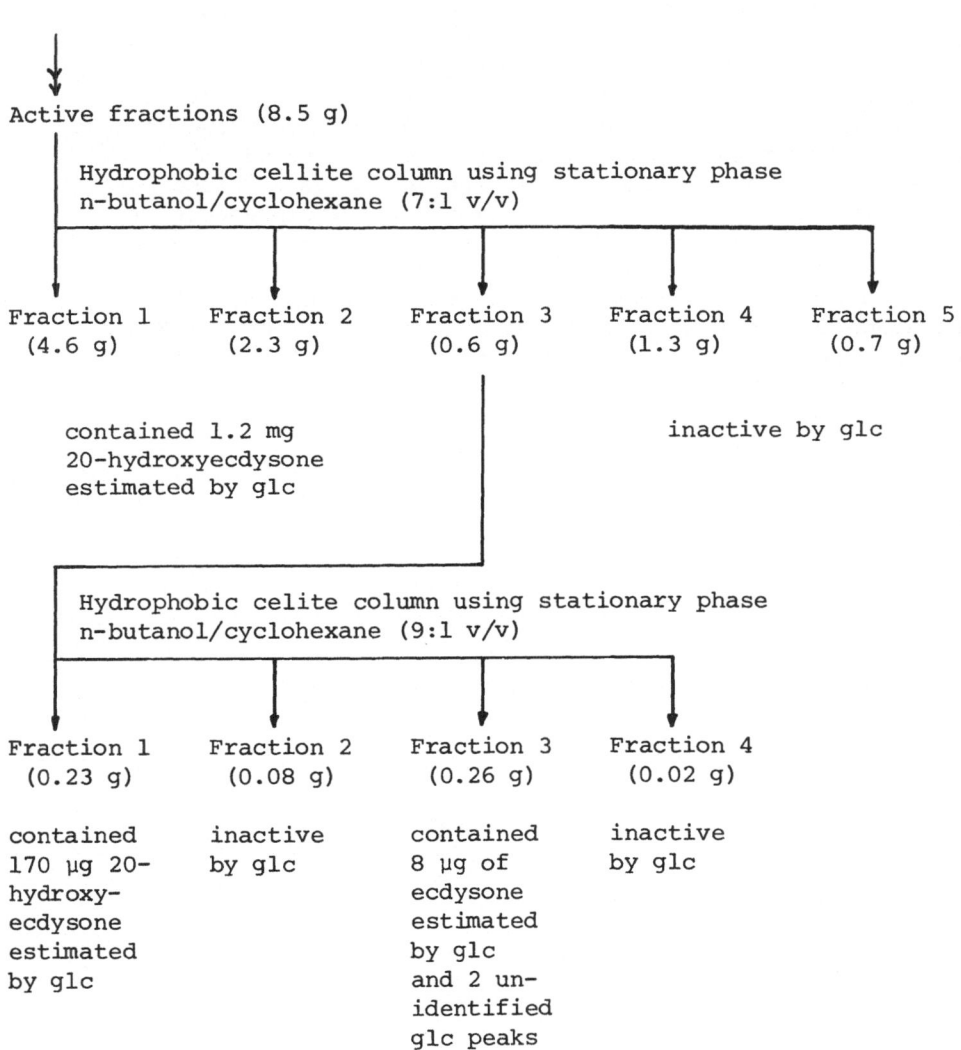

Recent work by Mizuno et al.[28] suggests two possible places in
the extraction process where considerable losses of moulting hormone may have occurred. Mizuno has shown that substances with
moulting hormone activity, or potential moulting hormone activity,
might be bound so firmly to the arthropod shell or cuticle that
they are not completely extracted by solvents commonly employed
for the extraction of moulting hormones at room temperature or
below. Mizuno et al. also found that ecdysone conjugates are partitioned into the aqueous phase of the n-butanol-water system.
This phase is almost totally neglected in moulting hormone extractions, since the partition coefficients of free ecdysones are in
favour of the n-butanol phase. Even if the aqueous phase were to
be assayed, moulting hormone activity would not be detected[28,29]
because ecdysone conjugates such as the sulphates or glucosides
are inactive in the bioassays in common use. The presence of
ecdysone would be detected only if the bioassay animal had the
ability to hydrolyse the conjugates. The only conjugate of ecdysone   showing biological activity is the β-glucoside of
ponasterone A: ponasteroside A[30] (Figure 5). The work of Mizuno
et al. suggests that a closer look must be taken at commonly used
moulting hormone extraction procedures.

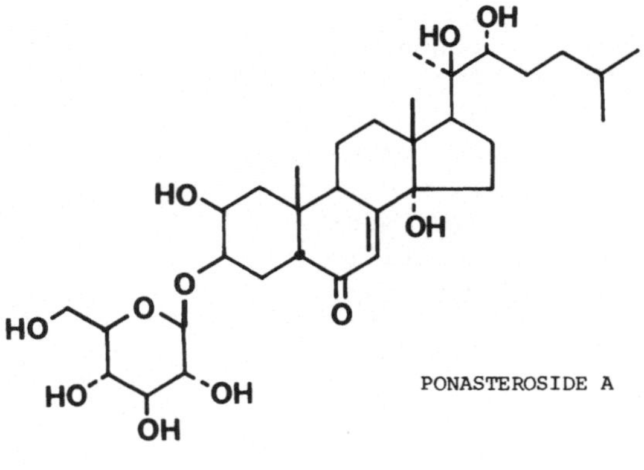

PONASTEROSIDE A

Figure 5.

BIOSYNTHESIS OF ECDYSONES IN CRUSTACEANS

Recent work, including our own, has thrown some light on the in vivo production of crustacean moulting hormones. Crustaceans cannot biosynthesise sterols from mevalonate and need a dietary sterol source from which to synthesise moulting hormones. Gagosian et al.[31] have shown that cholesterol is converted to 20-hydroxyecdysone in the lobster Homarus americanus. Spaziani et al.[32] have demonstrated a selective uptake of radio-labelled cholesterol by the Y-gland of the crab Hemigrapsus nudus, obtaining products which co-chromatographed with ecdysone. However, the route by which cholesterol is converted into ecdysone in crustaceans is largely unknown. More is known about later stages in the biosynthesis, and taking into account the work done with insects we have postulated a number of possible biosynthetic pathways for the later stages of moulting hormone synthesis in crustaceans (Figure 6).

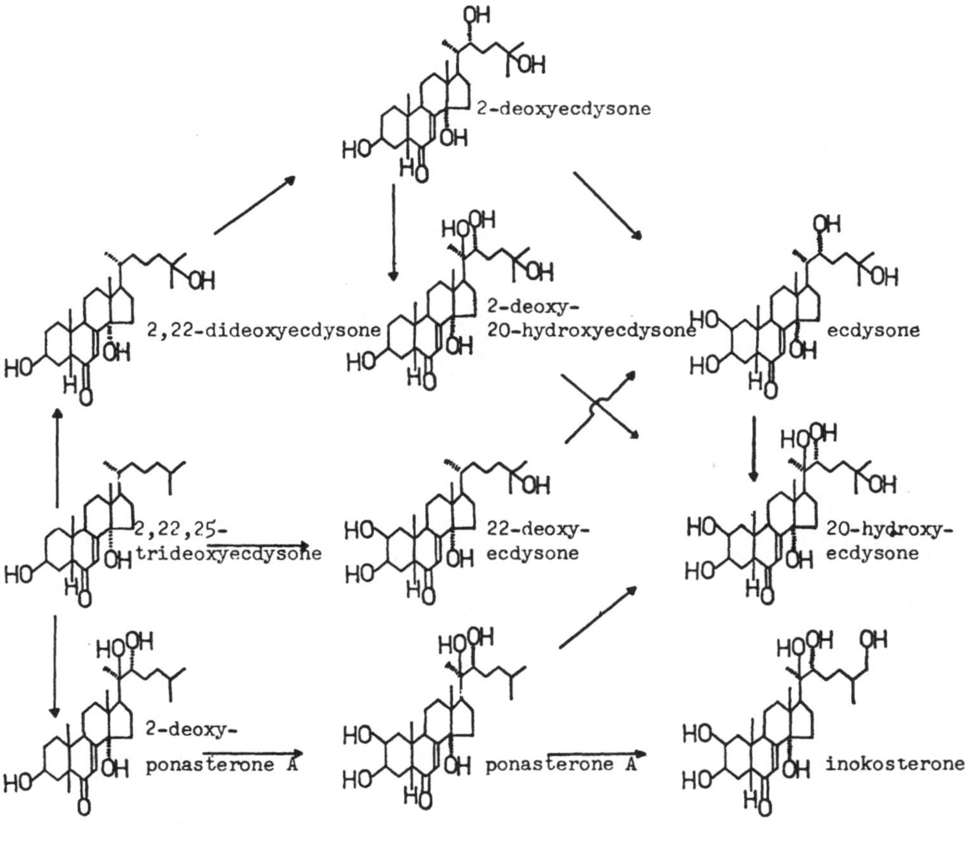

Figure 6.

Our proposed scheme starts with 2,22,25-trideoxyecdysone[33].
This, and 2,22-dideoxyecdysone, have not been detected in crustaceans, but 22-deoxyecdysone has been shown to be an ecdysone precursor in insects and to have moulting hormone activity in crustaceans[34,35]. 2-Deoxyecdysone has not been found in crustaceans,
but it has been found in the fern Blechnium minus, along with
ecdysone and 2-deoxy-20-hydroxyecdysone[36]. It is highly active in
the Calliphora bioassay, and it is reasonable to suppose that it can
be converted in crustaceans to either ecdysone or 2-deoxy-20-
hydroxyecdysone. Ecdysone has been detected in the barnacle
B. balanoides, along with 20-hydroxyecdysone[24]. The identification
of ecdysone in barnacles is supported by the work of Bollenbacher
et al.[37], who cultured the Y-glands of the crab Pachygrapsus
crassipes, and Willig et al.[38], who cultured the Y-gland of the
crayfish Orconectes limonus. Bollenbacher et al. detected only
ecdysone in the medium; Willig et al. detected ecdysone and 20-
hydroxyecdysone in the medium but suggested that the latter had
been formed by conversion of ecdysone in epidermal tissue attached
to the Y-gland. King et al.[39] have demonstrated that ecdysone is
rapidly converted to 20-hydroxyecdysone in crustaceans. 2-Deoxy-
20-hydroxyecdysone has been isolated, along with 20-hydroxyecdysone,
from J. lalandei, and it is reasonable to suppose that the former
ecdysone is readily converted into the latter.

2-Deoxyponasterone A has not been detected in either insects
or crustaceans, but ponasterone A has been detected in Calliphora
after injection of 22,25-dideoxyecdysone[40]. In the same insect,
ponasterone A has been shown to be converted at about equal rates
to 20-hydroxyecdysone and inokosterone. Inokosterone has been
found in the crab Callinectes sapidus during the first stage of
ecdysis[13]. At the later premoult stage, inokosterone is accompanied by a smaller amount of 20-hydroxyecdysone. After moulting,
20-hydroxyecdysone is the major hormone and is accompanied by a
small amount of makisterone A.

It is unlikely that there is just one biosynthetic route to
moulting hormones in crustacea, and indeed it is likely that there
are several routes within one crustacean which could be utilized
to provide different hormones at different stages of the animal's
life-cycle. With the advent of very sensitive methods for the
analysis of ecdysones, such as our GLC method, it is becoming
easier to investigate biosynthetic pathways of moulting hormones.
At present we are investigating ecdysone concentrations throughout
the life cycle of B. balanoides.

In the two crustaceans which have been the subject of mass
extractions, 20-hydroxyecdysone has been shown to be the major
ecdysone present. In B. balanoides it was present with very small
amounts of ecdysone, and in J. lalandei with very small amounts of
2-deoxy-20-hydroxyecdysone. Siddall et al.[41] found that in

<u>Manducta sexta</u> (the tobacco hornworm) injected ecdysone was rapidly converted to the conjugated form and 20-hydroxyecdysone was present mainly as the free hormone. If we take into account the findings of Mizuno et al.[28], the the reason that most workers find mainly 20-hydroxyecdysone in their extracts, with only traces of other ecdysones, could be that their extraction procedures are designed to concentrate the free hormone. There may be equally large amounts of other ecdysones present which are not detected, the trace amounts of some ecdysones observed being the products of conjugate hydrolysis during extraction. It is reasonable to suppose that, in order to have hormones available as and when required, an "inactive" store is kept of immediate precursors as conjugates. This view is supported by Hsiang et al.[42], who found moulting hormone conjugates in the crab <u>Uca pugilator</u> and showed that their concentration decreased with impending natural ecdysis or ecdysis induced by eyestalk or limb removal. We are presently investigating conjugate formation in <u>B. balanoides</u>.

ACKNOWLEDGEMENTS

Part of this work was supported by the Procurement Executive, Ministry of Defence. We are glad to acknowledge their aid and the help of Dr. D. J. Tighe-Ford, Mr. D. C. Vaile, and Mr. D. R. Houghton, of the Central Dockyard Laboratory, Portsmouth. We thank Dr. D. H. S. Horn for a sample of 20-hydroxyecdysone and Roche Products, Ltd. for ecdysone.

REFERENCES

1. J. P. Visscher, Bull. U. S. Bur. Fish., <u>43</u>, 193 (1928).

2. C. F. T. Young, The London Drawing Association <u>8</u>, 212 (1867).

3. H. A. Gardener, Circular No. 157, Education Bureau, Scientific Section, Paint Manufacturers' Association of United States, Washington, 1922.

4. E. Hentschel, International Revue der gesamten Hydrobiologie und Hydrographie, <u>11</u>, 238 (1923).

5. E. Hentschel, Mitteilungen aus dem Zoologischen Staatsinstitut und Zoologischen Museum zu Hamburg, <u>41</u>, 1 (1924).

6. A. Butenandt and P. Karlson, Z. Naturforsch., <u>9b</u>, 389 (1954).

7. P. Karlson, H. Hoffmeister, W. Hoppe, and R. Huber, Ann. Chem., <u>662</u>, 1 (1968).

8.   W. W. Doan, in <u>Developmental Systems: Insects</u>, Vol 2, eds. S.J. Counce and C. H. Waddington, Academic Press, New York and London, 1972, 291.

9.   M. J. Thompson, J. N. Kaplanis, W. E. Robbins, and R. T. Yamamoto, Chem. Comm., 650 (1967).

10. J. N. Kaplanis, W. E. Robbins, M. J. Thompson, and S. R. Dutky, Science, <u>180</u>, 307 (1973).

11. J. N. Kaplanis, S. R. Dutky, W. E. Robbins, M. J. Thompson, E. L. Lundquist, D. H. S. Horn, and M. N. Galbraith, Science, <u>190</u>, 681 (1975).

12. D. H. S. Horn, S. Fabbri, F. Hampshire, and M. E. Lowe, Biochem. J., <u>109</u>, 99 (1968).

13. A. Faux, D. H. S. Horn, E. J. Middleton, H. M. Fales, and M. E. Lowe, Chem. Comm., 175 (1969).

14. D. J. Tighe-Ford and D. C. Vaile, J. Exp. Mar. Biol. Ecol., <u>9</u>, 19 (1972).

15. P. J. Cheung, J. Exp. Mar. Biol. Ecol., <u>15</u>, 223 (1974).

16. P. Karlson and E. Shaaya, J. Insect. Physiol., <u>10</u>, 797 (1964).

17. D. W. Borst and J. D. O'Connor, Steroids, <u>24</u>, 637 (1974).

18. C. F. Poole, E. D. Morgan, and P. M. Bebbington, J. Chromatogr., <u>104</u>, 172 (1975).

19. C. F. Poole and E. D. Morgan, J. Chromatogr., <u>115</u>, 587 (1975).

20. N. Ikekawa, F. HaHori, J. Rubio-Lightbourn, H. Miyazaki, M. Ishibashi, and C. Mori, J. Chromatogr. Sci., <u>10</u>, 233 (1972).

21. H. Miyazaki, M. Ishibashi, C. Mori, and N. Ikekawa, Anal. Chem., <u>45</u>, 1164 (1973).

22. C. F. Poole and E. D. Morgan, J. Chromatogr., <u>116</u>, 333 (1976).

23. P. Karlson and P. Schmialek, Z. Physiol. Chem., <u>316</u>, 83 (1959).

24. P. M. Bebbington and E. D. Morgan, J. Comp. Biochem. Physiol., in press.

25. D. J. Tighe-Ford, Crustaceana, <u>15</u>, 15 (1968).

26. E. D. Morgan, A. P. Woodbridge, and P. E. Ellis, J. Insect.
     Physiol., $\underline{21}$, 979 (1975).

27. E. D. Morgan, A. P. Woodbridge, and P. E. Ellis, Acrida., $\underline{4}$,
     69 (1975).

28. T. Mizuno and E. Ohnishi, Devel. Growth and Diff., $\underline{17}$(3), 219
     (1975).

29. A. Sanwasi and P. Karlson, Zool. Jb. Physiol. Bd., $\underline{78}$, 378
     (1974).

30. T. Takemoto, S. Arihara, and H. Hikmo, Tetrahedron Lett.,
     4199 (1968).

31. R. B. Gagosian, R. A. Bourbonniere, W. B. Smith, E. F. Couch,
     C. Blanton, and W. Novak, Experientia, $\underline{30}$, 723 (1975).

32. E. Spaziani and S. B. Slater, Gen. Comp. Endocr., $\underline{20}$, 534
     (1973).

33. M. N. Galbraith, D. H. S. Horn, and J. A. Thomson, Experientia,
     $\underline{31}$, 873 (1975).

34. D. S. King, Amer. Zool., $\underline{12}$, 343 (1972).

35. A. Krishnakumaran and H. A. Schneiderman, Biol. Bull., $\underline{139}$,
     520 (1970).

36. Y. K. Chong, M. N. Galbraith, and D. H. S. Horn, Chem. Comm.,
     1217 (1970).

37. W. E. Bollenbacher and J. D. O'Connor, Amer. Zool., $\underline{13}$, 1274
     (1973).

38. A. Willig and R. Kellor, Experientia, $\underline{32}$, 936 (1976).

39. D. S. King and J. B. Siddall, Nature, $\underline{221}$, 955 (1969).

40. J. A. Thomson, J. B. Siddall, M. N. Galbraith, D. H. S. Horn,
     and E. J. Middleton, Chem. Comm., 669 (1969).

41. J. N. Kaplanis, S. R. Dutky, W. E. Robbins, and M. J. Thompson,
     in Invertebrate Endocrinology and Hormone Heterophylly,
     Springer-Verlag, Berlin, Heidelburg and New York, 1974, p. 172.

42. J. Hsiang and E. Premuzic, Amer. Zool., $\underline{15}$, 786 (1975).

* Current address: 31 Catharine Road, Chell Heath, Stoke on Trent,
   Staffordshire, ST6 6PT, England.

HORMONAL ASPECTS OF BARNACLE ANTIFOULING RESEARCH

D.J. Tighe-Ford

Central Dockyard Laboratory, HM Naval Base

Portsmouth, Hants, U.K.

Part of the current antifouling research programme at CDL has been aimed at understanding and attempting to exploit developmental processes of barnacles and algae [1]. Such work is in addition to the screening of candidate toxicants and the synthesis of biologically active compounds. Studies with barnacles have been carried out in the presumption that the control of development and growth of these cirriped crustaceans is similar to that of other crustaceans and insects, which are major Classes of the Phylum Arthropoda. Two characteristics dominate the activity of arthropods: i) a life cycle in which larval stages undergo a succession of changes of form and function before they enter the adult stage, and ii) the possession of an integument which is periodically moulted to allow such changes in form and increase in size. Post-embryonic growth, ecdysis and metamorphosis are under hormonal control and reviews have been made by many authors, eg.[2-6].

Most studies of arthropod hormonal systems have been carried out on insects, although an increasing amount of work has been done recently with crustaceans and other groups. Three major hormones control insect development and growth. The moulting process is initiated by an activation hormone produced by brain neurosecretory cells. Although its structure has been described as steroidal, it is generally considered to be a protein or, most likely, a polypeptide. Production is stimulated by a range of extrinsic factors such as light, temperature and wounding. Activation hormone passes _via_ cell axons to the corpora cardiaca, which act as a neurohaemal organ; from here it is liberated into the haemolymph and transported to the prothoracic glands (or their equivalents), which are stimulated to release directly and/or

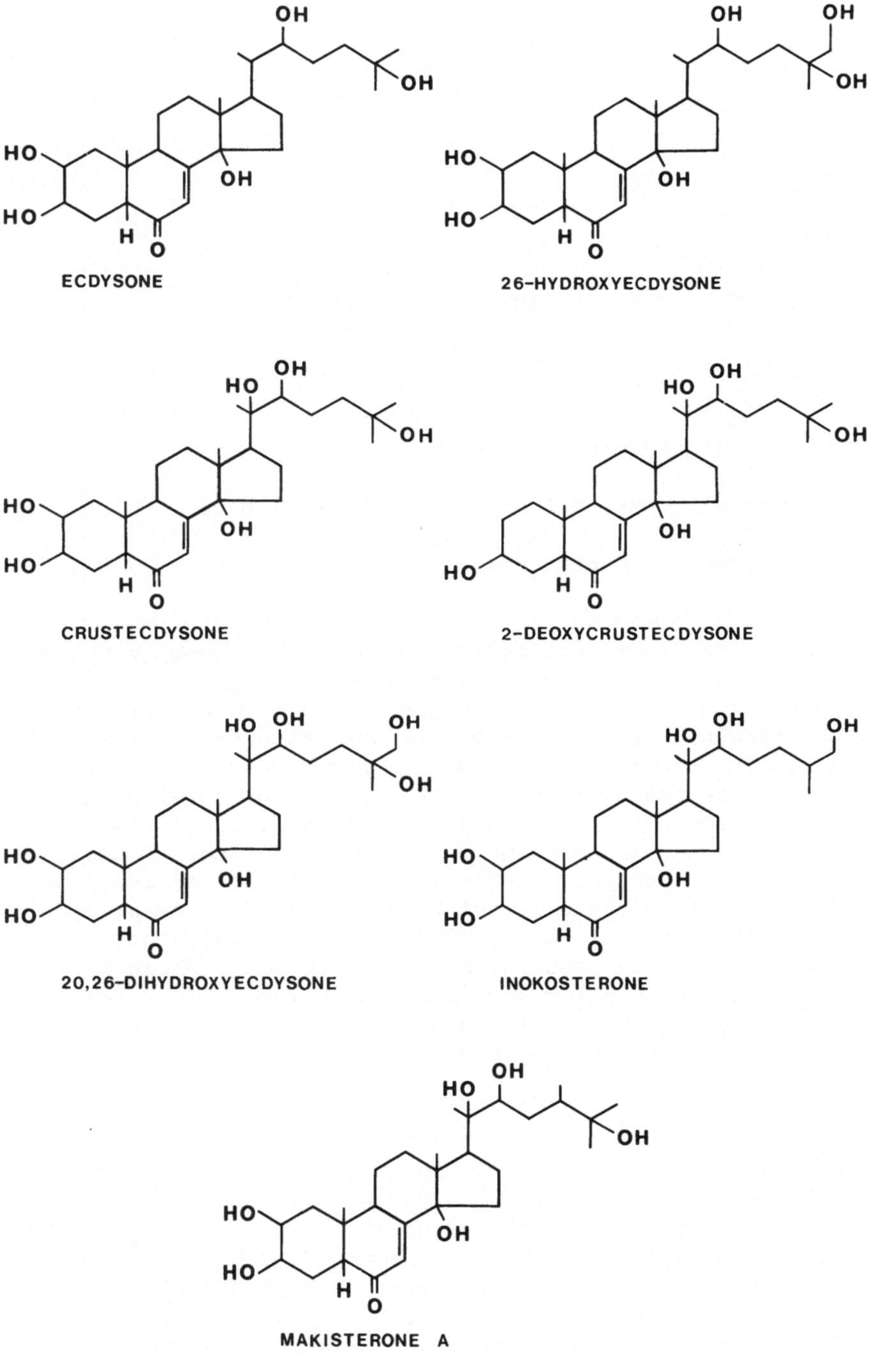

ECDYSONE

26-HYDROXYECDYSONE

CRUSTECDYSONE

2-DEOXYCRUSTECDYSONE

20,26-DIHYDROXYECDYSONE

INOKOSTERONE

MAKISTERONE A

Figure 1. Structure of ecdysones isolated from insects and crustaceans.

indirectly the hormones which stimulate the tissues involved in the moulting processes of larvae and adults. These are steroid hormones, termed ecdysones (Figure 1), in particular, ecdysone and crustecdysone; the latter has been variously termed β-ecdysone, 20-hydroxyecdysone and ecdysterone[7,8]. Activation hormone also controls the activity of another pair of glands, the corpora allata, which produce juvenile hormones (JH); these possess a sesquiterpene-like structure (Figure 2). During developmental stages the production of JH determines the form assumed after a moult; its action is to favour the expression of larval rather than adult characteristics. In holometabolous insects, relatively high levels of JH are produced during each larval instar except the last, and ensure the formation of another larval form; production is diminished in the last instar, and the moult to the pupal stage results. Little or no JH is secreted before the final moult and the adult emerges after ecdysis. In most insects, except the primitive Apterygota, which continue to moult throughout adult life, the prothoracic glands then degenerate and no further moulting occurs. The corpora allata, however, retain activity and their secretions are concerned with activities such as reproduction, ovarian development, embryogenesis, sex attractant production and diapause[6,9].

Endocrine control of crustacean moulting activities is similar to that of insects, although until recently information about the mechanisms involved was restricted to the Decapoda, particularly crabs (Brachyura). The moulting hormones (MH) are ecdysones[10] and are apparently produced directly or indirectly by the Y-organs in the thoracic region, which may be considered analogous to the insect prothoracic glands. The activity of these

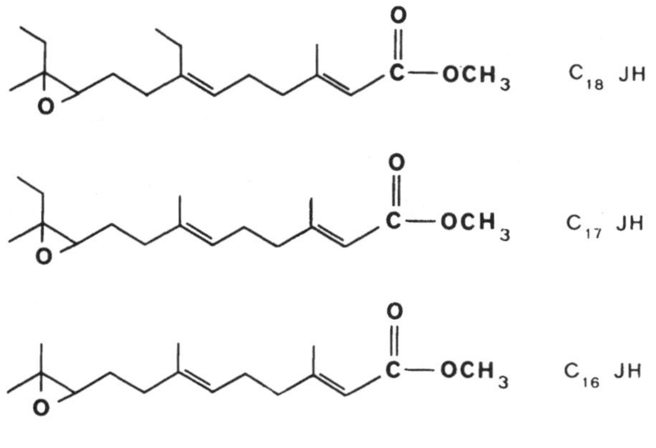

Figure 2. Structure of the insect juvenile hormones.

glands is, however, inhibited by neurosecretion from cells consti-
tuting the X-organ in the medulla terminalis of the eye-stalk;
this moult-inhibiting hormone (MIH) is most probably a polypeptide[3,11] and is released <u>via</u> or modified by the sinus glands,
which are neurohaemal organs. MH production is initiated when
secretion of MIH ceases - the reverse of the control in insects,
in which neurosecretion stimulates moulting hormone production.
Nevertheless, moulting in both Classes is controlled by a two-step
sequence (Figure 3). There is, however, some evidence of a moult-
accelerating hormone in crustaceans which may be produced by
neurosecretory cells of the brain or eye-stalk, depending upon
species[2]. Moulting continues in the adult form of many crustaceans, particularly decapods. In species such as the edible crab
<u>Cancer</u> <u>pagurus</u> and the lobster <u>Homarus</u> <u>vulgaris</u>, the Y-organs
retain function and thus moulting and growth continue until
death; the puberty moult in the spider crab <u>Maia</u> <u>squinado</u>, however,
is the final one, as the Y-organs then apparently atrophy[3]. Very
little is known of the mechanisms controlling the larval development of non-insect arthropods, although as early as 1958
Schneiderman and Gilbert[12] reported that extracts from a number of
crustaceans possessed JH activity when tested against insects and
suggested that this hormone may play a role in crustacean development. It is only recently that ecdysones have been shown to induce

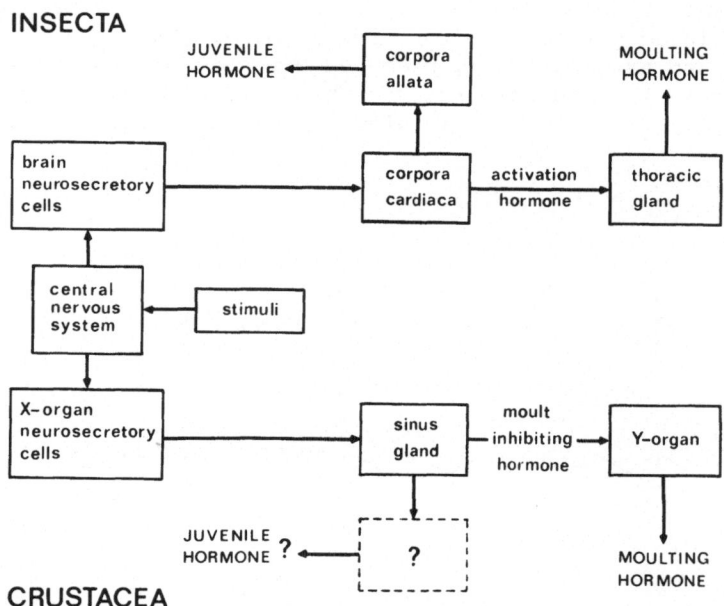

Figure 3. Hormonal systems controlling development and moulting
in insects and crustaceans.

ecdysis in larvae of crustacean and chelicerate arthropods such
as the barnacle <u>Balanus</u> <u>eburneus</u>[13] and the horseshoe crab
<u>Limulus polyphemus</u>[14], respectively.

The life cycle of a typical acorn barnacle comprises seven
free-swimming larval stages and an adult which is permanently
attached to a substratum. The first six stages are typical crustacean nauplius larvae, while the seventh is a cypris larva;
progression is by a succession of ecdyses followed by the cypris
moulting to the adult form after settlement. Metamorphosis may
be considered as comprising the changes between the stage VI
nauplius, the cypris larva and the adult, as described in detail
by Walley[15]. After exploration and selection of a substratum,the
cypris attaches itself by a proteinaceous cement[16,17] produced by
glands in the two antennules[16,18]. The hardening processes[16,19,20]
may be similar to those in the tanning of insect cuticle, in which
quinones cross-link with proteins to form structurally stable
macro-molecules. Metamorphosis to the young adult then occurs,
followed by the development of shell plates. Enlargement and
thickening of these plates continues throughout adult life,and the
cyclic activity of the shell-producing tissues follows that of the
moult[21]. In parallel, a cement-producing system of epidermal origin develops within the base of the young adult; new networks of
cement ducts are laid down during every intermoult cycle[22,23].
Moulting continues throughout adult life and entails the shedding
of the integument covering the body and mantle, within the shell.
The duration of the intermoult period varies widely, from around
one day to more than a month, depending upon factors such as
species, age, temperature and season. A mean duration of 8.3
days was reported[24] for adult <u>Balanus</u> <u>amphitrite</u> maintained in
the laboratory at 23°C; there was marked variation, not only between individuals but also between successive cycles in the same
animal.

Until recently very little was known of the systems controlling development and growth in barnacles. Evidence as to possible
hormonal function was limited to observations on the presence of
neurosecretory cells in the nervous system exhibiting cyclic
activity[25,26], the presence of nervous system factors which were
active upon crab chromatophores and whose production was associated
with the moult cycle[27,28], and the possible presence of endocrine
mechanisms for the control of breeding[29,30,31]. What appears to
be the first direct evidence was obtained by Carlisle (unpublished,
personal communication) who found that an extract of adult
<u>Elminius</u> <u>modestus</u> induced a resumption of moulting when injected
into Y-organ-ablated shore crabs, <u>Carcinus</u> <u>maenas</u>; the response
was apparently the same as that induced by ecdysone extracts from
other crustaceans and insects (Carlisle[32]), indicating the possible presence of such a hormonal system in barnacles. Later,
crustecdysone was shown to markedly increase moulting activity

when injected into adult <u>Balanus</u> <u>balanoides</u> and to induce an early
resumption of moulting during the anecdysis period which follows
the annual breeding[33,34]. A dose as low as 0.005 µg/animal resulted in a stimulation of moulting[35] representing a response at
0.05 µg/g (based upon a wet tissue weight of 0.1 g). This is
similar to the most sensitive of the responses reported in other
crustacean and chelicerate arthropods; ecdysis has been induced,
or intermoult period shortened, by doses of 0.125 µg/g in the crayfish <u>Procambarus</u> <u>simulans</u>[36], 0.0214 µg/g in the crayfish
<u>Orconectes</u> <u>obscurus</u>[37] and 0.06 µg/g in larvae of the horseshoe
crab <u>Limulus</u> <u>polyphemus</u>[38]. These responses compare very favour-
ably with those in insect species used for the assay of ecdysones; although a dose of 0.005 µg crystecdysone will induce
pupariation in 50% of larvae of the house-fly <u>Musca</u> <u>domestica</u> (the
most sensitive assay),the dose/weight response is 0.18 µg/g[39].
Activity towards barnacle larvae was confirmed by the ability of
crustecdysone to induce premature, albeit abnormal, metamorphosis
in cyprids of <u>B</u>. <u>eburneus</u> exposed to hormone dissolved in sea
water at concentrations down to 3×10^{-7} w/v[13]; although the adult
form was attained,the cyprid valves were not shed. It may be significant that no control cyprids metamorphosed, suggesting that
early stage larvae had been employed. Jegla, Costlow and
Alspaugh[38] observed that larvae of <u>L</u>. <u>polyphemus</u> injected with
ecdysones early in the moult cycle failed to complete ecdysis,
whereas over 80% of those injected late in the cycle did. Failure
to complete the moulting process has been reported in other crustacean and chelicerate arthropods, e.g. adult fiddler crabs <u>Uca</u>
<u>pugilator</u>[40] and adult <u>L</u>. <u>polyphemus</u>[41]. The natural occurrence of
ecdysones in barnacles has recently been confirmed[42]; the major
component in adult <u>B</u>. <u>balanoides</u> was identified as crustecdysone,
at a concentration of 1 µg/Kg. This level is comparable to that
reported in other species; 2 µg/Kg in the crayfish <u>Jasus</u>
<u>lalandei</u>[43], 6 µg/Kg in the lobster <u>Homarus</u> <u>americanus</u>[44] and 4-280
µg/Kg, depending upon moult stage, in the crab <u>Callinectes</u>
<u>sapidus</u>[45].

There is also evidence that barnacles possess a moult-
inhibiting hormone (MIH). Davis and Costlow[46] found that extracts
of central nervous systems removed from Stage C (interecdysis)
adult <u>Balanus</u> <u>improvisus</u> increased the time to ecdysis when injected into Stage D_O (early proecdysis) animals, whereas extracts
from Stage D_O animals injected at Stage C did not. The possibility that the overall hormonal control of moulting in barnacles
is functionally similar to that in other, particularly decapod,
crustaceans is supported by Davis and Costlow's report that eye-
stalk extracts from post- or inter-ecdysis crabs <u>Uca</u> <u>pugilator</u>
retard ecdysis in Stage D_O <u>B</u>. <u>improvisus</u>. The existence of a
further hormonal system, in larval stages, is indicated by studies
on the effects of insect JH. Development in barnacles is analogous to that of holometabolous insects, as in both there is a

gradual progression through the nauplius/larval stages, followed
by a marked transformation into a quite different form (the barnacle cypris and the insect pupa) before the adult is attained.
The cypris has been described[15] as a "locomotive pupa" because of
the tissue histolysis and morphogenetic activity during metamorphosis from the stage VI nauplius and to the adult form. In
insects the activity of JH is mimicked by a large number of
naturally occurring and synthetic compounds (e.g. 9, 47) which
exert "juvenilizing" effects upon immature stages. Their effects
were summarized by Slama[48] as follows: "An exogenous supply ....
to last instar larvae or pupae before a certain critical period
causes partial or complete inhibition of metamorphosis, manifested
by partial or complete retention of the old epidermal structures
on the next instar. The abnormal specimens thus formed are recognised as larval-pupal intermediates, larval-adult intermediates,
...... supernumary or extra larvae, secondary pupae, etc". Gomez
et al.[49] found that a JH analogue, ZR-512 (ethyl 3,7,11-trimethyl-
dodeca-2,4-dienoate), in ethanolic sea water solution stimulated
premature metamorphosis of Balanus galeatus cyprids at concentrations as low as 10^{-8} w/v. It was reported that the analogue
had no effects upon Stage IV nauplii, as might be expected from
analogy with holometabolous insects. Similar stimulation of metamorphosis was also induced by farnesol[50] and JH itself[51] . Although
Cheung[13] considered that such an effect was contrary to JH action
upon insects,such a view does not take into account the ability of
analogues to mimic the prothoracotrophic effects of natural hormone
by stimulating the activity of the insect prothoracid glands, e.g.
52-54. It was reported by Gomez et al.[49] and Cheung and Nigrelli[50]
that a proportion of the cyprids metamorphosing in analogue solution failed to attain the normal adult form. Both morphological
and size abnormalities were induced in larvae of Elminius modestus
exposed to farnesyl methyl ether and analogue Ro-8-4314 (ethyl,10,
11-epoxy-3,7,10,11 tetra methyl-2-cis-trans-6-cis-trans-dodecadie-
noate) at concentrations of 10^{-5} and 10^{-6} v/v in acetone solutions
in sea water[55]. Metamorphosis of Stage VI nauplii was retarded,
and those that did metamorphose formed morphologically abnormal
larvae, intermediate in size between the nauplius and cypris
stages; although these larvae were cypris-like,they apparently
retained some nauplius characteristics. Cyprids either metamorphosed to unattached adults, as was also reported by the earlier
authors, or formed larvae which were either larger or morphologically abnormal. There was evidence that the effects may have
been related to the physiological development of the larvae at time
of exposure. The results from these different studies thus suggest
that JH analogues exert effects upon barnacles which are analogous
to their juvenilizing and prothoracotrophic action upon insect
larvae and pupae. Although the interference with metamorphosis
may result from non-hormonal effects of agents which are biologically active, the nature of the responses is compatible with a
possible natural role for a JH-type of system. Evidence that JH

and analogue ZR-512 may interact with the same receptor site(s) in
barnacles was obtained by Ramenofsky, Faulkner and Ireland[51]: pre-
treatment of B. galeatus cyprids with certain sub-threshold concentrations of JH (possessing only 10^{-3} of the activity of ZR-512)
reduced the ability of subsequently added ZR-512 to stimulate premature metamorphosis.

     The effects of arthropod hormones and analogues upon barnacles
point to the conclusion that development and growth are regulated
by hormonal systems similar to those in other arthropods. Figure
4 illustrates the possible nature of such systems, based on the
information available to date. There is evidence to suggest that
the moulting hormone system may play a role in other processes.
Wounding of adult animals, by drilling through the shell, stimulates moulting, and the wound tissue formed within the mantle at
the site of damage is gradually enclosed by a ring of calcification extending inwards from the shell[56]. Interpretation of these
results in the light of the relationship between the activity of
shell-producing tissue and moulting[21] suggests that ecdysone(s)
may be associated with the control of calcification processes,
as it is in other crustaceans[57-59] and insects[60]. As cement gland
activity also follows the moulting cycle[21,61] and new cement
ducts are laid down during every inter-moult period[22,23], it may
be that cementation processes are under a similar control. The
effects of analogues of juvenile hormone upon barnacles indicate
the possible presence of a JH-type of hormonal system regulating
the form assumed at each larval moult. It remains to be seen
whether such a system could have a wider function and, as in insects, play a role in adult activities such as reproduction. The
apparent integration of JH and ecdysone systems in barnacles may
extend beyond the control of larval development; it has been
shown that both hormones are involved in the control of insect
ovarian development (e.g. [6]). Studies with crabs[62,63] suggest
that crustecdysone-like pheromones are involved in copulatory
activities of crustaceans. Isolation of the water-borne sex
pheromone yielded material with characteristics of crustecdysone,
and this hormone itself stimulated precopulatory activities in the
crab species Pachygrapsus crassipes, Cancer magister and C.
productus. It has been suggested that there is a copulatory
moult in the barnacle B. balanoides and that receptivity of a
functional female may be related to the stage of the moult
cycle[64-66]. Crustecdysone, however, had no effect upon the
penis searching activity of B. balanoides at known intervals
after ecdysis[35]. It has, however, been suggested[63] that there
may be a species-specificity for sex-pheromones and that some
species may respond to other ecdysones occurring naturally in
crustaceans.

     There is evidence to suggest that, as in other arthropods,
neurosecretory phenomena occur in barnacles and that these may

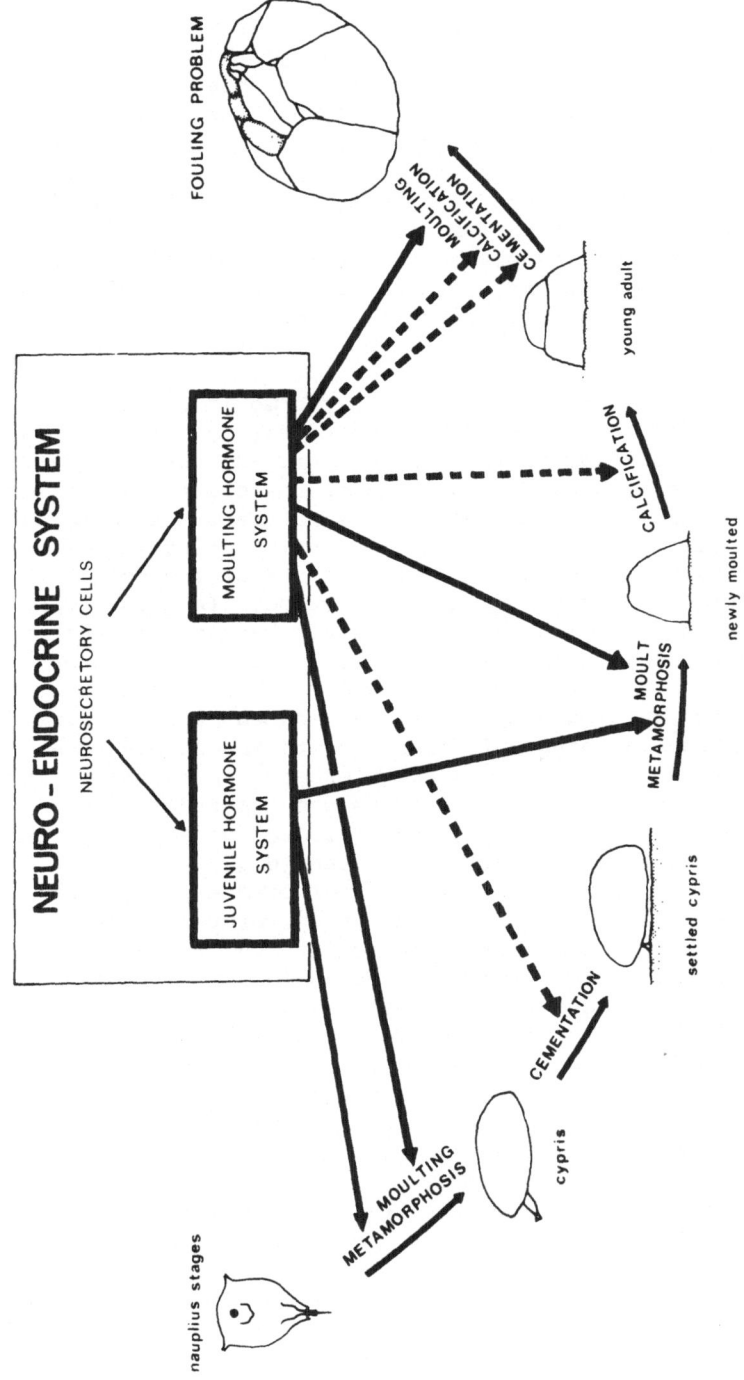

Figure 4. Postulated neuro-endocrine system for the control of development and growth in barnacles. Solid lines denote experimental evidence for a control of, or association with, an activity; dotted lines denote a hypothetical role.

mediate the relationship between activity and the external environment. Neurosecretory centres are present in the supra- and sub-
oesophageal ganglia[25] and the median photoreceptor[67] of adults:
the latter author reported the presence of a neurosecretion within
the median optic nerve to the supra-oesophageal ganglia. Light
acting upon the photoreceptors exerts an inhibitory effect upon
second-order neurones in these ganglia, and the "on/off" responses
are graded with the intensity of illumination[68,69]. There would,
therefore, appear to be a mechanism for neurosecretory control.
It would seem reasonable to suppose that stimulation of moulting
in adult B. balanoides by wounding and by continuous illumination
during winter/reproductive anecdysis[34,70] is mediated via neurosecretory activity.

The apparent presence of a MIH during the moult cycle of B.
improvisus[46] provides further evidence to suggest that the regulation of moulting is similar to that established in other crustaceans. If this is the case, MIH may be a neurosecretion inhibiting
the production of moulting hormone(s). The nature of reproductive
anecdysis in B. balanoides and the factors contributing to its
induction[71,72] raise the question of the nature of the possible
hormonal regulation operating during this period. In crustaceans
generally, MIH is most active during the winter anecdysis that
commonly occurs in many Brachyura and Astacura of northern seas;
this neurosecretion inhibits moulting hormone production and
apparently relates moulting activity to environmental conditions[73].
According to Jenkin[73], anecdysis is analogous to insect diapause,
which can occur without any associated quiescence in larvae[74] and
adults[75]. These then show characteristics such as reduced activity, feeding, and respiration; lack of reproductive activity; and
extreme cold-tolerance. Facultative diapause (e.g. [3]) depends for
its appearance upon appropriate environmental conditions; photoperiod and temperature are critical, and often related, factors.
Several characteristics of B. balanoides during the winter/
reproductive anecdysis period appear to be analogous to those of
diapause, eg. a marked reduction in feeding activity[76], decreased
oxygen consumption[77] and increased cold-tolerance[78]. Short
photoperiod and low temperatures induce the breeding condition in
this northern species, which results in the abrupt cessation of
moulting; absence of the appropriate environmental conditions
prevents fertilization and allows moulting and feeding to continue. The ability of light to induce an early resumption of
moulting during reproductive anecdysis indicates an effect upon a
hormonal system. Aiken[79] suggested that short day length favours
the production of MIH, whereas long day length allows moulting
hormone formation. The reason for lack of light effects at other
times of the year[70] is not known.

Except for the neurosecretory centres mentioned, virtually
nothing is known of the barnacle tissues which may be responsible

for the production and release of hormonal material. Kauri[80]
suggested that the frontal filament bases in the nauplii were
sensory-papilla-X-organs and may, therefore, represent neurohaemal
organs for the release of neurosecretory material. In the apparent
absence of adult storage organs, such material may be released
from the central nervous system into the perineural sinus[25]. However, no tissues have been identified as analogous to the endocrine
glands of other crustaceans and insects. Such a situation is not
peculiar to barnacles; the location, structure and origin of the
ecdysial glands of chelicerates, such as Limulus polyphemus, is
not known[41] and there has been some confusion as to the identity
of the Y-organs of macruran crustaceans[81,82].

The possible natural function and metabolism of the compounds involved in barnacle hormonal systems can be indicated by
analogy with other arthropods in which many common mechanisms
have been established. At least seven ecdysones (Figure 1) have
been isolated from the phylum[10,83], two or more of which can occur
in any one species: ecdysone, crustecdysone, 2-deoxycrustecdysone,
20,26-dihydroxyecdysone, 26-hydroxyecdysone, inokosterone and
makisterone A (= callinecdysone A & B, respectively[45]).
Bebbington[42] reported that, as well as the major occurrence of
crustecdysone, much smaller amounts of ecdysone were present in
adults of the barnacle B. balanoides; gas liquid chromatography
also indicated the possible presence of other ecdysones. It would
appear that ecdysone and crustecdysone have different functions in
both insects (e.g. 84-86) and crustaceans (e.g. 40); for example,
ecdysone may initiate the moulting cycle and crustecdysone may
regulate subsequent processes. Differing titres of three ecdysones
are present during different stages in the moult cycle of the crab
Callinectes sapidus[45]. It has been suggested[42] that as ecdysone
occurs at only a level of 6 µg/Kg (crustecdysone = 1 µg/Kg) in the
adult barnacle B. balanoides, comparable to that of 2-deoxycrus-
tecdysone (70 µg/Kg) in the crayfish Jasus lalandei, both perform
a similar function and may represent prohormones for crustecdysone. It is likely that barnacles share the dietary requirements
of other crustaceans[44,87-89] and insects (e.g. 90) in which
steroids, such as stigmasterol, β-sitosterol and ergosterol are
converted via desmosterol and cholesterol to ecdysones. Examination of the 'sterol' composition of eleven decapod crustaceans[91]
showed that cholesterol was the main sterol; desmosterol was a
minor constituent in marine species, whereas 24-methyl- and 24-
ethyl cholesterol were present in fresh water species. Both
cholesterol (59.8% of total sterols) and desmosterol (34.2%) are
present in barnacles[92]; although there is no evidence as to the
dietary requirements of these animals,it may be supposed that any
essential structures are met from their diet and/or by direct
uptake from the sea. Cholesterol is the principal sterol of zoo-
and phytoplankton, and sea water contains, in addition, β-sitosterol, campesterol, stigmasterol and fucosterol (see 93). Two

pathways have been postulated for the biosynthesis of crustecdysone in arthropods, involving conversion of 2-deoxyecdysone via
either 2-deoxycrustecdysone or ecdysone[94]. The first pathway has
been demonstrated in Jasus lalandei, and Thomson and co-workers
suggested that this may be the major pathway in other arthropods.
The second pathway may be present in the crab Uca pugilator and
the shrimp Crangon nigricauda, as they will convert ecdysone to
crustecdysone[95] ; the isolation of ecdysone in B. balanoides, the
first report for crustaceans, indicates that this may also be a
pathway in barnacles. Kaplanis et al.[83] suggested that in insects
different ecdysones could function at different stages of
development and that the qualitative nature of the hormones may
determine the type of moult. These authors also proposed that
there may be qualitative and quantitative differences in bio-
synthetic/metabolic pathways in different developmental stages.

Relatively little is known of the metabolism of juvenile
hormones in insects, although these animals have the ability to
synthesise a wide range of terpenoids (e.g. [96]). Three hormones
have so far been reported in insects (Figure 2) - C_{17} and C_{18} JH
from the silkmoth Hyalophora cecropia[97,98] and C_{16} JH from the
tobacco hornworm Mandura sexta[99]. Metzler et al.[100] suggested
that although acetate may contribute to the carbon skeleton of
the two H. cecropia hormones, biosynthesis probably does not occur
via mevalonate and may not follow the isoprenoid pathway. Schooley
et al.[101], however, reported that corpora allata of M. sexta, in
vitro, incorporated into mevalonate and acetate into both C_{16} and
C_{17} JH (propionate, also, was incorporated into the latter).
Their results suggest that C_{16} JH is derived from three molecules
of mevalonate (nine molecules of acetate); synthesis would thus
follow the accepted terpenoid pathway. With C_{17} JH, however, the
isoprenoid skeleton probably arises from mevalonate and the
homoisoprenoid units from two acetates and one propionate. The
ester methyl carbon is apparently derived from methionine[100]. Two
major enzymic pathways for the degradation of insect JH have been
established[102]; ester hydrolysis may either precede or succeed
cleavage of the epoxide ring, resulting in an epoxy acid and a
dihydroxyester, respectively, followed by the formation of a di-
hydroxy acid.

The apparent hormonal function in barnacles would seem to
offer the means of exploiting the activities which lead to the
establishment of these animals as a fouling problem. Such an
approach is analogous to that involved in the development of
third-generation insecticides (e.g. [103-105]). It is appropriate
at this stage to consider this approach in relation to fouling by
other animals and plants; the practical limitation for anti-
fouling purposes is that agents may be selective in their action.
There are, however, indications that a common line of approach may
be possible, through a sharing of some biologically active compounds

by plants and animals. More than 100 ecdysones or ecdysone-like compounds have been isolated from a wide range of plants (see 6, 10, 106); crustecdysone constitutes over 1% of the dry weight of rhizomes of the fern <u>Polypodium</u> <u>vulgare</u>[107]. Plant triterpenoids, such as meliantriol[108] and the related azadirachtin[109], are active towards insects, and there is evidence that the latter compound has some inhibitory effect upon moulting in adult barnacles[35]. Similarly, sesquiterpenoids (such as abscissic acid) and diterpenoids (such as the gibberellins) have profound effects upon plant development and growth (e.g. 110); gibberellin and indolyl-3-acetic acid activity has been reported from locusts[111,112]. Evaluation of such shared phenomena, coupled with a synthesis of compounds, may yield agents active towards a range of fouling species.

The approach to the exploitation of the role of hormones in barnacle development will depend upon the system selected for attack. Although the results obtained to date favour the use of JH analogues to interfere directly with the settlement and/or metamorphosis of the cypris stage, ecdysone is apparently associated with a wider range of developmental activities and would appear to be active throughout the processes leading to the established adult. There are, of course, good reasons for considering the prevention of settlement and/or metamorphosis as a primary target for exploitation. Nevertheless, inhibition of subsequent processes such as moulting, calcification and cementation may provide an equally valid approach and allow a longer time-scale for exploitation. Broadly, it would seem that the potential application of JH analogues lies in the activity of such compounds <u>per se</u>, whereas exploitation of the ecdysone system appears to depend upon an inhibition of the synthesis, release or action of hormone(s). Possible approaches are indicated by results from insect studies. Hypocholesterolemic agents such as triparanol and 22, 25-diazocholesterol, retard growth and induce abnormal larval development by the inhibition of steroid synthetic pathways. Svoboda and Robbins[113] suggested that these responses arose from reduced availability of ecdysones and/or the formation of minor steroid metabolites which act as growth inhibitors. Synthetic ecdysone analogues, some with only minimal structural features of natural ecdysones, apparently exert disruptive effects via hormonal activity[114], as was proposed for azadirachtin[115]. A range of steroids active towards arthropods has been reviewed by Herout[110]. It may be significant with respect to practical aspects of barnacle control that non-volatile solvents such as undecylenic acid, α-tocopherol and caprylic acid allow or facilitate penetration of the hydrophilic ecdysones through the insect cuticle[116]. Compounds with JH activity need not necessarily possess structures resembling that of the natural hormones; Bowers[47] reported that certain insecticide synergists with a methylenedioxyphenyl moiety, such as sesoxane and piperonyl

butoxide, possess JH activity in their own right, possibly by
inhibiting the enzymes which normally metabolize JH[102,118];
extremely effective compounds were formed by combining structural
features of such synergists with those of the natural hormones[119].
Reviews of a range of JH analogues and their activity have been
given by Bowers[9] and Slama[48].

It is apparent that knowledge of hormonal function in barnacles
lags behind that already established for many other arthropods.
Although interpretation by analogy must be carried out with caution,
it would seem that the general, common pattern which is emerging
from the phylum provides a basis for future investigations. What
is now required is a blending of studies of hormonal function with
investigations of the processes involved in individual activities,
such as cementation, calcification, etc., at tissue and cell level.

REFERENCES

1. D. R. Houghton, Unverwater Sci. Tech. J. June, 100 (1970).

2. L. M. Passano, in The Physiology of Crustacea, Vol. 1, ed.
    T. H. Waterman, Academic Press, London, 1960, p. 473.

3. K. C. Highnam and L. Hill, The Comparative Endocrinology of
    Invertebrates, Arnold, London, 1969.

4. V. B. Wigglesworth, Insect Hormones, Oliver & Boud, Edinburgh,
    1970.

5. M. Fingerman, Life Sci. 14, 1007 (1974).

6. V. J. A. Novak, Insect Hormones, Chapman & Hall, London, 1975.

7. D. H. S. Horn, E. J. Middleton, J. A. Wunderlich, and F.
    Hampshire, Chem. Comm., 339 (1966).

8. M. N. Galbraith, D. H. S. Horn, P. Hocks, G. Schulz, and
    M. Hoffmeister, Naturwissenschaften 59, 471 (1967).

9. W. S. Bowers, in Naturally Occurring Insecticides, eds.
    M. Jacobsen and D. G. Crosby, Dekker, New York, 1971, p. 307.

10. D. H. S. Horn, in Naturally Occurring Insecticides, eds.
    M. Jacobsen and D. G. Crosby, Dekker, New York, 1971, p. 333.

11. K. Rangarao, Experientia 21, 593 (1965).

12. H. A. Schneiderman and L. I. Gilbert, Biol. Bull, Mar. Biol.
    Lab., Woods Hole, 115, 530 (1958).

13. P. J. Cheung, J. Exp. Mar. Biol. Ecol. 15, 223 (1974).

14. T. C. Jegla and J. D. Costlow, Gen. Comp. Endocr. 14, 295
     (1970).

15. L. J. Walley, Phil. Trans. R. Soc. B. 256, 237 (1969).

16. G. Walker, Mar. Biol. 9, 205 (1971).

17. P. J. Cheung and R. F. Nigrelli, Zoologica 57, 79 (1972).

18. J. A. Nott and B. A. Foster, Phil. Trans. R. Soc. Ser. B.
     256, 115 (1969).

19. J. R. Saroyan, E. Lindner, C. A. Dooley, and H. R. Bleile,
     Ind. Eng. Chem. Prod. Res. Devel. 9, 122 (1970).

20. E. Lindner and C. A. Dooley, Proc. Third Int. Congr. Mar. Corr.
     Fouling, October 1972, National Bureau of Standards,
     Gaithersburg, Maryland, U.S.A., 1972.

21. J. Bocquet-Vedrine, Darwin. Archs. Zool. Exp. Gen. 105, 30
     (1965).

22. J. Bocquet-Vedrine, Archs. Zool. Exp. Gen. 111, 521 (1970).

23. J. R. Saroyan, E. Lindner, and C. A. Dooley, Biol. Bull. Mar.
     Biol. Lab., Woods Hole, 139, 333 (1970).

24. C. W. Davis, U. E. H. Fyhn, and M. J. Fyhn, Biol. Bull. Mar.
     Biol. Lab., Woods Hole, 145, 310 (1973).

25. H. Barnes and J. J. Gonor, Sowerby J. Mar. Res. 17, 81 (1958).

26. D. B. McGregor, J. Exp. Mar. Biol. Ecol. 1, 154 (1967).

27. M. I. Sandeen and J. D. Costlow, Biol. Bull. Mar. Biol. Lab.,
     Woods Hole, 120, 192 (1961).

28. J. D. Costlow, Biol. Bull. Mar. Biol. Lab., Woods Hole, 124,
     254 (1963).

29. H. Barnes, J. Mar. Biol. Ass. U. K. 43, 717 (1963).

30. H. Barnes and M. Barnes, J. Exp. Mar. Biol. Ecol. 1, 1 (1967).

31. D. J. Tighe-Ford, Nature 216, 920 (1967).

32. D. B. Carlisle, Gen. Comp. Endocr. 5, 366 (1965).

33.   D. J. Tighe-Ford and D. C. Vaile, J. Exp. Mar. Biol. Ecol.
      9, 19 (1972).

34.   D. J. Tighe-Ford and D. C. Vaile, in Proc. Third Int. Congr.
      Mar. Corr. Fouling, National Bureau of Standards, Gaithersburg,
      Maryland, U.S.A., 1972, p. 744.

35.   D. J. Tighe-Ford, Ph.D. Thesis, C.N..A.A., London, 1974.

36.   M. E. Lower, D. H. S. Horn, and M. N. Galbraith, Experientia
      24, 518 (1968).

37.   A. C. Warner and J. R. Stevenson, Gen. Comp. Endocr. 18, 454
      (1972).

38.   T. C. Jegla, J. D. Costlow, and J. Alspaugh, Gen. Comp. Endocr.
      19, 159 (1972).

39.   J. N. Kaplanis, L. A. Tabor, M. J. Thompson, W. E. Robbins,
      and T. J. Shortino, Steroids 8, 625 (1966).

40.   K. R. Rao, M. Fingerman, and C. Hays, Z. vergl. Physiol. 76,
      270 (1972).

41.   W. S. Herman, Gen. Comp. Endocr. 18, 301 (1972).

42.   P. M. Bebbington, Ph.D. Thesis, University of Keele, 1975.

43.   D. H. S. Horn, S. Fabri, F. Hampshire, and M. E. Lowe,
      Biochem. J. 109 ,399 (1968).

44.   R. B. Gagosian, R. A. Bournbonierre, W. B. Smith, E. F. Couch,
      C. Blanton, and W. Novak, Experientia 30, 723 (1974).

45.   A. Faux, D. H. S. Horn, E. J. Middleton, H. M. Fales, and
      M. E. Lowe, Chem. Comm. 175 (1969).

46.   C. W. Davis and J. D. Costlow, J. Comp. Physiol. 93, 85 (1974).

47.   F. Sehnal, in Chemical Zoology, Vol. 6, Arthropoda, Part B,
      eds. M. Florkin and B. T. Scheer, Academic Press, New York,
      1971, p. 307.

48.   K. Slama, A. Rev. Biochem. 40, 1079 (1971).

49.   E. D. Gomez, D. J. Faulkner, W. A. Newman, and C. Ireland,
      Science 179, 813 (1973).

50.   P. J. Cheung and R. F. Nigrelli, Am. Zool. 13, 1339 (1973).

51. M. Ramenofsky, D. J. Faulkner, and C. Ireland, Biochem. Biophys. Res. Comm. <u>60</u>, 172 (1974).

52. A. Krishnakumaran and H. A. Schneiderman, J. Insect Physiol. <u>11</u>, 1517 (1965).

53. L. I. Gilbert, S. Applebaum, T. A. Gorell, J. B. Siddall, and Y. C. Siew, Bull. World Health Org. <u>44</u>, 397 (1971).

54. L. M. Riddiford, Biol. Bull. Mar. Biol. Lab., Woods Hole, <u>142</u>, 310 (1972).

55. D. J. Tighe-Ford, in press.

56. D. J. Tighe-Ford and D. C. Vaile, J. Exp. Mar. Biol. Ecol. <u>14</u>, 295 (1974).

57. M. E. Lowe and D. H. S. Horn, in <u>Physiological Systems in Semi-Arid Environments</u>, eds. C. C. Hoff and M. L. Riedsel, University of New Mexico Press, Albuquerque, 1969, p. 155.

58. M. A. McWhinnie, M. O. Cahoon, and R. Johanneck, Am. Zool. <u>9</u>, 841 (1969).

59. F. Graf, C. r. hebd. Seanc. Acad. Sci., Paris <u>274</u>, 1731 (1972).

60. G. Fraenkel and C. Hsiao, J. Insect Physiol. <u>13</u>, 1387 (1967).

61. U. E. Fyhn and J. D. Costlow, Biol. Bull. Mar. Biol. Lab., Woods Hole, <u>150</u>, 47 (1976).

62. E. P. Ryan, Science <u>151</u>, 340 (1966).

63. J. S. Kittredge, M. Terry, and F. T. Takahashi, Fish. Bull. <u>69</u>, 337 (1971).

64. H. Barnes and M. Barnes, Arch. Soc. Zool. Bot. Fenn. "Vanamo," <u>11</u>, 11 (1956).

65. D. J. Crisp and B. S. Patel, Nature <u>181</u>, 1078 (1958).

66. B. Patel and D. J. Crisp, Crustaceana <u>2</u>, 89 (1961).

67. W. H. Fahrenbach, Z. Zellforsch. mikrosk. Anat. <u>46</u>, 233 (1965).

68. G. F. Gwilliam, Biol. Bull. Mar. Biol. Lab., Woods Hole, <u>125</u>, 470 (1963).

69. G. F. Gwilliam, Biol. Mar. Biol. Lab., Woods Hole, <u>129</u>, 244 (1965).

70. H. Barnes and R. L. Stone, J. Exp. Mar. Biol. Ecol., 15, 275
     (1974).

71. D. J. Crisp and B. S. Patel, Biol. Bull Mar. Biol. Lab., Woods
     Hole, 118, 31 (1960).

72. D. J. Crisp and B. Patel, Mar. Biol. 2, 283 (1969).

73. P. M. Jenkin, Control of Growth and Metamorphosis, Pergamon,
     Oxford, 1970.

74. M. E. Clay and C. E. Venard, Ann. Ent. Soc. Am. 64, 968 (1971).

75. W. S. Bowers and C. C. Blickenstaff, Science 154, 1673 (1966).

76. D. A. Ritz and D. J. Crisp, J. Mar. Biol. Ass. U. K. 50, 223
     (1970).

77. H. Barnes, M. Barnes, and D. M. Finlayson, J. Mar. Biol. Ass.
     U. K. 43, 185 (1963).

78. D. J. Crisp and D. A. Ritz, Helgolander wiss. Meeresunters.,
     15, 98 (1967).

79. D. E. Aiken, Science 164, 149 (1969).

80. T. Kauri, Crustaceana 11, 115 (1966).

81. J. B. Sochasky, D. E. Aiken, and N. H. F. Watson, Can. J.
     Zool. 50, 993 (1972).

82. D. B. Carlisle and R. O. Connick, Can. J. Zool. 51, 417
     (1973).

83. J. N. Kaplanis, W. E. Robbins, M. J. Thomson, and S. R. Dutky,
     Science 180, 307 (1973).

84. H. Oberlander, J. Insect Physiol. 18, 223 (1972).

85. U. Clever, I. Clever, I. Storbeck, and N L. Young, Devl.
     Biol. 31, 47 (1973).

86. H. Oberlander, C. E. Leach, and C. Tomblin, J. Insect Physiol.
     19, 993 (1973).

87. A. Kanazawa, N. Tanaka, S. Teshima, and K. Kashiwada, Bull.
     Jap. Soc. Scient. Fish. 37, 1015 (1971).

88. A. Kanazawa and S. Teshima, Bull. Jap. Soc. Scient. Fish. 37,
     891 (1971).

89. S. Teshima and A. Kanazawa, Mem. Fac. Fish., Kagoshima University, 22, 15 (1973).

90. W. E. Robbins, J. N. Kaplanis, J. A. Svoboda, and M. J. Thompson, A. Rev. Ent. 16, 53 (1971).

91. S. Yasuda, Comp. Biochem. Physiol. 44B, 41 (1973).

92. U. H. M. Fagerlund and D. R. Idler, J. Am. Chem. Soc. 79, 643 (1957).

93. A. Saliot and M. Barbier, J. Exp. Mar. Biol. Ecol. 13, 207 (1973).

94. J. A. Thomson, D. H. S. Horn, M. N. Galbraith, and E. J. Middleton, in Invertebrate Endocrinology and Hormone Heterophylly, ed. W. J. Burdette, Springer-Verlag, Berlin, 1974, p. 172.

95. D. S. King and J. B. Siddall, Nature 221, 955 (1969).

96. R. B. Clayton, J. Lipid Res. 5, 3 (1964).

97. H. K. von Roller, H. Dahm. C. C. Sweeley, and D. M. Trost, Angew. Chem. 79, 190 (1967).

98. A. S. Meyer, H. A. Schneiderman, E. Hanzmann, C.J.H.K.O., Proc. Natn. Acad. Sci., U.S.A., 60, 853 (1968).

99. K. J. Judy, D. A. Schooley, L. L. Dunham, M. S. Hall, B. J. Bergot, and J. B. Siddall, Proc. Natn. Acad. Sci., U.S.A., 70, 1509 (1973).

100. M. Metzler, K. H. Dahm, D. Meyer, and R. R. Roller, Z. Naturf. 26b, 1270 (1971).

101. D. A. Schooley, K. A. Judy, B. J. Bergot, M. S. Hall, and J. B. Siddall, Proc. Natn. Acad. Sci., U.S.A., 70, 2921 (1973).

102. M. Slade and C. F. Wilkinson, Science 181, 672 (1973).

103. C. M. Williams, Sci. Am. 217, 13 (1967).

104. P. E. Ellis, E. D. Morgan, and A. P. Woodbridge, Pest Art. News Serv. 16, 434 (1970).

105. R. C. Reay, Int. J. Environ. Studies 5, 93 (1973).

106. H. Hikino and T. Takemoto, Naturwissenschaften 59, 91 (1972).

107. J. Jizba, V. Herout, and R. F. Sorm, Tetrahedron Lett., 1689
      (1967).

108. D. Lavie, M. K. Jain, and S. R. Shpan-Gabrielith, Chem. Comm.,
      910 (1967).

109. J. M. Butterworth and E. D. Morgan, J. Insect Physiol. 17,
      969 (1971).

110. V. Herout, Prog. Phytochem. 2, 143 (1971).

111. G. V. Hoad and P. E. Ellis, Nature 237, 108 (1972).

112. S. D. Hendrix and R. L. Jones, Pl. Physiol. 50, 199 (1972).

113. J. A. Svoboda and W. E. Robbins, Science 156, 1637 (1967).

114. W. E. Robbins, J. N. Kaplanis, M. J. Thompson, T. J. Shortino,
      C. F. Cohen, and S. C. Joyner, Science 161, 1158 (1968).

115. C. N. E. Ruscoe, Nature 236, 159 (1972).

116. C. M. Williams, in Conference on Insect-Plant Interactions,
      Santa Barbara, California, Bioscience 18, 797 (1968).

117. W. S. Bowers, Science 161, 895 (1968).

118. G. T. Brooks, Nature 245, 382 (1973).

119. W. S. Bowers, Science 164, 323 (1969).

CHEMICAL INTERACTIONS IN LARVAL SETTLING OF A MARINE GASTROPOD

Michael G. Hadfield

Kewalo Marine Laboratory, Pacific Biomedical Research

Center, University of Hawaii, Honolulu, Hawaii 96813

INTRODUCTION

Prior to the 1930s, it was a generally accepted notion that larvae of benthic marine invertebrates were, in the timing of their metamosphosis, at the mercy of chance. If the ocean's currents carried them over substrata suitable for adult life when the time for metamorphosis arrived, they survived; if the substrate were not suitable, the larvae perished. Beginning with observations of Mortensen[1] and Day and Wilson[2], students of marine ecology began to see that the situation relative to larval settling was more controlled, that larvae could execute a "choice" of substratum. It was additionally recorded that larvae could actually delay metamorphosis until suitable substrata were found[3].

After noting that larvae could, in the setting of a laboratory culture, choose between substrata, Wilson and others began to question the nature of the "cues" perceived by the larvae in selecting particular substrata. Position, texture, colour and chemical factors have all been investigated in a variety of species-substratum interactions. In the history of such studies, it has been noted that larvae can respond selectively to both non-living (i.e. particular sands or muds) and living substrata (generally the prey or fodder of the settling species). A most valuable summary of such interactions has been presented by Crisp[4].

Generally, three observational/experimental steps lead to the conclusion that a chemical interaction occurs between metamorphically-competent larvae and substrata. These are: (1) field observations indicate that species A is always found on or near species B or a particular substratum; (2) in laboratory tests,

larvae are shown to choose to settle on species or substratum B
when a number of different substrata are available; and (3) a
chemical stimulus is separated from its usual source and isolated
or transferred to an otherwise inactive substratum and shown to
stimulate larval metamorphosis (adapted from Crisp[4]). Crisp referred to these situations as "associative settling" of the larvae,
and while he meant by this term two-species interactions such as
predator settling in response to prey, the term can serve equally
well for describing specific settlng on non-living substrata[5,6].

The specific problem which we have addressed is the induction
of larval metamosphosis in the nudibranch mollusc Phestilla
sibogae by its prey, the stoney coral Porites compressa[7,8,9,10].
Having completed investigative stages 1-3, outlined above, we
learned to prepare lyophilized distilled-water extracts of the
coral which retain the metamorphosis-inducing capacity of living
coral (seawater in which coral has stood for 18-24 hours is also
effective). This "crude inducer" forms the basis for our studies
on identification of the inducer (not to be reported here) and on
the nature of the larval response. Specifically, we report here
on investigations into: (1) the relationship between inducer concentration and larval response (i.e. % of larvae responding);
(2) response as a function of larval age; (3) exposure time vs.
larval response (as % metamorphosing); (4) the fate of the inducer
during induction; and (5) specificity of response for the particular coral named above.

                        MATERIALS AND METHODS

Methods for maintaining Phestilla sibogae in laboratory culture are given in Bonar and Hadfield[7]. Experiments on metamorphic
induction are carried out in replicated runs wherein: 40 ml
lidded Stender dishes are filled with 20-25 ml of Millipore-
filtered seawater (MPF-SW); the seawater is a solvent for various
test substances including the "crude inducer" (see below); 20
larvae are added to each dish. Controls consist of larvae placed
in MPF-SW or other solutions as appropriate.

Unless otherwise stated, larvae are 12 days old at the start
of each experiment (time course: egg laying = day 0; larvae are
hatched on day 7; larvae are maintained in dark bowls of MPF-SW
until day 12). We have empirically determined that 12 days is the
youngest larval age at which a given batch of larvae is maximally
competent to metamorphose. Unless otherwise stated, counts to
determine percent metamorphosis are made 24 hours after larvae are
placed in a test situation. Maximum response is never 100%, and
while variation is considerable from experiment to experiment,
maximum is usually considered to have occurred when metamorphosis
is over 75% for 12-day or older larvae.

"Crude inducer" is prepared by collecting fresh <u>Porites</u>
<u>compressa</u> from local coral reefs, rinsing the freshly collected
coral briefly in distilled water and then placing it in clean distilled water for one hour. The latter is usually accomplished by
filling a 1 litre beaker with small coral "fingers" and filling
the jar with distilled water (usual ratio = 600 ml/800 g). After
one hour the coral is removed and the distilled water extract is
lyophilized. The tan-coloured powders obtained are accumulated
and homogenized so that large, uniform batch of "crude inducer"
is available for a large number of subsequent experiments.

Partial purification of the "crude inducer" shows it to be a
basic, water-soluble compound with a molecular weight under 1000.

RESULTS

1.   The basic experiment for quantifying the response of larvae
to inducer is one which tests the response as a function of inducer
concentration. The percentages of larvae which metamorphose in
response to varying concentrations of crude inducer are shown in
Figure 1. Data are given for both the first and second day after
continuous exposure to the inducer. The fact that the curves for
these two summaries are parallel indicates that qualitative differences in the response are probably not occurring.

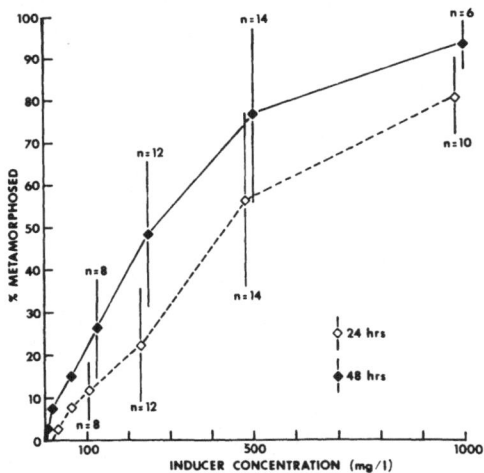

Figure 1. Response of larvae to varying inducer concentrations;
shown as percent metamorphosed (ordinate) vs. inducer concentration (abscissa). Results are shown after 24 and 48 hours
exposure. Each test consisted of 20 larvae; thus, n=12, for
instance, is 12 tests of 20 larvae each, or a total of 240
larvae tested at 0.25 g inducer/litre. Bars show standard
deviations.

2.   Utilizing knowledge of maximum effective inducer concentration
derived from Figure 1, experiments were conducted to quantify metamorphic response as a function of larval age. Results, shown in
Figure 2, indicate that responsiveness -- or sensitivity -- to
inducer increases with larval age and reaches a maximum around the
12th day. No larvae are competent to respond on days 6 and 7;
such larvae have normally not yet hatched. To be certain that
larvae younger than 12 days old were not simply sensitive to different, higher, inducer concentrations, 9-day-old larvae were
tested at 0.5, 1.0, 2.5, 5.0 and 10.0 g/l crude inducer concentration and the respective, mean responses were:

Concentration	Mean Response
0.5 g/l	10.0%
1.0 g/l	22.5%
2.5 g/l	15.0%
5.0 g/l	22.5%
10.0 g/l	dead
control (0 g/l)	0.0%

These results indicate that increasing inducer concentrations
above 1 g/l (shown to be maximally effective in Figure 1) do not
increase the metamorphic response of 9-day-old larvae.

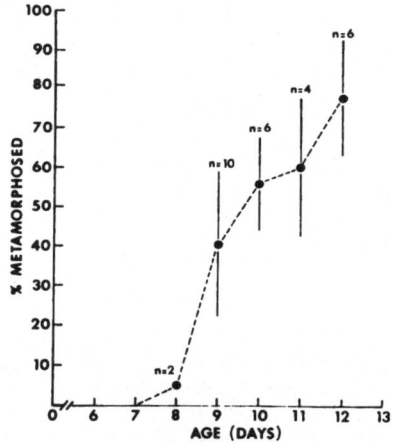

Figure 2. Age-dependent variation in larval metamorphosis. Larvae
aged 6 to 12 days from egg deposition (0-6 days after hatching)
were exposed to inducer (1.0 g/l) and the percent metamorphosing
was determined after 24 hours. 6- and 7-day larvae were artificially hatched. n is determined as in Figure 1. Bars represent
standard deviations.

3.   Knowing (Figure 1) that inducer concentration determines percentage of larvae metamorphosing, we tested to see if varying
duration of exposure to inducer (maximum concentration from
Figure 1) produced similar variation in response. The results,
illustrated in Figure 3, indicate that metamorphic response in a
given batch of larvae is a function of duration of exposure to
inducer. That this variation is not simply another expression of
that seen in Figure 1 (i.e. response is a function of inducer concentration), is demonstrated in Figure 4 where responses to various
inducer concentrations are plotted over succeeding days of exposure. Note that batches of larvae continuously exposed to 125 mg
inducer/l (=12.5% of maximum effective dose) never achieve response levels of batches exposed to greater concentrations. That
is, increased length of exposure does not compensate for reduced
inducer concentration.

     An alternate hypothesis to explain the results shown in
Figure 4 would be that the inducer molecule breaks down during the
course of the experiment. That this is clearly not the case is
shown in the next section.

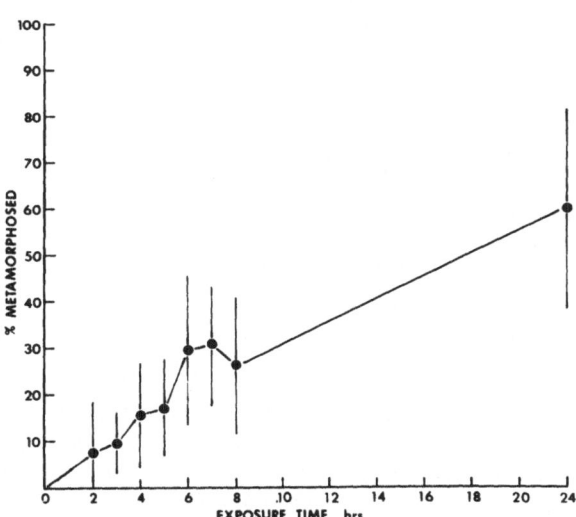

Figure 3. Response of 12-day-old larvae to varying duration of
exposure to inducer; shown as percent metamorphosing (ordinate)
vs. duration of exposure (abscissa). Larvae were passed through
3 changes of filtered seawater after exposure to inducer and the
results were determined 24 hours after initial exposure to the
inducer. n derived as in Figure 1. Bars indicate standard
deviation.

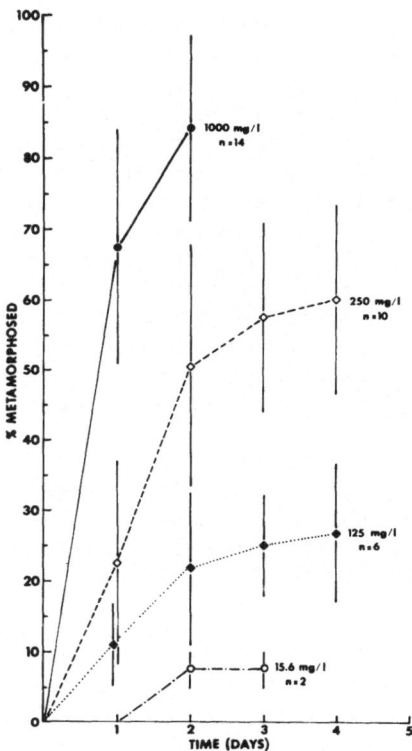

Figure 4. Response of larvae to various inducer concentrations
over time. 12-day-old larvae were introduced to different
inducer concentrations,and the total number having completed
metamorphosis was counted on succeeding days until it ceased to
increase. All experiments were started at "time 0"; the effect
shown at 15.6 mg/l is due to the total absence of metamorphoses
during the first 24 hours. n is calculated as in Figure 1.
Bars indicate standard deviations.

4.   The mode of action of the inducer was studied by examining
the fate of the inducer molecule during metamorphic induction.
The basic question here is: is the inducer "used up" or irreversibly bound to larvae during the induction process? This was
studied by exposing batches of larvae to 50% and 75% of the maximuu
effective concentration of inducer, removing all larvae and metamorphosed juveniles after 24 hours and noting percent metamorphose
Then new larvae were added to the same solutions, the percent
metamorphosis was noted after 24 hours and the animals removed.
The whole process was repeated two more times so that each single
inducer solution was used in the induction of four separate batche
of larvae.

Table 1. Responses of successive batches of larvae to the same
inducer solutions at two concentrations.

Inducer: 0.75 g/l		Inducer: 0.50 g/l	
Test	Mean % Metamorphosing	Test	Mean % Metamorphosing
1	67.5	1	62.5
2	72.5	2	60.0
3	62.5	3	65.0
4	72.5	4	57.5

The results, summarized in Table 1, fall very close to the
percent of metamorphosis predicted for each concentration by
Figure 1. The fact that approximately the same, though submaximal,
numbers of larvae metamorphose on each succeeding day clearly
indicates that the inducer is not used up, irreversibly bound, or
destroyed during the induction process and thus further substan-
tiates that the time-dependent variation in induction response is
qualitatively different from the concentration-dependent variation.

5.   Specificity of response of larvae of Phestilla sibogae to
various coral species was checked in two ways. First, three species
of coral were allowed to stand in dishes of seawater for 24 hours,
the coral was removed, the water filtered and larvae were added.
Secondly, all coral species were extracted in distilled water, the
solutions lyophilized and seawater test solutions made up of equal
concentrations of "crude inducer" powder equal to 1.0 g/l. Results
of both tests are presented in Figure 5. The usual inducer is the
coral Porites (P. compressa); the others, both common Hawaiian
reef corals, are Pocillopora damicornis and Montipora verrucosa.
Phestilla is not known to eat either of the latter two species.
It is important to note that some larvae will respond to presumably inappropriate corals.

DISCUSSION

The data presented in Figures 1-5 indicate that the interactions between larvae and the soluble coral product which brings
about metamorphosis are complex. While each larva is presented
with only a single "choice" in the presence of the metamorphic
inducer --to metamorphose or not-- the likelihood of any given
larva, in a batch of larvae, undergoing metamorphosis varies with
larval age, concentration of inducer, duration of exposure to
inducer, and the coral species extracted.

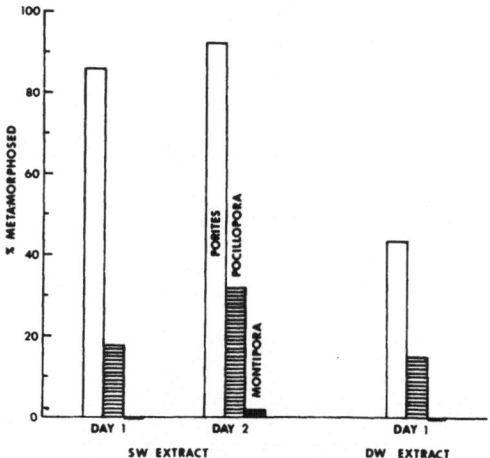

Figure 5. Specificity of the metamorphic response of <u>Phestilla</u>
<u>sibogae</u> larvae. Histograms to the left show percent metamorphosis
of larvae exposed to seawater extracts of the corals <u>Porites</u>
<u>compressa</u>, <u>Pocillopora</u> <u>damicornis</u>, and <u>Montipora</u> <u>verrucosa</u> after
24 and 48 hours exposure. Histogram on the right shows response
of larvae to equal concentrations (1 g/l) of lyophilized distilled
water extracts of the same three corals.

At a given, high inducer concentration (1 g/l) the probability
of metamorphosis increases from about 0.4 on day 9 to about 0.75
on day 12. As seen in Figure 2, variation from one batch of larvae to the next is quite high. The probability of 8-day-old
larvae undergoing metamorphosis is nearly 0. At the same time,
only about 18% as many 9-day-old larvae will metamorphose as
12-day-old larvae when both are introduced to inducer concentrations about half maximal (0.5 g/l). It is important to recognize
that day 8 is the typical hatching day (younger larvae tested were
artificially hatched) and that developmental progression is apparently not highly synchronized between larvae. Age-dependent
variation in response to inducer thus probably represents variation in the rate at which larvae achieve competence to respond to
the inducer. In each age class, the group of larvae which is meta-
morphically competent still shows a range of individual thresholds
to inducer concentration.

At first I suspected that the variation seen in the response
of larvae to different durations of exposure to "crude inducer"
might be only another expression of the concentration curve (i.e.
Figure 3 could be derived from Figure 1). However, this possibility was ruled out by further experiments in which batches of
larvae, exposed to different inducer concentrations, were allowed
to remain in the inducer solutions and the number having

metamorphosed counted daily until it no longer increased (Figure 4).
Since these larvae were all 12 days old at the outset and allowed
to remain in the inducer indefinitely, the possibility that reduced exposure time to inducer represents the same phenomenon as
reduced concentration of inducer is discounted. That is, individual
larvae have separate thresholds for minimal inducer concentration
and minimal exposure time to inducer. Means of both thresholds
vary widely, and probably independently, among larval populations.

   The fact that a given inducer concentration is capable of
producing, in a given larval population, a predictable number of
metamorphoses which does not increase in time (Figure 4), implies
that larvae in some way sequester or inactivate the inducer. This
possibility is dubious to begin with because of the probable
extreme disparity between the small number of larvae and the
great number of inducer molecules likely to be present. The ability of inducer solutions with 75% and 50% of maximum induction
capacity to repeatedly induce metamorphosis at the same level in
successive groups of larvae placed in the solutions (Table 1)
affirms that the inducer concentrations are not significantly decreased during the induction process. These experiments also
demonstrate the temporal stability of the inducer in seawater.
These findings substantiate the conclusion that each individual
larva is in some manner sensitive to an absolute minimum inducer
concentration.

   Specificity of the metamorphic response of larvae of
Phestilla sibogae to a particular coral, Porites compressa, is
high. The fact that it is not absolute (Figure 5) may be due to
either (1) the other coral species produce the inducer in much
lower concentrations, or (2) other coral species produce compounds
whose structures are sufficiently similar to the "true" inducer
that they produce metamorphosis in some larvae. Without further
knowledge of the molecular identity of the inducer, we cannot
distinguish between these two possibilities.

   We cannot, at present, make good comparison between the
activity of the metamorphic inducing agent studied here and those
studied in other species. The large molecular entities which induce barnacle settling act only as adsorbed layers; concentrations
have not been varied and quantified, nor have exposure times and
larval age been studied as variables (see summary in Crisp[4]).
Characteristics of biological films-- presumably fungi, bacteria,
algae, and their products-- have been studied for their effective-
ness in inducing metamorphosis in hydroid planula larvae[11,12] and
sea urchin larvae[13]. Only in the former were studies somewhat
comparable to ours conducted. These included construction of
dose-response curves which have, in fact, shapes similar to those
obtained in the present study.

Workers in Japan[14] have isolated and identified an algal product which stimulates settling of larvae of the hydrozoan _Coryne uchidai_. They have not, unfortunately, analyzed the nature of the effect of the compound(s) on the larvae.

Crisp[4] postulated that chemical settlement-stimulating agents must all act as absorbed surface components. Our data, along with those cited in the previous paragraphs, tend to refute this notion. All larvae studied in the experiments represented by Figure 3, for example, were exposed to inducer solutions for brief time periods and transferred through several changes of filtered seawater. Only several hours later did they undergo metamorphosis in complete absence of inducer. The importance of an adsorbed inducer, as stressed by Crisp, is that continuous contact with it would lead to metamorphosis at that site only. Obviously, we are observing a different phenomenon.

The biological significance of the variability in metamorphic responses of _Phestilla_ larvae to coral extract lies in the adaptive value of the variability. With offspring leaving the pelagic realm and entering the juvenile phase at mandatorily different ages, inducer concentrations, inducer exposure durations, and possibly other parameters than those investigated here, the species is obviously in no danger of placing its entire reproductive effort in a single, perhaps precarious, location.

ACKNOWLEDGEMENTS

The success of the experiments presented here is due in no small part to the excellent technical assistance of Wm. Van Heukelem. My sincere thanks go to Drs. Van Heukelem and Marilyn F. Dunlap for their valuable suggestions during the course of the experiments and the writing of this paper. Parts of these researches were supported by NSF research grant GB-36702 and NIH research contract NO1-RR-4-2168.

REFERENCES

1. T. Mortensen, Studies of the Development and Larval Forms of Echinoderms, G. E. C. Gad, Copenhagen, 1921.

2. J. H. Day and D. P. Wilson, J. Mar. Biol. Ass. U. K., 19, 655 (1934).

3. G. Thorson, Reproduction and Larval Development of Danish Marine Bottom Invertebrates, Medd. Komm Danmarks Fisk-. Havunders., Ser. Plankton, No. 4, Copenhagen, 1946.

4. D. J. Crisp, in Chemoreception in Marine Organisms, eds. P. T. Grant and A. M. Mackie, Academic Press, New York, 1974, p. 177.

5. R. S. Scheltema, Biol. Bull., 120, 92 (1961).

6. D. P. Wilson, Annls. Oceanogr. Monaco., 27, 49 (1952).

7. D. B. Bonar and M. G. Hadfield, J. Exp. Mar. Biol. Ecol., 16, 227 (1974).

8. M. G. Hadfield, Amer. Zool., 12, 721 (1972).

9. M. G. Hadfield and R. H. Karlson, Amer. Zool., 9, 317 (1969).

10. L. G. Harris, Biol. Bull., 149, 539 (1975).

11. W. A. Müller, Wilhelm Roux' Arch., 173, 107 (1973).

12. W. A. Müller, Wilhelm Roux' Arch., 173, 122 (1973).

13. R. A. Cameron and R. T. Hinegardner, Biol. Bull., 146, 335 (1974).

14. T. Kato, A. S. Kumarireng, I. Ichinose, Y. Kitahara, Y. Kakinuma, M. Nishihara, and M. Kato, Experientia, 31, 433 (1975).

INTERSPECIFIC RELATIONSHIPS IN THE FIELD OF BACTERIA AND PHYTOPLANKTON IN A MARINE ENVIRONMENT

M. Aubert, M. J. Gauthier, and J. M. Gastaud

C.E.R.B.O.M. (I.N.S.E.R.M.)

1, avenue Jean Lorrain, 06300 Nice, France

Over the past few years our laboratory has conducted a systematic investigation into the interspecific relationships that take place in the microbiological and planktonic fields.

The results of our studies have suggested a close relationship between the various organisms, such connections being based on the transmission and reception of messages carried throughout the medium by specific chemical substances. Some researchers, such as Lucas, Nigrelli and Fontaine, have already expressed a similar assumption, but our experiments have made it possible to isolate a few of the messages and provide evidence of their functions. Based on a sufficiently large number of such functional mechanisms, it is possible to get an idea of the overall processes which control the biological equilibrium of the sea, at least as far as microorganisms are concerned. Considering the major functions of the marine species, it can be thought that nutrition, reproduction, defense and motion may be accomplished through chemotaxis. Such functions involve biological mechanisms including nutritional attraction, sexual attraction, attractions generating commensalism and symbiosis, migratory attraction, and the capability to synthesize substances that bring about various metabolisms and aggressive or defensive behaviours. It appears that these biological mechanisms require the action of chemical substances which "warn" certain organisms of the presence of other organisms which are necessary for their reproduction and possibly their nutrition, or will monitor their migratory activities. We have called such substances "telemediators," with the following definition: "Chemical telemediators are substances synthesized by marine animal or plant species which, when released into the environment, act remotely upon the behaviour or the biological

functions of the same species or other species." Such a definition
eliminates any physical or chemical action that is connected with
the natural environment or artificially created. For instance, we
shall ignore the biological actions resulting from the introduction
of chemical pollutants into the sea, as well as the biological
action of nutrients introduced by rivers, or, as another example,
the alteration of certain substances of bacterial origin under the
influence of solar radiation. This concept of function control
in the marine environment might be compared to the way a pluri-
cellular organism works, with its hormonal balance depending on
pacemaking mechanisms generated by chemical mediators. This is an
imperfect example because, in that case, the type of action is
entirely limited to the organisms in which it takes place; the
chemical mediators do not move outside the organism but are con-
veyed through the various parts in the circulatory system. In the
marine environment, however, the similar mechanism which we describe
includes an external stage, when active substances are released
into the oceanic medium, and the receiver is an organism other than
the sender.

This may be extended to all marine organisms, although it is
difficult to identify the facts and to isolate the "responsible"
substances where the large pelagic species are concerned. Yet it
is possible to envisage experimental protocols to investigate
these phenomena in the marine microorganisms of the planktonic
species. Our experiments were carried out on the latter, and the
results obtained are summarized in this paper.

TELEMEDIATION MECHANISMS BETWEEN TWO SPECIES

In this field, the best known example concerns the biological
equilibrium which takes place between certain phytoplanktonic
species, for instance, the antagonism observed between populations
of diatoms and dinoflagellates. As was shown by M. Aubert in a
series of experiments illustrating the growth of these mixed species
with time, the antagonism is quite significant. More recently,
J. M. Pincemin, through a series of experiments carried out
in vitro, provided clear evidence of this interspecific opposition.
In his thesis, he described the collapse of a population of the
diatom Asterionella japonica after it was placed in the presence
of Glenodinium monotis; he also demonstrated that the presence of
the culture medium in which this diatom had lived would enhance
the growth of the dinoflagellate. In view of the slow growth of
the dinoflagellate compared with that of the diatom, the substances
released by each species into the environment inhabited by the other
provide a means to control the proliferation of the diatom, so that
biological equilibrium is maintained between the two species. This
example shows that, in this case, the mechanism is two-fold, as it

involves two types of organisms that are linked by at least two chemical mediators. The latter have not yet been isolated and their chemical nature is not known.

Other authors have reported similar facts. Thus, D. M. Pratt observed an antagonism in the growth of a diatom, Skeletonema costatum, and a Xanthophycea, Olisthodiscus luteus, although it was not possible to isolate a chemical mediator. Similar observations were made in fresh-water environments.

Let us mention another example of the mechanism of dynamic equilibrium between populations of diatoms and terrestrial bacteria released into the marine environment. Since 1961 our studies have provided evidence of the antagonism which takes place between a number of diatoms and a great number of bacteria of terrestrial origin which reach the marine environment in runoff waters, rivers, and urban sewage waters. Because of the many papers published on this subject, we shall not dwell on a description of these phenomena although, contrary to the stage previously described, we have been able to isolate and analyze certain chemical mediators that are responsible for the antagonism. Our experiments on this subject have revealed the presence of a fatty acid and a low molecular weight nucleosidic compound, given off by a few diatom species, which have an antibiotic action toward a great number of terrestrial bacteria. This is, therefore, a biologically and chemically proven fact. Here again, it involves two stages of the ocean biomass.

In the past decade, other scientists, such as Sieburth, have demonstrated independently, through various substances which they analyzed (tannin, acrylic acid, phenolic derivatives, etc.), that other species of phytoplankton or algae possess antibacterial activity.

In the same connection, as another example of the two-stage mechanism involved in marine microbiology, we shall mention the antagonism which takes place between marine bacteria and some terrestrial bacteria. This was reported by ZoBell and Rosenfeld and, later, by Krassil'Nikova. More recently, M. J. Gauthier demonstrated the interest of such interbacterial antagonism, providing evidence of the existence of a chemical mediator produced by some alteromonads: it is an acidic polyoside. Here again, it is a two-stage process that is proven both biologically and chemically.

Similar observations were made by other authors. Sieburth, for instance, showed the antagonism taking place in situ between pseudomonads and arthrobacters due to two substances, the chemical determinations of which have not yet been possible.

TELEMEDIATION MECHANISMS BETWEEN MULTIPLE SPECIES

When looking more closely into the phenomena, one finds
processes that are more complex and involve more species, with
actions upon highly specific metabolic functions. We shall give
two examples.

From a number of experiments carried out either in situ or
in vitro, we were able to demonstrate that Asterionella japonica,
a diatom, stops synthesizing its antibiotic when placed close to
Prorocentrum micans, a dinoflagellate. This results in the annihilation of the antibacterial activity of the diatom. Aubert and
Pesando were able to identify the nature of the chemical mediator,
which is released by the dinoflagellate and acts remotely. Spectroscopic studies, titrations, and the use of enzymes made it
possible to determine that the telemediator involved was a protein
found both in the cells of Prorocentrum micans and in the medium
where it had been living. This specific action, the various mechanisms of which we have demonstrated, is a three-stage biological
process, each mediator having been analyzed and chemically identified. This three-stage process, as compared to the two-stage
process, provides evidence for the existence of secondary telemediators, whose action is indirect, in addition to the primary
telemediators, whose action is direct.

In the experiments we have just mentioned, we demonstrated
the successive phases of the phenomena by investigating the origin
of the initial producer. Other actions, either analogous or
opposite, can be identified. Thus we have demonstrated that,
rather than the blocking of antibiotic production by a diatom, it
is the action of a mediator which induces such a synthesis. We
discovered that for synthesizing its antibiotic the diatom
required in its environment the presence of a number of substances
which, using spectrographic techniques, we assume to be nucleo-
proteins, although we do not know their chemical structure or the
way they are emitted.

These phenomena, which involve both plankton and bacteria, are
fairly close to the ones demonstrated by M. J. Gauthier concerning
the relationships between marine and terrestrial bacteria. It
appears that certain of the latter, although sensitive to the antibiotic polyanions produced by some marine alteromonads, are
protected against their lethal activity by several enzymes
(catalase, peroxidase, glucuronidase). It also appears that, as
in the antibacterial action of diatoms, induction of release may
be achieved through the action of telemediators from other marine
bacteria of still different species.

TELEMEDIATION MECHANISMS INVOLVING FEEDBACK PROCESSES

On reviewing the examples mentioned, it appears that there are two types of telemediating actions. One is a direct, "primary" type, where a telemediator synthesized by a species directly controls the metabolism of another species. The other is an indirect, "secondary" type in which the telemediator synthesized by a species triggers a metabolic action in a second species, which, in turn, controls a third species. A combination of the two mechanisms may result in cycles and, eventually, feedback mechanisms which control the interspecific biological equilibrium. A number of closed-chain processes were thus identified. For instance, the terrestrial bacteria which abound in sewage waters release products such as Vitamin B_{12} which enhance the growth of certain antibiotic-producing diatoms, which, in turn, stop the proliferation of the bacteria.

On the whole, these experimental phenomena demonstrate two types of action, one where the telemediator acts directly on the receiver and another in which, owing to the very low activity of the telemediator, it acts as a signal, in the cybernetic sense of the word. For instance, the very low concentrations at which we recovered the protein responsible for the annihilation of the antibiotic-producing function in Asterionella japonica suggest that, in this case, it was a "message," as defined in genetics.

Thus, the idea of the life of organisms being controlled by an exchange of messages between species results in a data-processing-like concept of marine biology.

BIBLIOGRAPHY

J. Adler, Science 153, 708 (1966).

J. Adler, Science 166, 1588 (1969).

M. Alexander, in Microbial Ecology, Wiley, 1971, p. 361.

J. B. Allison and W. H. Cole, Mt Desert I. Biol. Lab. Bull., 24 (1935).

M. Aubert, Rev. Int. Oceanogr. Med. 21, 5 (1971).

M. Aubert and M. J. Gauthier, Rev. Int. Oceanogr. Med. 5, 63 (1967).

M. Aubert, J. Aubert, M. J. Gauthier, and D. Pesando, Rev. Int. Oceanogr. Med. 6-7, 43 (1967).

M. Aubert and D. Pesando, Rev. Int. Oceanogr. Med. 15-16, 29 (1969).

M. Aubert, D. Pesando, and M. J. Gauthier, Rev. Int. Oceanogr. Med.
   28-29, 69 (1970).

M. Aubert, D. Pesando, and J. M. Pincemin, Rev. Int. Oceanogr. Med.
   17, 5 (1970).

M. Aubert and D. Pesando, Rev. Int. Oceanogr. Med. 21, 17 (1971).

M. Aubert, D. Pesando, and J. M. Pincemin, Rev. Int. Oceanogr.
   Med. 25, 17 (1972).

M. Aubert, M. J. Gauthier, B. Donnier, D. Pesando, J. M. Pincemin,
   and M. Barelli, Rev. Int. Oceanogr. Med. 28, 129 (1972).

M. Aubert and D. Pesando, Rev. Int. Oceanogr. Med. 35-36, 195 (1974).

M. Aubert, M. J. Gauthier, and D. Pesando, Rev. Int. Oceanogr. Med.
   37-38, 69 (1975).

J. Aubert, D. Pesando, and H. Thouvenot, Rev. Int. Oceanogr. Med.
   10, 259 (1968).

J. Aubert and J. P. Gambarotta, Rev. Int. Oceanogr. Med. 25, 39
   (1972).

J. Aubert, J. P. Belaich, F. Ferneix, J. Pouthier, and D. Pesando,
   in Marine Pollution and Marine Wastes Disposal, eds. Pearson
   and Frangipane, Pergamon Press, Oxford and New York, 1975,
   p. 111.

H. Augier, Ph.D. Thesis, Marseilles, 1972.

L. G. M. Gaas-Becking, Ann. Bot. 39, 613 (1925).

W. Bell and R. Mitchell, Biol. Bull. Mar. Biol. Lab. 143, 265 (1972).

J. A. Bentley, Nature 181, 1499 (1958).

J. A. Bentley, J. Mar. Biol. Ass. U. K. 39, 433 (1960).

B. R. Berland and S. Y. Maestrini, Mar. Biol. 3, 334 (1969).

B. R. Berland, D. J. Bonin, and S. Y. Maestrini, Mar. Biol 12, 189
   (1972).

B. R. Berland, D. J. Bonin, and S. Y. Maestrini, Thetys 4, 339
   (1972).

B. R. Berland, D. J. Bonin, A. L. Cornu, S. Y. Maestrini, and J. P.
   Marino, J. Phycol. 8, 383 (1972).

B. R. Berland, D. J. Bonin, and S. Y. Maestrini, Mar. Biol. Mar. Oceanogr. 3, 1 (1973).

P. R. Burkholder, in International Oceanographic Congress Preprints, ed. M. Sears, Amer. Ass. Adv. Sci, 1959, p. 912.

G. Cahet, Vie et Milieu 16, 917 (1965).

G. Cahet, C. R. Acad. Sci. Paris 263, 691 (1966).

B. M. Chassy, L. L. Love, and M. I. Krichevsky, Proc. Acad. Sci. U.S.A. 64, 296 (1969).

I. Chet, S. Fogel, and R. Mitchell, J. Bacteriol. 106, 863 (1971).

A. Collier, S. M. Ray, A. W. Magnitski, and J. O. Bell, Fish. Bull. U.S. 54, 167 (1953).

H. C. Davis and R. R. Guillard, Fish. Bull. U.S. 58, 293 (1958).

L. E. Ericson and L. Lewis, Ark. Kemi. 6, 247 (1953).

S. Fogel, I. Chet, and R. Mitchell, Bacteriol. Proc. G. 31 (1971).

M. Fontaine, Conférence Congrès de l'A.F.A.S., Paris, 1970.

M. J. Gauthier, Rev. Int. Oceanogr. Med. 15-16, 103 (1969).

M. J. Gauthier, Rev. Int. Oceanogr. Med. 15-16, 41 (1969).

M. J. Gauthier, in Atti del 5e Colloquio Internazionale di Oceanographia Medica, ed. S. Genovese, 1973, p. 623.

M. J. Gauthier, Antimicrob. Agents Chemother. 9, 361 (1976).

M. J. Gauthier, J. P. Breittmayer, and M. Aubert, Proc. 10e Europ. Mar. Biol. Symp, Ostende, in press.

M. J. Gauthier, J. M. Shewan, D. Gibson, and J. V. Lee, J. Gen. Microbiol. 87, 211 (1975).

M. J. Gauthier, M. Aubert, and P. Bernard, Rev. Int. Oceanogr. Med. 43, 13 (1976).

L. G. Gutvieb, A. G. Benzhitski, and M. N. Lebedeva, in Atti del 5e Colloquio Internazionale di Oceanographia Medica, ed. S. Genovese, 1973, p. 161.

A. C. Hardy, Discovery Rep. 11, 511 (1935).

J. A. Hellebust, Limnol. Oceanogr. 10, 192 (1965).

S. H. Hutner, A. Cury, and H. Baker, Anal. Chem. 30, 849 (1958).

E. G. Jørgensen, Physiol. Plant. 15, 530 (1962).

S. H. Larsen, R. W. Reader, E. N. Kort, W. W. Tso, and J. Adler, Nature 249, 74 (1974).

J. Legall and J. R. Postgate, in Advances in Microbial Physiology, Vol. 10, eds. A. H. Rose and T. W. Tempest, 1974, p. 81.

C. E. Lucas, Cons. Perm. Explor. Mer., J. Cons. 13, 309 (1938).

C. E. Lucas, Biol. Rev. 22, 270 (1947).

C. E. Lucas, Deep-Sea Res. 3 (suppl.), 139 (1955).

C. E. Lucas, Symp. Soc. Exp. Biol. 15, 190 (1961).

J. J. A. McLaughlin, J. Protozool. 5, 75 (1958).

R. Margalef, L'Annee Biologique, Ser. 4, 2, 3 (1963).

R. Margalef, in Fundacion la Salte de Ciencias Naturales, Caracas, 1967, p. 377.

R. Mitchell, Nature 230, 257 (1971).

R. Mitchell and Z. Nevo, Nature 205, 1007 (1964).

R. Mitchell and Z. Wirsen, J. Gen. Microbiol. 52, 335 (1968).

C. Nalewajko, Limnol. Oceanogr. 11, 1 (1966).

R. F. Nigrelli, Trans. N. Y. Acad. Sci., Ser. II, 20, 248 (1958).

R. F. Nigrelli, Ann. N. Y. Acad. Sci. 90, 1 (1958).

A. Paoletti, in Proc. Symp. Pol. Mar. Microorg. Prod. Petr., C.I.E.S.M.M., 1964, p. 133.

D. Pesando, Rev. Int. Oceanogr. Med. 25, 49 (1972).

J. M. Pincemin, Rev. Int. Oceanogr. Med. 22-23, 165 (1971).

D. M. Pratt, Limnol. Oceanogr. 11, 447 (1966).

L. Provasoli, in The Sea, Vol. 2, ed. M. N. Hill, Interscience, New York, 1963, p. 165.

S. Samuel, N. M. Shaw, and G. F. Fogg, J. Mar. Biol. Ass. U. K. 51, 793 (1971).

A. K. Saz, S. Watson, S. R. Brown, and D. L. Lowert, Limnol. Oceanogr. 8, 63 (1963).

J. McN. Sieburth, Limnol. Oceanogr. 4, 419 (1959).

J. McN. Sieburth, Science 132, 676 (1961).

J. McN. Sieburth, J. Bacteriol. 82, 72 (1961).

J. McN. Sieburth, Honors lecture, Univ. Rhode Island, 1962.

J. McN. Sieburth, J. Bacteriol. 93, 1911 (1967).

J. McN. Sieburth, Bull. Misaki Mar. Biol. Inst. 12, 49 (1968).

J. McN. Sieburth, Mar. Biol. 11, 98 (1971).

J. McN. Sieburth and D. M. Pratt, Trans. N. Y. Acad. Sci., Ser. II, 24, 498 (1962).

J. P. Thomas, Mar. Biol. 11, 311 (1971).

S. Ulitzur and M. Shilo, Biochem. Biophys. Acta 201, 350 (1970).

E. F. J. Wood, Oceanogr. Mar. Biol., Ann. Rev. 1, 197 (1963).

C. E. ZoBell, J. Bacteriol. 33, 86 (1937).

C. E. ZoBell, J. Bacteriol. 46, 39 (1943).

GENUS AND SPECIES INDEX

SUBJECT INDEX

431